Adoption of New Technologies and Practices for Sustainable and Smart Agriculture

Adoption of New Technologies and Practices for Sustainable and Smart Agriculture

Guest Editors

Yang Li
Jing Nie

MDPI

Basel • Beijing • Wuhan • Barcelona • Belgrade • Novi Sad • Cluj • Manchester

Guest Editors

Yang Li
College of Mechanical and
Electrical Engineering
Shihezi University
Shihezi
China

Jing Nie
College of Mechanical and
Electrical Engineering
Shihezi University
Shihezi
China

Editorial Office
MDPI AG
Grosspeteranlage 5
4052 Basel, Switzerland

This is a reprint of the Special Issue, published open access by the journal *Sustainability* (ISSN 2071-1050), freely accessible at: https://www.mdpi.com/journal/sustainability/special_issues/U4X11AD63T.

For citation purposes, cite each article independently as indicated on the article page online and as indicated below:

Lastname, A.A.; Lastname, B.B. Article Title. *Journal Name* **Year**, *Volume Number*, Page Range.

ISBN 978-3-7258-7142-1 (Hbk)
ISBN 978-3-7258-7143-8 (PDF)
https://doi.org/10.3390/books978-3-7258-7143-8

Contents

About the Editors

Yang Li

Yang Li is an Associate Professor at the College of Mechanical and Electrical Engineering, Shihezi University, China. He is an IEEE Senior Member and the director of the Department of Electronic Information and Automation. He received an M.S. in Electrical Engineering from Dalian University of Technology, China, and a Ph.D. in Information and Communication Engineering from Tianjin University, China. His main research interests include image processing, pattern recognition, data quality assessment, and smart agriculture. He is an Associate Editor for several reputable journals, including *Information Processing in Agriculture, Precision Agriculture, IEEE Journal of Selected Topics in Applied Earth Observations and Remote Sensing*, etc. He has managed many special issues as lead Guest Editor for reputable journals, such as *Computers and Electronics in Agriculture, IEEE Transactions on Consumer Electronics*, etc. He has published over 100 research papers in reputable journals, including COMPAG, JIII, ASOC, IEEE TII, IEEE TITS, IEEE IoTJ, etc. He won the 2024 ICT Express Contribution Award and 2023 ICT Express Best Paper Award. He was included in the World's Top 2% Scientists by Stanford University from 2023 to 2025.

Jing Nie

Jing Nie is a Professor at the College of Mechanical and Electrical Engineering, Shihezi University, China. He received an M.S. in Mechatronic Engineering from Xinjiang University, China, and a Ph.D. in Mechanical Engineering from Shihezi University, China. His main research interests include deep learning, image processing, pattern recognition, and intelligent control. He is an Associate Editor for several reputable journals, such as *Plant Methods, Turkish Journal of Agriculture and Forestry, CABI Agriculture and Bioscience*, etc. He has published over 70 research papers in reputable journals, including COMPAG, JIII, ASOC, IEEE IoTJ, IEEE JSTARS, IEEE TCE, etc. He won the 2025 ICT Express Contribution Award.

sustainability

MDPI

Systematic Review

Towards Sustainable Vertical Farming: A Systematic Review of Energy Return on Investment Efficiency and Optimization Strategies

Abdulaziz Aborujilah

Department of Management Information System, College of Commerce & Business Administration, Dhofar University, Salalah 211, Oman; aaborujilah@du.edu.om

Abstract: Vertical farming offers one potential sustainable solution to food production in cities as populations increase and available farmland decreases. However, the large-scale adoption of vertical farms is still impeded by high energy requirements and costs. This research attempts to assess how energy optimization strategies that improve the sustainability and feasibility of vertical farming systems are applied systematically. Following the PRISMA guidelines, a systematic review of 52 articles published between 2014 and early 2024 was conducted on four major academic databases: ScienceDirect, Scopus, Web of Science, and Google Scholar. These reviews revealed that some modern technologies like high-efficiency LED lights, smart HVAC control, and IoT-based smart irrigation systems provided great advancements in reducing electricity consumption. However, even with these innovations, energy savings were heavily impacted by factors such as crop variety, climate, facility layout, system design, and geography. Other critical factors like high upfront spending, limited access to qualified personnel, inconsistent reporting standards, and a lack of real-world data further impede widespread adoption of the technology. This review emphasizes the need for multidisciplinary longitudinal field studies, standardized metric definitions, strategic integration of renewable resources, and supplementary training for operators. These analyses provide a foundation which can assist policymakers, researchers, and investors in developing energy-efficient, low-cost, and eco-friendly vertical farming systems.

Keywords: vertical farming; energy optimization; sustainability; smart irrigation; LED lighting; renewable energy integration; urban agriculture; automation and IoT

1. Introduction

Vertical farming has emerged as a forward-looking answer to rapid urban growth, shrinking farmland, and the urgent need for greener food systems. Cities like Singapore now showcase how stacked gardens can boost local vegetable supplies and lighten environmental footprints. Rooftop farms planned for the city's public housing alone could lift harvest volumes by 700%, satisfying more than a third of demand while cleverly using scarce land [1]. Seeking to make production smoother and cheaper, Singapore and similar metropolises are quickly rolling out high-tech growing methods such as hydroponics, aeroponics, and aquaponics. Despite these advancements, challenges such as limited available land, regulatory complexity, and other operational barriers persist. Innovative startups—including Plenty, AeroFarms, and Bowery Farming—are advancing enabling technologies for space- and resource-efficient production, leveraging automation and data analytics to control growing conditions and maintain a steady supply of high-quality produce [2,3]. National initiatives worldwide increasingly promote smart agriculture and urban sustainability. For instance, South Korea has launched major plans to commercialize convergence technologies and implement vertical smart farms, even in unconventional locations such as subway stations [4,5]. However, the adoption of AI, big data, and other advanced technologies in agriculture remains limited in both scope and effectiveness [6]. Similarly, China's use of artificial intelligence and plant factories with artificial lighting is promising, yet hindered by climate variability and a fragmented landscape of smallholder farms [7,8].

In Sweden, integrating controlled environment agriculture (CEA) with cultural and educational initiatives has fostered innovative urban food systems, but significant challenges remain regarding greenhouse gas emissions and skilled labor shortages [9,10]. Urban agriculture policies in cities such as New York and London have sought to address space constraints and infrastructure gaps, though progress is tempered by policy limitations and persistent skepticism toward new farming models [11,12]. In Budapest researchers are actively examining how IoT-enabled smart farming systems can operate within city settings, highlighting the ongoing demand for hands-on studies that fine-tune these fledgling technologies [13,14]. A consistent observation across these investigations is that vertical farms still consume a lot of energy, mainly because of the heat and light needed to keep crops growing. Indeed, recent estimates show that it takes between 10 and 18 kilowatt-hours to grow one kilogram of lettuce in a stacked system—a figure that dwarfs the less-than-1 kilowatt-hour needed for traditional field production. Such a gulf makes it clear that any sustainable future for controlled-environment agriculture hinges on smarter lighting, sharper climate controls, and a reliable mix of renewable power. Even so, key knowledge gaps persist. Most urgently, the field still lacks large-scale, real-world data and agreed-upon benchmarks to judge whether specific energy-saving measures really work. Many current studies overlook the practical integration of renewable energy and rarely address solutions accessible to small-scale operators. Additionally, workforce training and public acceptance are underexplored, ultimately constraining the real-world sustainability and scalability of vertical farming systems.

To bridge these gaps and establish a foundation for further progress, this review systematically evaluates current energy optimization strategies employed in vertical farming, critically examining their inherent limitations. Furthermore, it delineates prospective research directions by integrating technological innovations, operational practices, and policy frameworks, thereby contributing to the development of scalable and energy-efficient vertical farming systems. This review aims to systematically synthesize current research on energy efficiency and optimization in vertical farming, focusing on technological innovations, smart energy management, and sustainability practices. The objective is to identify strategies that reduce energy consumption without compromising productivity, providing a framework for sustainable vertical farming development. The following key research questions guide this review:

1. How can energy optimization strategies improve sustainability in vertical farming?
2. How do smart irrigation systems contribute to energy efficiency in vertical farming?
3. What are the main barriers to adopting advanced energy solutions in vertical farming?

The structure of this paper is as follows. First, it presents an overview of how vertical farming can help address food security challenges in urban areas. The next section outlines the review's methodology. This is followed by a summary of key studies, focusing on energy use and smart irrigation in vertical farming. The paper then analyzes the main challenges in applying energy-saving strategies, including issues related to advanced models and system complexity. Finally, the study highlights the potential of smart energy systems to improve sustainability in vertical farms and offers suggestions for future research and policymaking.

2. Materials and Methods

To address the research questions and synthesize evidence on energy optimization in vertical farming, this review employed the PRISMA framework for systematic reviews.

The systematic search identified a total of 270 records from databases. Before screening, 100 duplicate records and 40 records deemed ineligible by automation tools were removed, leaving 130 records for screening. Of these, 50 records were excluded, and 80 reports were sought for retrieval. However, 28 reports could not be retrieved, resulting in 52 reports assessed for eligibility. During eligibility assessment, 14 reports were excluded for not focusing on vertical farming systems, 7 for not addressing energy efficiency, and 7 as opinion papers without empirical data. Ultimately, 52 studies were included in the review. A comprehensive search was conducted across four major academic databases—ScienceDirect, Scopus, Web of Science, and Google Scholar—to ensure broad and relevant coverage. The initial search strategy focused on identifying studies related to energy challenges and solutions in vertical farming, using keywords such as "vertical farming," "energy optimization," "energy efficiency," and "sustainability."

Figure 1 illustrates the PRISMA flow diagram used in the systematic review process. It visually represents the steps taken to identify, screen, and select relevant studies for inclusion in the final analysis.

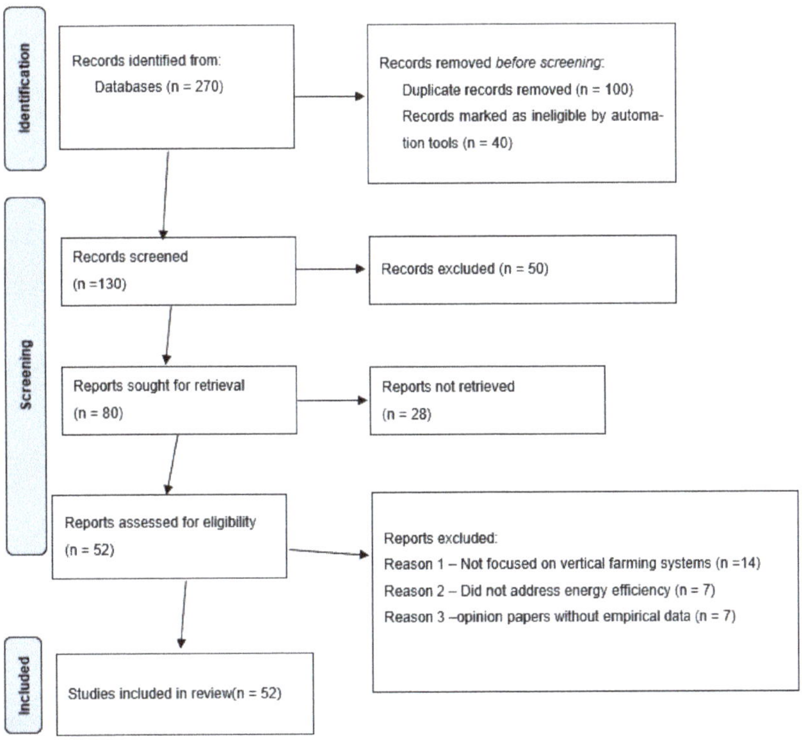

Figure 1. Review process flow.

2.1. Eligibility Criteria

Studies were eligible for inclusion if they (1) focused on energy optimization and/or efficiency in vertical farming within the context of sustainable agriculture, (2) were published in peer-reviewed journals between January 2014 and March 2024, (3) were written in English, and (4) provided full-text access with sufficient methodological detail.

Exclusion criteria included duplicate records, books, conference proceedings, non-peer-reviewed industry/government reports, studies loosely related to vertical farming without energy-related analysis, and studies lacking quantitative or qualitative data on energy use or sustainability impacts. Studies were grouped into thematic categories (e.g., technological innovations, renewable integration, energy management systems, circular economy practices) during synthesis.

2.2. Information Sources

The literature search was conducted in the following databases: ScienceDirect, Scopus, Web of Science, Google Scholar, IEEE Xplore, and SpringerLink. Searches were completed on: ScienceDirect (10 March 2024), Scopus (12 March 2024), Web of Science (13 March 2024), Google Scholar (14 March 2024), IEEE Xplore (15 March 2024), and SpringerLink (15 March 2024). Reference lists of included studies were also screened to identify additional eligible records.

2.3. Search Strategy

Search strings were adapted to each database but maintained consistent Boolean logic. The core query was:

("vertical farming" OR "plant factory" OR "indoor farming") AND ("energy optimization" OR "energy efficiency" OR "power management" OR "sustainable energy" OR "energy-saving" OR "renewable energy") AND ("sustainability" OR "sustainable agriculture" OR "sustainable farming" OR "smart agriculture")

2.4. Selection Process

All retrieved records were exported to EndNote for duplicate removal. Two reviewers independently screened titles and abstracts against the eligibility criteria. Full texts of potentially relevant studies were retrieved and assessed independently by both reviewers. Disagreements were resolved through discussion; if unresolved, a third reviewer acted as arbiter. No automation tools were used for screening or selection.

Table 1 categorizes the reviewed literature into focus areas and sub-themes relevant to urban agriculture, vertical farming, smart technologies, sustainability, policy, and emerging innovations. References are provided for each sub-theme to guide further study.

Table 1. Thematic analysis—References by Focus Area.

Focus Area	Sub-Theme	References
Urban Agriculture	Land Use Planning	[1,15]
	Urban Horticulture	[16–18]
	Case Studies	[19,20]
Vertical Farming	Environmental Impact	[21–23]
	Business Models	[24,25]
	Technological Innovations	[7,26,27]

Table 1. *Cont.*

Focus Area	Sub-Theme	References
Smart Farming Technologies	IoT and Automation	[10,28–34]
	AI and Data-Driven Farming	[4,35]
	Energy Efficiency	[5,9,36]
Sustainability	Sustainable Practices	[37–40]
	Climate Change Mitigation	[41,42]
Policy and Social Aspects	Policy Frameworks	[43,44]
	Social Infrastructure	[14,45]
Emerging Technologies	Alternative Energy	[6,11,46,47]
	Innovative Practices	[3,48]

3. Vertical Farming Background

Vertical farming relies heavily on advanced actuator systems to achieve precise control over key environmental conditions. These actuators regulate devices responsible for managing airflow, lighting, and irrigation, thereby creating optimal environments for plant growth. By automating the operation of pumps and valves within irrigation systems, actuators minimize resource waste and ensure that plants receive consistent and appropriate hydration [25,41,49]. This technological foundation is central to maximizing both resource efficiency and crop yields in vertical farming.

3.1. Smart Systems for Energy Optimization

Building on this background, the following section reviews core energy optimization strategies that enable sustainable operation and address the substantial energy demands inherent in these systems.

Figure 2 illustrates a smart aquaponic vertical farming system enhanced by Internet of Things (IoT) technologies for real-time monitoring and automated control. The system integrates multiple sensors to measure critical parameters such as soil moisture, pH, water temperature, and ammonia levels. Data collected from these sensors are transmitted via a communication gateway to a microcontroller, which processes the signals and relays them to a cloud platform for storage, analytics, and decision-making.

The cloud-based system facilitates remote monitoring through web or mobile applications, enabling users to track environmental conditions and system performance in real time. An LED lighting system supports photosynthesis, while a pump circulates nutrient-rich water from the fish tank to the grow tray. The return flow, post-filtration, ensures efficient water reuse and nutrient delivery. This closed-loop configuration exemplifies sustainable resource use, optimizing energy, water, and nutrient cycles.

Another Smart System for Energy Optimization is schematic of integrating RESs in a smart GH; Figure 3 presents a conceptual model of a smart greenhouse system that integrates renewable energy technologies and Internet of Things (IoT)-enabled automation to enhance energy efficiency and sustainability. The system utilizes multiple renewable energy sources, including photovoltaic solar panels, geothermal heating, and wind turbines, to power critical greenhouse operations. These energy sources are connected to an inverter, heat pump, and battery energy storage unit to ensure stable and efficient energy supply. Inside the greenhouse, environmental conditions such as temperature, humidity, and soil moisture are monitored using air and soil sensors. These data are transmitted via gateways and processed by central controllers through a wired MODBUS communication protocol. The controllers activate specific actuators—including pumps, heaters, sprinklers, and LED lights—based on real-time sensor inputs to maintain optimal growing conditions. Remote monitoring and control are facilitated through a cloud-connected platform that enables users to access system data and adjust parameters via mobile or web applications. This integrated approach supports precision agriculture by optimizing resource use, reducing energy waste, and enabling autonomous operation under varying climatic conditions.

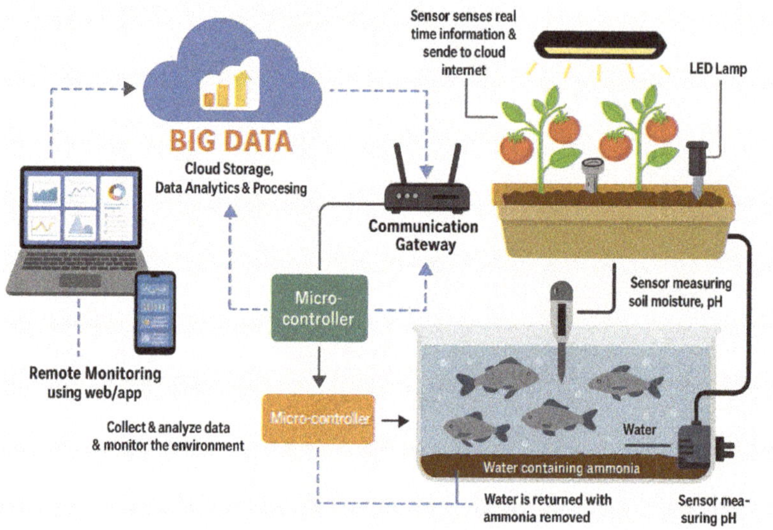

Figure 2. Schematic representation of implementation of IoT in vertical farming. (Adapted from Sowmya et al. [40]).

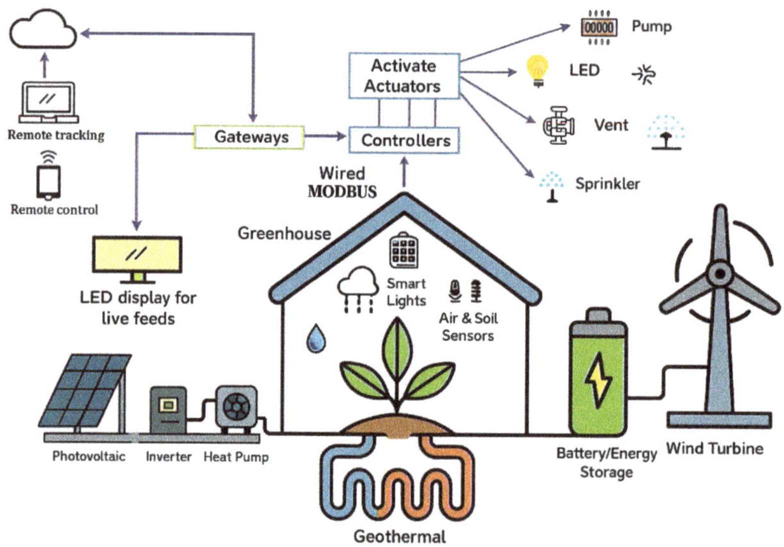

Figure 3. A schematic of integrating RESs in a smart GH (Adapted from Jamshidi et al. [8]).

3.2. *Life Cycle Assessment of Energy Use in Vertical Farming Systems*

Vertical farming—the cultivation of crops in stacked layers or on vertically inclined surfaces—has emerged as a promising solution to food insecurity in urban environments. Despite its notable advantages, vertical farming is characterized by substantial energy requirements, prompting ongoing research into strategies for optimizing energy use without compromising productivity. The advent of "Vertical Farming 2.0" marks a significant technological advancement, defined by the integration of advanced automation, artificial intelligence, sensor networks, and real-time environmental controls. These innovations enable the precise regulation of light, temperature, humidity, and CO_2 levels, allowing crops to be cultivated under optimal, energy-efficient conditions while minimizing waste [50]. A particularly effective approach for reducing energy consumption involves synchronizing LED lighting schedules with off-peak electricity hours, thereby lowering operational costs and enhancing crop yields [33]. Beyond on-farm technological improvements, vertical farms are increasingly collaborating with municipal infrastructure to harness organic waste, capture CO_2, and utilize surplus local energy, thus supporting regional sustainability and cost reduction [51]. Flexibility in energy use, such as adjusting operations according to real-time electricity prices or renewable energy availability, further increases operational efficiency [16]. Renewable energy-rich farms now employ solar panels and wind turbines alongside AI and IoT tools that automatically swap loads between the grid and on-site generation, fine-tuning power use to current supply and market tariff signals [30]. Achieving low-power draw remains a core requirement for the long-term financial and planetary viability of vertical horticulture. Integrating hydroponics or aquaponics within a circular economy model then enables precise nutrient recycling while cutting downstream organic waste [17]. Further gains materialize through next-generation LED lighting and crop biostimulants such as spirulina phycocyanin, both of which lift produce quality and allow milder climate settings, curbing the burden on energy-hungry controls [28,29]. To quantify the overall gains, researchers performed a life-cycle assessment on a typical facility yielding around 185,000 plants, equivalent to close to 5500 kg of edibles each year. The cradle-to-gate life-cycle assessment covered every operational stage, including cultivation, energy use, material transport, and local distribution, and centered on leafy greens and herbs while also accounting for on-farm waste management practices [35]. Figure 3 lays out the LCA findings and thereby supplies a comprehensive view of the energy-consumption pattern seen in contemporary vertical farms. With that background in place, the next section turns to the enduring obstacles that still hinder both the broad uptake and the meaningful execution of energy-saving techniques across vertical-farming systems. Figure 4 illustrates a comprehensive life cycle assessment (LCA) framework for vertical farming systems, outlining the flow of materials, energy, and processes from input to end-use. The system boundary, indicated by the red dashed line, captures all stages within the vertical farm's operational lifecycle. Material inputs such as pots, seeds, growing medium, fertilizers, water, and packaging are transported into the farm and contribute directly to plant cultivation. Supporting this process is the infrastructure backbone, which includes pumps, LED lighting systems, structural components, electronics, and nutrient tanks. These components enable controlled environmental conditions necessary for optimal crop growth. Energy inputs—comprising heating, ventilation, lighting, and pumping—are central to maintaining the internal climate and are among the most energy-intensive operations. Once crops are cultivated, they move to the harvest and packaging stage, where they are prepared for retail and distribution. Simultaneously, waste handling processes manage byproducts such as unused plant material and packaging waste. The system also accounts for heat recovery and avoided energy, where excess heat and energy are recaptured for reuse, contributing to improved energy efficiency. External processes, including intermediate processing and supply logistics, are represented outside the system boundary and connect the vertical farming unit to broader industrial networks. This LCA model enables the identification of high-impact stages within the production cycle and supports the development of strategies for reducing energy consumption, minimizing waste, and enhancing sustainability across the entire vertical farming operation.

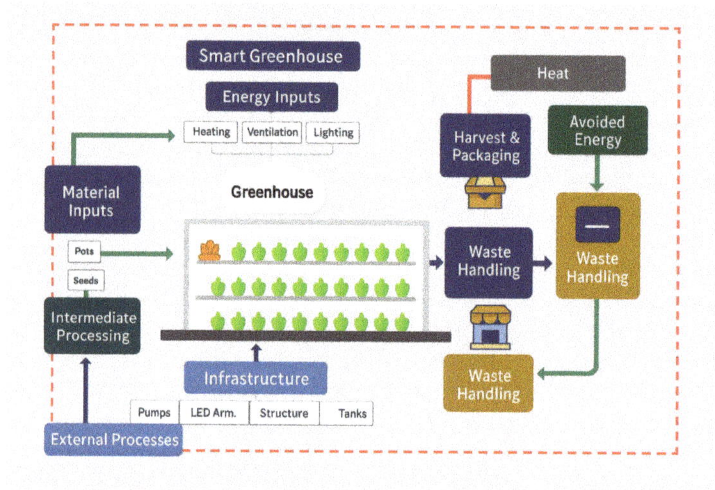

Figure 4. Life Cycle Assessment for Energy Use in Vertical Farming (Adapted from Martin et al. [36]).

3.3. *Model-Based Energy Optimization in Vertical Farming*

Optimization in vertical farming increasingly relies on advanced modeling and simulation techniques to navigate the complexities of sustainable production. By employing coupled heuristic approaches, vertical farms are able to fine-tune critical variables such as water use, radiation, and temperature throughout the growth cycle, thereby maximizing yield while minimizing energy consumption [52]. The adoption of sophisticated optimization tools—such as the P-Graph Framework—enables vertical farms to address multifaceted operational challenges by formulating and solving complex problems. These methods support the identification of strategies that are both economically and energetically optimal, guiding farms toward more sustainable and cost-effective operations. This emphasis on computational optimization highlights the growing importance of data-driven decision-making in contemporary vertical farming. The upcoming section reviews concrete obstacles—engineering, organizational, and economic—that can limit how far and how quickly these cutting-edge optimization methods can actually be used in the field. Table 2 lists several energy-saving practices for vertical farms, including smart sensors, spectral lighting, and system coupling, clearly showing how each one cuts power use and boosts overall sustainability.

Table 2. Model-Based Energy Optimization in Vertical Farming.

Technique	Description	Energy Impact	Sustainability Impact	Reference
Smart Sensors and AI-based systems	Real-time control of light, temp, humidity, CO_2.	Cuts energy Waste via automation.	Boosts efficiency, lowers emissions.	[44]
Dynamic Lighting control system	Shifts lighting to low-cost energy periods.	Reduces lighting costs and load.	Cuts carbon footprint.	[6]
Municipal Integration	Taps into local waste, CO_2, and surplus energy.	Less reliance on external energy.	Promotes reuse and local synergy.	[11]
Price & Renewable Response technique	Adapts energy use to market rates and green sources.	Optimizes energy timing and source.	Encourages clean energy use.	[3,5,36,53]

Table 2. *Cont.*

Technique	Description	Energy Impact	Sustainability Impact	Reference
Renewable Energy Switching technique	Alternates between solar, wind, and grid power.	Decreases fossil fuel use.	Enhances clean energy adoption.	[35]
Circular Resource Use technique	Recycles hydroponic/aquaponic waste as nutrients.	Lowers resource and energy waste.	Minimizes farm- generated waste.	[35]
Advanced LED technique	Uses high-efficiency LEDs for growth.	Slashes lighting energy use.	Less energy- intensive lighting.	[10]
Biostimulants technique	Enhances plant growth naturally (e.g., Spirulina).	Reduces energy via better growth.	Lowers input needs, boosts efficiency.	[28]
Growth Optimization technique	Simulates ideal growth conditions.	Balances yield with minimal energy.	Limits overuse of water and power.	[29]
P-Graph Optimization technique	Plans most efficient energy/ cost operations.	Identifies optimal energy strategies.	Informs sustainable decision-making.	[11]

3.4. Optimizing Water and Energy Use Through Intelligent Irrigation Systems

Integrating advanced energy-conservation strategies with smart irrigation systems is pivotal for boosting the efficiency of vertical farms. Smart irrigation uses sensors, automation, and IoT to fine-tune both water and power consumption, directly addressing two key hurdles in sustainable agriculture. Central to this setup is the automated water-delivery network, guided by probes that track soil moisture, weather trends, and plant-growth phases. It applies exactly what each crop needs when it needs it, cutting down both over- watering and wasted resource. By curbing excessive pumping, the system also slashes unwanted energy use and trims operating costs [32,36].

Real-time updates to the irrigation schedule, based on live soil and climate data, close the efficiency gap that plagues many conventional approaches. Integration with IoT permits round-the-clock surveillance and self-correcting operation, sparing growers from constant manual checks and lifting accuracy still higher [24,39]. In this way, automation not only streamlines management but also shields the system from human error and guarantees uniform water distribution across the growing cycle. Energy-saving sensors and microcontrollers substantially cut power use, delivering precise data that guides resource decisions without draining the system [44]. A key benefit of smart irrigation lies in its capacity to stop crops from being overwatered, a process that trims the energy needed to pull water from wells, treat it, and move it around, ultimately saving both energy and water while softening many side effects on the environment [23,42,54]. Because these tools keep soil moisture within a narrow, target range they can also cut the number of watering events, letting plants grow steadily while freeing energy for other critical tasks in the farm [22,44]. Taken together, these advantages show that smart irrigation is vital to making vertical farms precise, efficient, and kinder to the planet. The next section therefore turns to the hurdles—hardware costs, limited data-sharing, and knowledge gaps—that still slow the broad rollout of energy-optimizing systems across the industry. To build upon the broad energy optimization strategies outlined in the previous table, it is important to recognize how these overarching approaches are complemented by targeted interventions within specific subsystems. Irrigation stands out as one of the most critical subsystems in vertical farming, where the adoption of smart technologies further enhances overall energy efficiency and sustainability. Table 3 focuses on smart irrigation systems, detailing their unique contributions and illustrating how they function synergistically with other energy optimization techniques to drive sustainable outcomes in vertical farming operations. Table 3 outlines various smart irrigation strategies in vertical farming and their specific effects on energy efficiency and sustainability. Each approach is briefly described, with supporting references, to demonstrate its role in reducing both energy and resource use.

Table 3. Smart Irrigation Strategies for Energy Efficiency and Sustainability in Vertical Farming.

Smart Irrigation Systems	Description	Impact on Energy Efficiency	Impact on Sustainability	References
Sensors	Monitors soil, weather, and plant growth for precise water control.	Cuts water pumping, saving energy.	Conserves water and energy, supporting sustainability.	[30,31,51]
Real-time Schedule Adjustment	Adjusts irrigation based on data (weather, soil moisture).	Reduces pump run time, lowering energy use.	Ensures efficient water use, reducing waste.	[30,31,51]
IoT Integration	Continuous monitoring and auto-adjustments via IoT.	Minimizes manual input, optimizing energy use.	Enhances resource management, supporting sustainable farming.	[12,32,33]
Automation	Automates irrigation, reducing human error.	Cuts energy by ensuring precise application.	Boosts efficiency, reduces errors, and promotes sustainability.	[30,33,49]
Energy-efficient Sensors	Uses low-energy sensors and Microcontrollers for data collection.	Lowers overall energy consumption of the system.	Enables better decision-making, reducing resource use.	[33]
Preventing Over-irrigation	Ensures water is used only when needed, preventing excess.	Reduces energy for extraction, treatment, and distribution.	Minimizes water and energy waste, promoting efficiency.	[22,55,56]
Optimal Soil Moisture	Maintains ideal moisture for healthy crops.	Reduces energy spent on excess watering.	Ensures efficient water and energy use, boosting crop health.	[32,51]
Sustainable Practices	Focuses on conserving water and energy in vertical farms.	Enhances efficiency, cutting energy and water costs.	Supports long-term sustainability with reduced environmental impact.	[30,49]

4. Challenges in Integrating Energy Optimization Approaches into Vertical Farming

Implementing optimized energy strategies in vertical farming presents significant complexity, particularly when these systems are proposed as solutions for food security. Vertical farming is inherently capital-intensive, requiring substantial upfront investment and ongoing operational expenditures [37]. Reliance on artificial lighting and HVAC (heating, ventilation, and air conditioning) systems contributes to high energy consumption and sustainability concerns [19,52]. The high day-to-day expenses tied to vertical farming, including lighting and climate control, sharpen the financial strain and make it harder for operators to break even [15,19,45]. Adding artificial intelligence, the Internet of Things, and automated equipment can improve yields, yet the price of new hardware and the cost of keeping it running usually outweigh immediate savings [26,29,45]. Dealing with these sophisticated setups also demands trained personnel fluent in both agriculture and tech, and the shortage of such specialists slows broader industry growth [26,29]. Adjusting energy use on the fly during price spikes or when wind and sunlight vary is still a tough puzzle for engineers and managers alike [16,27,50]. Although solar panels and wind turbines cut carbon emissions, their fickle output forces growers to rely on back-up power or battery stores, pulling more money into system design [16,27,50]. Because of the heavy power draw, critics maintain that vertical farms—even those using renewables—must take further steps if they want to claim truly sustainable credentials.

Making vertical farming truly sustainable requires that operators fine-tune energy use without sacrificing either crop quantity or quality [18,37,55]. Widespread doubts about the technology, compounded by limited public experience, slow its adoption, underscoring the urgent need for educational campaigns that clearly explain the system's benefits [18]. When officials attempt to scale vertical farms as a way to bolster food security in crowded cities, they quickly run into economic and logistical hurdles, because local differences in energy grids, climate, and wealth shape what is practical. For that reason, small-scale studies based in each community are critical; they provide the localized data needed to craft workable solutions and guide larger rollouts [29].

By addressing these interconnected economic, technical, social, and regional challenges, future research and policy initiatives can pave the way for more sustainable, efficient, and widely adopted vertical farming systems. The concluding section summarizes key findings and presents recommendations for advancing the field. Table 4 highlights major challenges to energy efficiency in vertical farming, including high energy demand, costs, and technical complexity. Each challenge is briefly described, along with its specific impacts and supporting references.

Table 4. Challenges Impacting Energy Efficiency in Vertical Farming.

Challenge	Description	Energy Efficiency	References
High Energy Demand for Lighting and HVACD	Relies on energy-intensive lighting and climate control systems.	High energy use limits efficiency.	[27,29,41,48]
High Energy Costs	Maintaining optimal conditions is costly.	Rising costs hinder energy optimization.	[27,41,42,48]
High Initial Investment in Technology	Upfront costs for AI, IoT, and automation.	Large investment slows energy-efficient tech adoption.	[23,27,39,41]
Need for Specialized Skills	Requires skilled operators for complex systems.	Lack of skills may lead to inefficient energy use.	[23,39]
Complexity of Demand Response Systems	Requires complex systems for real-time data and renewable energy integration.	Complexity may reduce efficiency and optimization.	[5,46,57]
Variability in Renewable Energy Supply	Solar and wind energy supply is variable.	Variability complicates energy management.	[21,46,57]
Environmental Impact Despite Resource Savings	Energy use still impacts despite reduced land and water use.	High energy Consumption affects efficiency.	[42,54,58]
Unfavorable Public Perception	Public skepticism slows adoption compared to traditional agriculture.	Negative views hinder investment in energy- efficient tech.	[58]

5. Discussion

As detailed in Sections 3.1 and 3.2, energy optimization strategies for vertical farming—including high-efficiency LED lighting, advanced HVAC controls, and smart irrigation—present promising opportunities for reducing both costs and resource consumption. However, their effective implementation is often complex and highly context-dependent. Energy optimization is widely acknowledged as fundamental to achieving the operational and financial sustainability of vertical farming systems. For example, adopting high-efficiency LED lighting can reduce lighting energy consumption by 30–50% compared to traditional high-pressure sodium (HPS) lamps, as confirmed by commercial-scale trials and modeling studies [7,13]. Ref. [13] reports that replacing high-pressure sodium lamps with LEDs cuts energy expenses by nearly $35,692 each year and trims maintenance costs by an additional $6580, yielding an overall payback period of roughly two to three years under standard operating circumstances. Despite the higher capital investment required, these savings render LED technology economically viable for many operators, though such upfront costs may remain prohibitive for small-scale or under-capitalized farms.

Integrating smart HVAC controls represents another substantial advance in energy-efficient vertical farming. Modeling and field studies suggest that smart, sensor-driven HVAC systems can lower overall energy use by 15–25%, although actual savings depend on climate, facility insulation, and system configuration [44,59]. Although these systems entail significant initial outlays for automation, sensors, and controls, the resulting operational cost reductions often lead to a payback period of three to five years for mid-sized facilities.

Smart irrigation systems—featuring real-time soil moisture sensors, automated valves, and IoT integration—are increasingly used to optimize water and energy consumption [12,25,36]. Empirical studies show that such systems can reduce unnecessary water pumping by 20–30%, directly lowering energy consumption and operating costs. The cost of sensor deployment is typically recovered within one or two growing seasons, especially in regions where water or energy is scarce.

Despite these technological advancements, contradictions and limitations persist in the literature. Reported energy savings from light-emitting diodes (LEDs) or intelligent heating, ventilation, and air-conditioning (HVAC) controls vary markedly between investigations, principally because crop species, environmental parameters, and facility layouts differ [13,44].

Studies based exclusively on simulation routinely inflate estimated savings relative to those measured in operating farms, underscoring the demand for extended, real-world trials. Moreover, many published analyses omit consistent reporting units—such as kilowatt-hours per kilogram—thereby hindering across-study comparison and meta-analysis. Economic questions persist alongside technical ones. While capital-intensive, advanced energy systems usually yield satisfactory returns in large vertical farms, small-scale growers often find the up-front investment prohibitive and the requisite skill scarce, leaving a research void in affordable, broadly applicable optimization techniques. Integrating renewable photovoltaics and wind can further curtail grid reliance, but their inherent intermittency complicates continuous energy supply. Batteries and demand–response schemes address part of the issue, yet field data on their durability and long-run cost-effectiveness remain sparse, limiting confidence in widespread adoption. In locations where powering the farm still means burning a lot of coal or gas, the green promise of vertical agriculture hangs on solid electricity management; sloppy energy use can wipe out the gains from saving soil and redirecting water. Nearly everyone now agrees that a serious education push and skilled workforce are essential. Running these dense, high-tech farms calls for operators who blend deep technical know-how with sharp management instincts. Tailored apprenticeships, step-by-step knowledge-sharing, and partnerships with trade schools are therefore critical if growers are to wield the new tools wisely. Meanwhile, doubts from the public and wavering investors dog the sector, especially fears that the energy crunch will sink the farms economically and ecologically. Openness in reporting, eye-catching pilot farms, and clear, honest outreach can calm skepticism and pave the way for larger-scale deployment.

6. Future Research Directions

Despite notable advances in energy optimization for vertical farming, several unresolved issues require targeted investigation. High operational costs and variability in reported energy savings—highlighted by studies such as [13,44]—underscore the need for research that extends beyond simulations and controlled environments to include large-scale, real-world deployments. Prioritizing field-based, longitudinal studies is essential to rigorously validate the economic and sustainability claims associated with technologies such as LED retrofitting, demand response strategies, and AI-driven environmental controls. There is also a pressing need for comparative research that evaluates the performance of energy optimization methods across diverse facility types, climate conditions, and crop varieties. Particular attention should be given to small-scale and resource-constrained vertical farms, where high upfront costs and skills gaps may inhibit technology adoption. Understanding the scalability and cost-effectiveness of advanced systems in these settings is critical for ensuring broader sustainability and accessibility. Given the technical challenges inherent in integrating renewable energy and storage solutions—especially the intermittency of solar and wind power—future work should focus on developing system-level solutions and enhancing operational resilience. The creation of standardized, universally accepted metrics for energy efficiency and environmental impact is equally important, as current inconsistencies hinder robust cross-study comparisons. A more comprehensive evaluation of the long-term environmental impacts of vertical farming is warranted, particularly in regions that rely on carbon-intensive energy grids. This includes the need for detailed life-cycle assessments and environmental modeling to capture indirect effects and potential trade-offs. Finally, addressing persistent workforce and educational gaps will require research on effective training programs, curriculum development, and knowledge transfer initiatives. Developing a skilled labor force is essential for fully realizing the potential of advanced energy management in vertical farming.

7. Conclusion

This review has demonstrated that the sustainability and operational success of vertical farming are fundamentally dependent on the strategic implementation of energy optimization technologies. While solutions such as LED lighting, smart HVAC systems, and automated irrigation show substantial promise for reducing energy demand and enhancing efficiency, the realized benefits are highly context-dependent, influenced by factors such as farm scale, local energy infrastructure, and available capital. Persistent inconsistencies in the literature—stemming from differences in study design, reliance on simulations, and the lack of standardized metrics—highlight the need for more robust empirical evidence, particularly in real-world settings. A major limitation of this review is that much of the available research is based on short-term or simulated data rather than large-scale, longitudinal field studies. The diversity in methodologies and reporting practices further complicates the ability to draw direct comparisons or synthesize outcomes across studies. In addition, there remains limited insight into the performance and adoption of energy optimization strategies within small-scale or resource-constrained vertical farms, as well as a lack of comprehensive analyses on the integration of renewable energy and storage technologies. To address these gaps, future research should prioritize rigorous, field-based, and longitudinal studies that evaluate the real-world energy and economic impacts of core optimization technologies. Comparative investigations across farm sizes, crop types, and geographic contexts are recommended, alongside the development of standardized metrics for energy and sustainability reporting. Further, dedicated research is needed to explore practical approaches for integrating renewables, advancing storage solutions, and overcoming the workforce and financial barriers that limit technology adoption—especially for smaller producers.

Funding: This paper was funded by the Ministry of Higher Education, Research and Innovation, Sultanate of Oman, Block Funding Program (BFP) No: BFP/RGP/ICT/24/113.

Institutional Review Board Statement: Not applicable.

Informed Consent Statement: Not applicable.

Data Availability Statement: No new data were created or analyzed in this study. Data sharing is not applicable to this article.

Acknowledgments: The authors gratefully acknowledge the valuable contributions of Mohammed Al-Sarem (College of Computer Science and Engineering, Taibah University, Saudi Arabia; Energy, Industry, and Advanced Technologies Research Center, Taibah University, Saudi Arabia), Fahad Ghabban (College of Computer Science and Engineering, Taibah University, Saudi Arabia; Energy, Industry, and Advanced Technologies Research Center, Taibah University, Saudi Arabia), Ibraheem Shayea (Electronics and Communication Engineering Department, Istanbul Technical University, Türkiye; Department of Intelligent Systems and Cybersecurity, Astana IT University, Kazakhstan), Nasser Tabook (Department of Computer Science, College of Arts and Applied Sciences, Dhofar University, Oman), Rawad Abdulghafor (Faculty of Computer Studies, Arab Open University—Oman), and Mohammad Nassef (Department of Computer Science, Cairo University, Egypt; Department of Computer Science and Artificial Intelligence, University of Jeddah, Saudi Arabia). Their dedicated efforts and collaborative work were instrumental in conducting this research and completing this paper.

Conflicts of Interest: The authors declare no conflict of interest.

References

1. Erekath, S.; Seidlitz, H.; Schreiner, M.; Dreyer, C. Food for future: Exploring cutting-edge technology and practices in vertical farm. *Sustain. Cities Soc.* **2024**, *106*, 105357. [CrossRef]
2. Büyüközkan, G.; Göçer, F.; Uztürk, D. A novel Pythagorean fuzzy set integrated Choquet integral approach for vertical farming technology assessment. *Comput. Ind. Eng.* **2021**, *158*, 107384. [CrossRef]
3. Martin, M.; Weidner, T.; Gullström, C. Estimating the potential of building integration and regional synergies to improve the environmental performance of urban vertical farming. *Front. Sustain. Food Syst.* **2022**, *6*, 849304. [CrossRef]
4. de Jong, A.; Milestad, R.; Bustamante, M.J.; Martin, M. Analyzing the divergence and development of business models for urban farming. *Urban Agric. Reg. Food Syst.* **2024**, *9*, e70004. [CrossRef]
5. Mannan, J.M.; Suguna, S.K.; Dhivya, M.; Parameswaran, T. Smart scheduling on cloud for IoT-based sprinkler irrigation. *Int. J. Pervasive Comput. Commun.* **2020**, *17*, 3–19. [CrossRef]
6. Langendahl, P.-A. The Politics of Smart Farming Expectations in Urban Environments. *Front. Sustain. Cities* **2021**, *3*, 691951. [CrossRef]
7. Ardakani, S.P.; Xie, H.; Liu, X. Smart Technologies for Urban Farming and Green Infrastructure Development: A Taxonomy. In *Urban Sustainability*; Springer: Singapore, 2022; pp. 379–397. [CrossRef]
8. Jamshidi, F.; Ghiasi, M.; Mehrandezh, M.; Wang, Z.; Paranjape, R. Optimizing Energy Consumption in Agricultural Greenhouses: A Smart Energy Management Approach. *Smart Cities* **2024**, *7*, 859–879. [CrossRef]
9. Kaya, C. Intelligent Environmental Control in Plant Factories: Integrating Sensors, Automation, and AI for Optimal Crop Production. *Food Energy Secur.* **2025**, *14*, e70026. [CrossRef]
10. Miserocchi, L.; Franco, A. Benchmarking energy efficiency in vertical farming: Status and prospects. *Therm. Sci. Eng. Prog.* **2025**, *58*, 103165. [CrossRef]
11. Lee, D.; Kim, K. National Investment Framework for Revitalizing the R&D Collaborative Ecosystem of Sustainable Smart Agriculture. *Sustainability* **2022**, *14*, 6452. [CrossRef]
12. Saad, M.H.M.; Hamdan, N.M.; Sarker, M.R. State of the art of urban smart vertical farming automation system: Advanced topologies, issues and recommendations. *Electron.* **2021**, *10*, 1422. [CrossRef]
13. Ágoston, G.; Pongrácz, F.; Horváth, K.G.; Bukodi, Z. Vertical farms and smart cities—Identification of common research areas, Tungsram's experience and vision in Central Europe. In Proceedings of the 2022 Smart City Symposium Prague (SCSP), Prague, Czech Republic, 26–27 May 2022; pp. 1–5. [CrossRef]
14. Augustine, C.; Balaji, K.; Dharanikumar, S.V.; Anand, A.J. Urban farming: Case study. In *Advanced Technologies for Smart Agriculture*; River Publishers: Aalborg, Denmark, 2023; pp. 321–338. Available online: https://www.taylorfrancis.com/books/edit/10.1201/9781032628745/advanced-technologies-smart-agricultâĂę (accessed on 5 June 2025).
15. Ahamed, M.S.; Sultan, M.; Monfet, D.; Rahman, M.S.; Zhang, Y.; Zahid, A.; Bilal, M.; Ahsan, T.M.A.; Achour, Y. A critical review on efficient thermal environment controls in indoor vertical farming. *J. Clean. Prod.* **2023**, *425*, 138923. [CrossRef]
16. Arabzadeh, V.; Miettinen, P.; Kotilainen, T.; Herranen, P.; Karakoc, A.; Kummu, M.; Rautkari, L. Urban vertical farming with a large wind power share and optimised electricity costs. *Appl. Energy* **2023**, *331*, 120416. [CrossRef]
17. Coon, D.; Lindow, L.; Boz, Z.; Martin-Ryals, A.; Zhang, Y.; Correll, M. Reporting and practices of sustainability in controlled environment agriculture: A scoping review. *Envrion. Syst Decis* **2024**, *44*, 301–326. [CrossRef]
18. Daniels, A.; Fink, M.; Leibold, M.; Wollherr, D.; Asseng, S. Optimal Control for Indoor Vertical Farms Based on Crop Growth. *IFAC- Pap.* **2023**, *56*, 9887–9893. [CrossRef]
19. Debdas, S.; Reddy, Y.P.K.; Das, D.; Das, S.; Hazra, S.; Chatterjee, S. Vertical Agriculture in the IoT Era. In Proceedings of the 2023 IEEE 3rd International Conference on Smart Technologies for Power, Energy and Control (STPEC), Bhubaneswar, India, 10–13 December 2023; pp. 1–6. [CrossRef]
20. Jeongsun, C.; Ji-In, C. Proposal for Improving Underground Smart Farm Spaces in Seoul Metro Stations-Focusing on Sangdo, Dapsimni, Cheonwang, Chungjeong-ro, and Euljiro3-ga Station Sangdo-. *Archit. Inst. Korea* **2024**, *40*, 25–33. [CrossRef]
21. Angotti, T. Urban agriculture: Long-term strategy or impossible dream?: Lessons from prospect farm in brooklyn, New York. *Public Health* **2015**, *129*, 336–341. [CrossRef]
22. Soofi, A.F.; Manshadi, S.D. Carbon-Aware Operation of Resilient Vertical Farms in Active Distribution Networks. *IEEE Trans. Smart Grid* **2024**, *15*, 431–443. [CrossRef]

23. Takagaki, M.; Hara, H.; Kozai, T. Micro- and mini-PFALs for improving the quality of life in urban areas. In *Plant Factory*, 2nd ed.; Academic Press: Cambridge, MA, USA, 2019; pp. 117–128. [CrossRef]
24. Awasthi, A.; Rangare, A.; Kaushik, R.; Immanuel, J. Live Demonstration: IoT based smart vertical farming framework with sensor network and mobile application for real-time monitoring. In Proceedings of the 2023 IEEE SENSORS, Vienna, Austria, 29 October–1 November 2023; p. 1. [CrossRef]
25. Devendiran, R.; Turukmane, A.V.; Sathiyaraj, A.; Srinivasa Rao, P.; Prasad, B.; Pulipati, S. Smart Irrigation: Revolutionizing Water Management in Agriculture for Sustainable Practices and Improved Crop Yield. In Proceedings of the 2023 6th International Conference on Recent Trends in Advance Computing (ICRTAC), Chennai, India, 14–15 December 2023; pp. 651–656. [CrossRef]
26. Ghiasi, M.; Wang, Z.; Mehrandezh, M.; Paranjape, R. A Systematic Review of Optimal and Practical Methods in Design, Construction, Control, Energy Management and Operation of Smart Greenhouses. *IEEE Access* **2024**, *12*, 2830–2853. [CrossRef]
27. Suhail, A.; Hasteer, N. Use of smart farming techniques to mitigate water scarcity. In *Artificial Intelligence and IoT-Based Technologies for Sustainable Farming and Smart Agriculture*; IGI-GLOBAL: Hershey, PA, USA, 2021; pp. 140–150.
28. Nwanojuo, M.A.; Anumudu, C.K.; Onyeaka, H. Impact of Controlled Environment Agriculture (CEA) in Nigeria, a Review of the Future of Farming in Africa. *Agriculture* **2025**, *15*, 117. [CrossRef]
29. Pimentel, J.; Friedler, F. Synthesis of Integrated Vertical Farming Systems with Multiperiodic Resource Availability. *Chem. Eng. Trans.* **2022**, *94*, 1039–1044. [CrossRef]
30. Rathore, T.; Gupta, D.K.; Kumar, N. Smart Irrigation system using IoT. In Proceedings of the 2023 Third International Conference on Secure Cyber Computing and Communications (ICSCCC), Jalandhar, India, 26–28 May 2023; pp. 605–610. [CrossRef]
31. Reynolds, K.; Gottfried, C.; Thomas, T. Racial equity and the USDA's Office of Urban Agriculture granting program and urban offices. *J. Agric. Food Syst. Community Dev.* **2024**, *14*, 129–139. [CrossRef]
32. Robbiani, G.; Törn, E. Intermittent Light Scheduling for Energy Cost Reduction in Vertical Farming. *IEEE Technol. Soc. Mag.* **2024**, *43*, 81–90. [CrossRef]
33. Sanjuan-Delmás, D.; Llorach-Massana, P.; Nadal, A.; Ercilla-Montserrat, M.; Muñoz, P.; Montero, J.I.; Josa, A.; Gabarrell, X.; Rieradevall, J. Environmental assessment of an integrated rooftop greenhouse for food production in cities. *J. Clean. Prod.* **2018**, *177*, 326–337. [CrossRef]
34. Wu, W.; Feng, X.; Lu, C. The rise of smart agriculture in China: Current situation and suggestions for further development. *Exp. Agric.* **2024**, *60*, e28. [CrossRef]
35. Milestad, R.; Carlsson-Kanyama, A.; Schaffer, C. The Högdalen urban farm: A real case assessment of sustainability attributes. *Food Secur.* **2020**, *12*, 1461–1475. [CrossRef]
36. Martin, M.; Bustamante, M.J. Growing-Service Systems: New Business Models for Modular Urban-Vertical Farming. *Front. Sustain. Food Syst.* **2021**, *5*, 787281. [CrossRef]
37. Al-Kodmany, K. Mitigating Climate Change: The Potential of Vertical Farming to Feed High-Density Cities. In *The Routledge Handbook on Greening High-Density Cities: Climate, Society, and Health*, 1st ed.; Routledge: Abingdon, UK, 2024; pp. 50–70. [CrossRef]
38. Ali, A.; Hussain, T.; Tantashutikun, N.; Hussain, N.; Cocetta, G. Application of Smart Techniques, Internet of Things and Data Mining for Resource Use Efficient and Sustainable Crop Production. *Agriculture* **2023**, *13*, 397. [CrossRef]
39. Takagaki, M.; Hara, H.; Kozai, T. Micro- and Mini-PFALs for Improving the Quality of Life in Urban Areas. In *Plant Factory: An Indoor Vertical Farming System for Efficient Quality Food Production*; Academic Press: Cambridge, MA, USA, 2015; pp. 91–104. [CrossRef]
40. Sowmya, C.; Anand, M.; Indu Rani, C.; Amuthaselvi, G.; Janaki, P. Recent developments and inventive approaches in vertical farming. *Front. Sustain. Food Syst.* **2024**, *8*, 1400787. [CrossRef]
41. Stanghellini, C.; Katzin, D. The dark side of lighting: A critical analysis of vertical farms' environmental impact. *J. Clean. Prod.* **2024**, *458*, 142359. [CrossRef]
42. Swetha, K.R.; Akash, K.R.; Beerendra, P.N.M.; Manoj, M.R.; Prajwal, B.N.; Thejaswini, H.B. An Automated Irrigation System for Agriculture Using IoT. In Proceedings of the 2023 International Conference on Applied Intelligence and Sustainable Computing (ICAISC), Dharwad, India, 16–17 June 2023; pp. 1–9. [CrossRef]
43. Brinks, H.; Kool, S.D. Farming with future: Implementation of sustainable agriculture through a network of stakeholders. In *Changing European Farming Systems for A Better Future: New Visions for Rural Areas*; Wageningen Academic: Wageningen, The Netherlands, 2006; pp. 299–303. [CrossRef]
44. Diehl, J.A.; Sweeney, E.; Wong, B.; Sia, C.S.; Yao, H.; Prabhudesai, M. Feeding cities: Singapore's approach to land use planning for urban agriculture. *Glob. Food Secur.* **2020**, *26*, 100377. [CrossRef]
45. Despommier, D. Vertical farms, building a viable indoor farming model for cities', Field Actions Science Reports. *J. Field Actions* **2019**, *Special Issue 20*, 68–73.
46. Kim, S.-J.; Yoe, H. Trend and Standardization of Smart Farm Technology. *Korean Inst. Commun. Inf. Sci.* **2022**, *47*, 1965–1973. [CrossRef]

47. Broad, G.M. Know Your Indoor Farmer: Square Roots, Techno-Local Food, and Transparency as Publicity. *Am. Behav. Sci.* **2020**, *64*, 1588–1606. [CrossRef]
48. Sumalatha, K.; Harshdeep, K. Smart Monitoring and Irrigation Regulation via IoT and Cloud. In Proceedings of the 2024 8th International Conference on I-SMAC (IoT in Social, Mobile, Analytics and Cloud)(I-SMAC), Kirtipur, Nepal, 3–5 October 2024; pp. 88–94. [CrossRef]
49. Sashika, M.A.N.; Gammanpila, H.W.; Priyadarshani, S.V.G.N. Exploring the evolving landscape: Urban horticulture cropping systems–trends and challenges. *Sci. Hortic.* **2024**, *327*, 112870. [CrossRef]
50. Kim, D.-I.; Kim, H.-J. A Study on Integrating Social Infrastructure through Program Convergence. *Archit. Inst. Korea* **2024**, *40*, 91–100. [CrossRef]
51. Popkova, E.G. Model of Uninterruptible and Highly Efficient Operation of Vertical Farms Based on Alternative Energy in the Interests of Sustainable Agriculture. In *Geo-Economy of the Future: Sustainable Agriculture and Alternative Energy: Volume II*; Springer: Cham, Switzerland, 2022; pp. 879–884. [CrossRef]
52. Debdas, S.; Chatterjee, S.; Das, S.; Das, D.; Hazra, S.; Shah, P.B. IoT Edge Based Vertical Farming. In Proceedings of the 2023 World Conference on Communication & Computing (WCONF), RAIPUR, India, 14–16 July 2023; pp. 1–5. [CrossRef]
53. Xie, S.; Martinez-Vazquez, P.; Baniotopoulos, C. Wind Aerodynamics and Related Energy Potential of Urban High-Rise Vertical Farms. In *Proceedings of the 4th International Conference "Coordinating Engineering for Sustainability and Resilience" & Midterm Conference of CircularB "Implementation of Circular Economy in the Built Environment"*; CESARE 2024. Lecture Notes in Civil Engineering; Springer: Cham, Switzerland, 2024; Volume 489, pp. 187–196. [CrossRef]
54. Wijnands, F.G.; Brinks, H.; Schoorlemmer, H.; de Bie, J. Integrated Pest Management Adoption in the Netherlands: Experiences with Pilot Farm Networks and Stakeholder Participation. In *Integrated Pest Management*; Springer: Dordrecht, The Netherlands, 2014. [CrossRef]
55. Singh, J.; Neeraj; Srinivas Reddy, P.; Hasan, D.S.; Alsahlanee, A.T.R.; Shaik, N. IoT-Enabled Automated Smart Irrigation System Incorporating Sensor Parameters. In *Proceedings of the Fifth Doctoral Symposium on Computational Intelligence*; DoSCI 2024; Lecture Notes in Networks and Systems; Springer: Singapore, 2024; Volume 1095, pp. 33–43. [CrossRef]
56. Sivamani, S.; Bae, N.; Cho, Y. A smart service model based on ubiquitous sensor networks using vertical farm ontology. *Int. J. Distrib. Sens. Netw.* **2013**, *9*, 161495. [CrossRef]
57. Tariq, B.; Amina, B.; Touhami, A.; Khelifa, B.; Chaimaa, S. Greenhouse Irrigation System Based IoT. In Proceedings of the 2024 4th International Conference on Embedded & Distributed Systems (EDiS), BECHAR, Algeria, 3–5 November 2024; pp. 285–290. [CrossRef]
58. Varia, J.; Kamaleson, C.; Lerer, L. Biostimulation with phycocyanin-rich Spirulina extract in hydroponic vertical farming. *Sci. Hortic.* **2022**, *299*, 111042. [CrossRef]
59. Chowdhury, H.; Asiabanpour, B. A circular economy integration approach into vertical farming with computer-based simulation model for resource optimization and waste reduction. *J. Clean. Prod.* **2024**, *470*, 143256. [CrossRef]

sustainability

Review

Integration of Plant Electrophysiology and Time-Lapse Video Analysis via Artificial Intelligence for the Advancement of Precision Agriculture

Maria Stolarz

Department of Plant Physiology and Biophysics, Institute of Biological Sciences, Maria Curie-Skłodowska University, Akademicka 19, 20-033 Lublin, Poland; maria.stolarz@mail.umcs.pl

Abstract: Biological research and agriculture are increasingly benefiting from the use of artificial intelligence algorithms, which are becoming integral to various areas of human activity. Fundamental knowledge of the mechanisms of plant germination, growth/development, and reproduction is the basis for plant cultivation. Plants provide food and valuable biochemicals and are an important element of a sustainable natural environment. An interdisciplinary approach involving basic science (biology and informatics), technology (artificial intelligence), and farming practice can contribute to the development of precision agriculture, which in turn increases crop and food production. Nowadays, a progressive elucidation of the mechanisms of plant growth/development involves studies of interrelations between electrical phenomena occurring inside plants and movements of plant organs. Recently, there have been increasing numbers of reports on methods for classifying plant electrograms using statistical and artificial intelligence algorithms. Artificial intelligence procedures can identify diverse electrical signals—signatures associated with specific environmental abiotic and biotic factors or stresses. At the same time, a growing body of research shows methods of precise and fast analysis of time-lapse videos via automated image analysis and artificial intelligence to study the movement and growth/development of plants. In both research fields, scientists introduce modern and promising methods of studying plant growth/development. Such basic research along with technological innovations will contribute to the development of precision agriculture and an increase in yields and production of healthier food in future.

Keywords: artificial intelligence; electrome; electrical signals; electrodes; electrophysiology; machine learning; phenomics; plant electrograms; plant movements; precision agriculture; time-lapse video

1. Introduction

Plants are part of the natural environment, from where they receive and process nutrient elements and stimuli and as a result show induced and spontaneous activity that changes the environment. Physiological processes occurring in plants interact with each other. The subject of my research and the following review are processes related to plant growth/development related to movements of organs and electrical changes (Figure 1). These processes exhibit a mutual relationship, which is currently being studied in greater detail. Plant movements also inspire modern architectural designs that support environmental sustainability [1]. Studying the influence of environmental factors on plant growth and development is essential for considerations of a sustainable natural and agricultural environment [2–5].

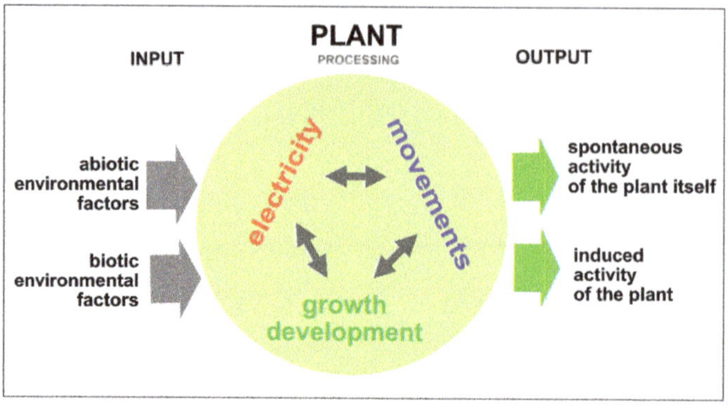

Figure 1. Schematic representation of the research field that includes investigations of the interrelationships between electricity, movements, and growth/development in plant. The influence of abiotic and biotic environmental factors and the induced and spontaneous activity of the plant are also taken into account.

1.1. Plant Electricity

Plant electrophysiology explores the role of electricity in plant physiology, including organ movements and growth/development processes. Electrical phenomena occurring in living organisms are generated by the presence of ions and electrons and are related to the properties of the cell membrane and photosynthetic and respiratory electron carrier chains. Changes in the trans-membrane electrical potential are essential in triggering the motor activity in animals, i.e., they exhibit stimulus-induced and spontaneous electrical activity. Electro-motive phenomena in animals rely on essential structural elements: the nervous system and the muscular system. Currently, comparative analyses of animal and plant behaviour are being carried out, and some processes that are important in animal and human biology are being studied in plants [6–12].

1.2. Plant Organ Movements

Organ movements, ubiquitous in growing plants, are studied at the level of cells, tissues, organs, and the whole plant. The whole plant exhibits rapid and slow endogenously and exogenously driven movements of organs. Endogenous ultradian nutations and circadian-regulated movements are a background for various exogenously induced tropisms and nastic movements. The time scale of plant movements ranges from milliseconds to hours. Plants can twine, bend towards and against stimuli, pull leaves down and up, and track the sun. The common classification of plant movement distinguishes tropisms, nastic movements, and nutations [13]. The motor behaviour is a collection of all types of movements performed by the plant. In a single plant, movements of various organs appear simultaneously and their combination results in its complex motor behaviour. Charles Darwin tried to explain the origin of organ movement in terms of the evolution of behaviour. He explained how plants achieved the ability to climb in his famous work "The Power of Movement in Plants " [14]. Nowadays, plant movements are an inspiration for biomimetic and biomechanic investigations, search for actuators, and robotic studies [15,16]. In animal studies, the connections between electrical and motor activity are obvious. In this work, we present both these activities in plants. A broad collection of videos of plant movements, especially nutations, and examples of electrical activity are available on my website: http://circumnutation.umcs.lublin.pl, accessed on 6 April 2025.

1.3. Method

This review was written based on literature searched in Google Scholar using a combination of such keywords as artificial intelligence, electrome, electrical signals, electrodes, electrophysiology, machine learning, phenomics, plant electrograms, plant movements, precision agriculture, and time-lapse video. Logical operators (and, or) were used to narrow down the results. The work focuses on articles published between 2000 and 2025, with an emphasis on the latest research from the last 5 years (2020–2025) to reflect the current state of knowledge and technological advances. Older primary publications by Charles Darwin (1880) [14] and Burdon-Sanderson and John Scott (1873) [17] and two reviews from 1975 and 1976 are also included. This review includes 68 literature items. Papers on image analysis of only single biological images were excluded. The focus was placed on time-lapse video sequences of plant movements analysed using special software and AI. Papers on general changes in electric potential and electrical changes related to microelectrode measurements were also excluded. The latest works from extracellular research with additional AI analysis in plants were the main focus. In this review, the possibility of systematic errors, which may affect the presented results, should be taken into account. The systematic errors that may have occurred in this work include publication bias, selection bias, and methodological heterogeneity of the studies, which makes direct comparisons difficult. The publication bias may result from the greater frequency of publication of studies with positive results. The selection bias is related to the limitation of searching for publications in English and using Google Scholar. Additionally, interpretation errors related to the assessment of the quality of the studies and their results cannot be excluded.

The aim of this paper is to show electrical and motor phenomena in plants, which are currently being analysed anew by modern statistics and artificial intelligence methods. New methods of recording and analysing electrical and motor changes in plants will be presented on the basis of the literature review. The potential application of these methods in the development of precision agriculture will be considered.

The paper presents an interdisciplinary approach, highlighting the contribution of research on plant electrophysiology and movement to fundamental plant sciences. These basic sciences, combined with technological and applied research, support the development of precision agriculture (Figure 2). Figure 2 illustrates converging interdisciplinary pathways that advance precision farming and promote environmental sustainability. The role of studies on plant electricity and movement in enhancing our understanding of plant physiology and biology is particularly noteworthy. This foundational knowledge may, in turn, find future applications in precision agriculture.

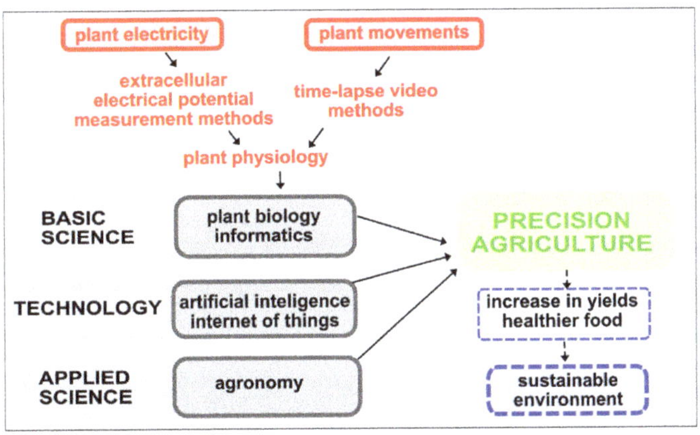

Figure 2. Interdisciplinary convergent pathways for development of precision agriculture leading to environment sustainability. Contribution of research on plant electricity and movements to plant physiology and basic science-biology (red colour).

2. Artificial Intelligence Procedures in Plant Phenomics

Artificial intelligence procedures increasingly support the plant biology research [18,19]. Studies at the level of the whole organism, i.e., the phenome, are currently much less developed than studies on the genome, proteome, or transcriptome. The studies described below mainly concern research on whole plants. Both the extracellular method of measuring electrical voltage and the time-lapse video technique are methods describing the plant phenome, contributing to the development of methods of plant phenotyping.

Subtle patterns of plant movements and electrical signals/signatures that are not visible in traditional statistical analyses can be detected by machine learning algorithms. This approach opens up new perspectives in the study of plant movements and behaviour [19]. To study the kinematic behaviour of plants, the following algorithms can be used: Unsupervised Learning (Anomaly Detection, Isolation Forest, One-Class Support Vector Machine, k-nearest neighbours), Supervised Learning (Decision Trees, random forest, support vector machines, Logistic regression), Ensemble Learning (Bagging, Boosting), and deep learning [19]. Similarly, various artificial intelligence algorithms are used in analyses of plant electrograms [20].

3. Plant Electrograms and Electrome Concept

Plant electrical activity is a set of all types of spontaneous and induced transmembrane potential changes, having different kinetics and occurring in different plant organs and tissues. The types of electrical changes include resting potential, action potential (AP), stimulus-induced long-lasting depolarisation and hyperpolarisation, and ultradian and circadian oscillations. All these changes in electrical activity, usually measured separately, create a system of changes in the electrical potential in the plant. It is an integral part of metabolic processes, in which flows of electric charges measured by electrodes take place. This system of diverse and dynamic distribution of electrical potential in the plant may perform signalling and coordination functions, similarly to animal systems. Recently, the concept of the electrome has appeared in reference to such concepts in biology as the proteome, transcriptome, and genome. The name electrome was proposed by De Loof (2016), and the plant electrome concept is currently being developed [21]. The set of ions, ion channels, and pumps located in the cell membrane determines the ionic flows taking place both in the cells and at the level of the entire organism, creating different levels of the electrome [21]. The electrome refers to the total electrical activity in plants, which plays a key role in their physiology, signalling, and adaptive capacity [20]. The dynamic development of the electrome concept is evidenced by the work addressed to young people [22], which is intended to emphasise its importance in plant life.

Mathematical modelling is also used in analyses of some electrophysiological data. An example is the mathematical model derived from the Hodgkin–Huxley equation for predicting the extracellular potential based on intracellular activity [23]. Such modelling allows predicting the value of extracellular signals. It was determined that extracellular signals are proportional to the transmembrane current and depend on the distance between cells and the resistance of the environment [23]. Mathematical modelling is also used to investigate changes in the AP response to stimuli. This work has been performed in *Nitellopsis* cells using salinity stimuli and modelling using the Thiel–Beilby model, which has previously been applied to *Chara australis* [24–26]. Most researchers have used the extracellular measurement method. Importantly, extracellular measurements reflect the total electrical activity of many cells rather than the exact intracellular potential [23]. Recent work using metal electrodes and extracellular measurements applies AI analyses to extracellular recordings, i.e., plant electrograms.

4. Effects of Environmental Factors and Stresses on Plant Electricity

Abiotic and biotic environmental factors affect the electrical state of the plant by inducing characteristic electrical signals (Figure 3, Table 1). Below are the types of environmental factors, discussed in the recent literature, that affect the electrical state of the plant and are detected by statistical and AI methods.

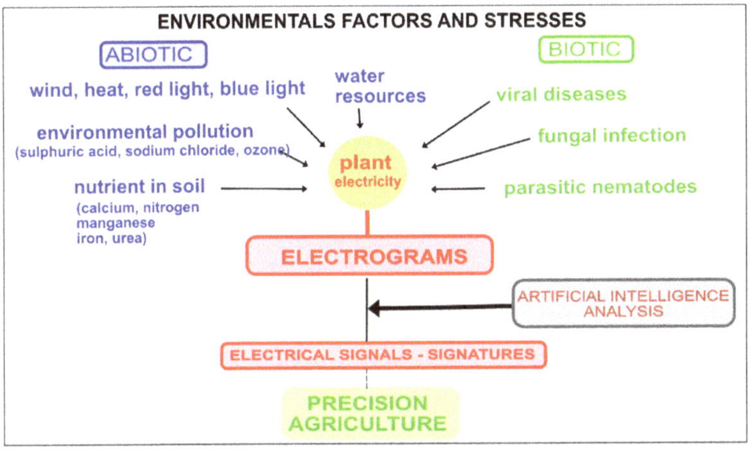

Figure 3. Environmental factors influencing plant electricity. Electrical signals/signatures identified via artificial intelligence analysis can be potentially introduced into precision agriculture in future (based on current literature shown in Table 1).

4.1. Abiotic Factors and Stresses

Deep learning will play an increasingly important role in development of methods of precision fertilisation in agriculture. Work is already underway to use AI to analyse soil and plant nutrients to optimise plant growth [27].

Plant yield and health are largely dependent on the availability of nutrients in the nutrient solution [28]. To date, the assessment of the plant's nutrient supply has been based on the visual observation of the plant. Recently, studies have been conducted to assess whether electrical signals from plants can indicate specific nutrient deficiencies. An attempt was made to use electrophysiological signals to detect calcium (Ca), nitrogen (N), manganese (Mn), and iron (Fe) deficiencies in tomato plants [28]. The analysis was based on signal decomposition using empirical mode decomposition and statistical feature extraction. The analysis of electrical signal patterns showed their specificity in relation to a specific deficiency. The development of this method may support real-time diagnosis of deficiencies as well as effective and precise intervention in the future agriculture [28].

Another paper presents studies in which the growth of peppers was monitored with different urea applications in the soil (low and high urea fertilisation) in relation to induced electrical signals [29]. A relationship was demonstrated between the electrical activity of the pepper plants and the degree of fertilisation and pepper growth. The advantages of the method were indicated as non-invasive, and the fertilisation status was presented in real time [29].

Electrical signals were also studied in tomato and cabbage plants after exposure to such chemicals as sulphuric acid (H_2SO_4), sodium chloride (NaCl), and ozone (O_3) [30]. An analysis of fifteen statistical features of the electrical signal was performed using eight types of algorithms in statistics and AI. The research aimed to develop more precise agriculture and check the level of environmental pollution based on electrical signalling in plants [30].

Table 1. Environmental factors, electrogram parameters, and statistics/AI analysis used in investigation of plant electricity.

Plant	Environmental Factors	Electrogram Parameters	Statistics/ AI Analysis	Literature (Year)
		Abiotic factors		
tomato	calcium (Ca), nitrogen (N), manganese (Mn), and iron (Fe) deficiencies	silver-coated copper wire electrodes continuous 3 weeks, 500 Hz	signal decomposition sample space reduction feature extraction	[28] (2022)
peppers	low and high urea fertilization	three stainless steel needle electrodes stem electrical resistance many days greenhouse and field experiment	average value of triplicates one-way ANOVA Duncan's multi-range test principal component analysis (PCA)	[29] (2023)
tomato cabbage	exposure to chemicals such as sulfuric acid (H_2SO_4), sodium chloride (NaCl) and ozone (O_3)	two stainless steel needle electrodes laboratory, Faraday cage 10 Hz	fifteen statistical features eight different classification algorithms, PCA	[30] (2023)
Hedera helix	ozone exposure	laboratory, Faraday cage 300 Hz	generic automatic toolchain machine learning models	[31] (2024)
Zamioculcas zamiifolia *Solanum lycopersicum* (tomato)	wind, heat, red light blue light	electrical potential and impedance many minutes lasting measures laboratory, 0.58 Hz	discriminant analysis deep-learning methods	[32] (2023)
grapevine	water status	two silver-plated needle electrodes many days climate chambers, 256 Hz.	two machine learning approaches based on classification and regression the prediction models	[33] (2024)
bean	water status	needle electrodes two hours measurements Faraday's cage 62.5 Hz.	arithmetic average of voltage variation, skewness, kurtosis, probability density function (PDF), autocorrelation, power spectral density (PSD), approximate entropy (ApEn), fast Fourier transform (FFT), and multiscale approximate entropy (ApEn(s), machine learning (ML)	[34] (2024)
Clivia	water gradient	patch electrodes Faraday's cage in the thermostatic and humidified incubator 60 min measurements 30 sec samples, 30 Hz	lightweight convolutional neural network (CNN) model (PlantNet)	[35] (2024)
		Biotic factors		
barley (*Hordeum vulgare*)	fungal infection (*Blumeria graminis, Bipolaris sorokiniana*)	pair of needle electrodes Faraday cage, 48 h measurements	descriptive statistics, Bayesian change point (BCP) analysis, ApEn, autocorrelation, ML (cluster analysis)	[36] (2023)
tobacco	viral diseases (alfalfa mosaic virus)	two fine needle electrodes field conditions, 8 s at a sampling rate of 250 Hz	median, autoregressive coefficients, autocorrelation—ML models (support vector machine, k-nearest neighbours, random forest)	[37] (2024)
tomato	parasitic nematodes	many days measurements	ML model	[38] (2024)

Ozone exposure was studied in *Hedera helix* plants, where the electrophysiological response was measured and analysed using machine learning methods [31]. The plant was used here as a sensor of ozone in the air. The tool chain presented in this work automates the development of algorithms for monitoring plants as air quality sensors [31]. The authors believe that a network of such sensor plants could monitor air quality in sustainable cities in the future [31].

Further studies were performed on *Zamioculcas zamiifolia* and *Solanum lycopersicum* (tomato) plants exposed to four different stimuli: wind, heat, red light, and blue light. The electric potential and impedance signals of the tissues were measured and analysed with a wide range of methods from discriminant statistical analysis to deep learning [32].

Another important problem for farmers worldwide is the real-time assessment of water status in grapevine plants [33]. This study proposed the use of plant electrophysiology as a new approach to assess water status. Experiments were conducted in a climatic chamber with potted grapevines under different irrigation regimes. Various morphological and physiological assessments were performed in parallel with electrophysiological measurements to correlate classical water status assessment methods with plant electrophysiological signals [33]. A binary classification model was used, and a regression model that effectively distinguishes between weak and strong irrigation signals was developed. These results therefore represent a promising new tool for future real-time monitoring and remote irrigation of grapevines. In the future, the development of this method will improve irrigation management and agronomic strategies [33].

In modern sustainable agriculture, saving water in crop planning is extremely important. Optimisation models balancing irrigation needs equipped with soil moisture and plant hydration sensors are currently being developed [27].

The study of water availability and accompanying electrical signals was the subject of experiments on common beans [34]. Extensive statistical methods and machine learning techniques were also used to classify electrical signals. This work indicated the possibility of using the electrome as a physiological indicator of water status in bean plants [34].

Another way to monitor water use by a plant is through the use of Clivia to determine the water gradient in the environment [35]. In this study, specific electrical signals in Clivia were documented under different soil moisture gradients. A lightweight convolutional neural network model was used for the analysis. Such studies demonstrate the potential of plants to be used as environmental sensors [35].

4.2. Biotic Stresses

In barley (*Hordeum vulgare*), bioelectric signals are produced as a response to fungal infection before visible disease symptoms appear [36]. Two fungal pathogens were studied: *Blumeria graminis*—a biotrophic fungus causing powdery mildew—and *Bipolaris sorokiniana*—a hemibiotrophic fungus causing brown leaf spot. The signals obtained after infection with both pathogenic fungi differed, giving a specific electrome signature. *Blumeria graminis* caused a decrease in entropy (more orderly signals), while *B. sorokiniana* increased entropy (more chaotic signals) [36]. In the study of barley signals, machine learning was applied to detect and classify bioelectrical changes in this plant. The combination of time series analysis (Bayesian Change Point, approximate entropy, autocorrelation), and AI clustering (K-means, K-modes, Balanced Iterative Reducing, and Clustering using Hierarchies) helped to create algorithms for early detection of fungal infection [36]. Interestingly, these algorithms can detect signals of infection within minutes of contact with the pathogen.

A very important problem in plant productivity today is posed by viral plant diseases [37]. It seems necessary to develop fast and non-invasive methods for early detection of virus infection to mitigate losses and limit the spread of viruses. Alfalfa mosaic virus infection of tobacco plants and accompanying electrical signatures were the subject of the latest research [37]. Electrical signals in uninfected and infected plants were examined. Precise techniques for feature selection (median, autoregressive coefficients, and autocorrelation) and machine learning models, including support vector machine, k-nearest neighbours, and random forest, were selected [37]. Computational techniques facilitated extraction of electrical signals indicating virus infection even before visual symptoms appeared. Moreover, an easy-to-use tool was built which, after improvement, will contribute to fast and inexpensive diagnosis of virus infection in the future, enabling timely interventions in the crop [37].

Another problem in agriculture is caused by parasitic nematodes that invisibly parasitize in the soil [38]. Analysis of electrical signals generated by uninfected, infected, and timely treated tomato plants was carried out. It was shown that the analysis of electrical signals effectively indicates the condition of plants in real time and facilitates therapeutic intervention [38].

As shown above, studying the electrical signature allows detection of early-stage infections by fungi [36], viruses [37], and parasitic nematodes [38]. In the future, this may help to reduce the use of chemicals, such as fungicides and nematicides, for pest control. Therefore, understanding the plant electrome may aid in making adequate decisions about the management of chemicals and support the production of healthier food in future.

5. Software for Time-Lapse Video Analysis for Investigation of Organ Movements in Plants

The studies on plant electricity presented above show that the phenomenon is related to plant responses to various environmental stimuli. In animals, electrical activity regulates many physiological processes and is inextricably linked to the induction and modulation of motor activity. Comparative analyses between the electrical activity of plants and their ability to move have been conducted for many years [7,8]. The best studied plants in this area are model plants *Dionaea muscipula* [17,39–41] and *Mimosa pudica* [42,43], which exhibit easily observable rapid movements of the trap and complex leaf. Recently, thanks to the rapidly developing techniques of recording and analysing digital video images, new possibilities have emerged in studying the biology of such slow processes as plant movements and growth/development (Table 2).

The brilliantly developing time-lapse video recording technique is currently at the forefront of research into the movements of plant organs and video imaging of growth and development. Digital video image processing techniques contribute to the objectivity, repeatability, and acceleration of scientific research [44]. The basic steps of the bioimage analysis workflow include preprocessing, segmentation, feature extraction, object tracking, and classification—these algorithms use machine learning and deep learning to streamline the operations [44].

Investigations of plant movement and growth/development start with the study of cell biology and video imaging [45–48]. Then, increasingly complex physiological processes can be imaged including organ (leaves, flower stem, roots) movement and growth/development [19]. The time-lapse video method, which records and visualises plant slow movements and growth, is invaluable in these studies. Specifically, there is a close link between elongation growth and slow, endogenous, rhythmic movements called nutations [49,50].

The slow nutational movements of various plant organs are being studied increasingly often. One of the first software applications for analysing nutational movements was the *Circumnutation Tracker* built by our team [51]. The software was built for semi-automatic analysis of standardised parameters of circumnutation movements. It was tested mainly on sunflower plants (*Helianthus annuus*) but has applications for analysis of the movements of various organs in different plant species [51]. In previous studies, rhythmic leaf movement and growth in *Arabidopsis thaliana*, *Petunia hybrida*, and *Solanum lycopersicum* plants were analysed by an inexpensive infrared vision system called OSCILLATOR [52]. Characteristic patterns of movement and growth were determined for day/night conditions [52].

Table 2. Software for time-lapse video analysis for investigation of organ movements in plants.

Software Name	Plant	Types of Movement	Literature (Year)
OSCILLATOR	*Arabidopsis thaliana* *Petunia hybrida* *Solanum lycopersicum*	rhythmic leaf movements and growth	[52] (2012)
Circumnutation Tracker, https://circumnutation.umcs.lublin.pl/ct, accessed on 6 April 2025	*Helianthus annuus* *various plant species*	analysis of the movements of various organs	[51] (2014)
3D stereovision machine system	beans	nutation movements of climbing plants	[53] (2024)
PALMA (Plant Leaf Movement Analyzer), https://sourceforge.net/projects/palma-leafmov/ accessed on 6 April 2025	*Arabidopsis thaliana*	periodic movements of leaves	[54] (2017)
TRiP (Tracking Rhythms in Plants), http://github.com/KTgreenham/TRiP, accessed on 6 April 2025	*Arabidopsis thaliana* *Brassica rapa* *Glycine max* *Cleome violacea* *Solanum lycopersicum* *Mimulus guttatus*	whole-plant images and periodic movements of cotyledons and leaves	[55] (2015)
Oskam et al. system, https://github.com/Pierik-Lab, accessed on 6 April 2025	*Arabidopsis thaliana*	nastic movements and growth of leaf as a part of the shade avoidance response	[56] (2024)
Rehman et al. system	*Arabidopsis thaliana* mutants	movements of plant leaves	[57] (2020)
Mao et al. tracker	*Arabidopsis thaliana*	circumnutation of flowering shoot apex	[58] (2023)
SLEAP (Social LEAP Estimates Animal Poses), https://zenodo.org/record/5764169#.YbCK0_FBxqt https://doi.org/10.5281/zenodo.5764169, accessed on 6 April 2025	*Arabidopsis thaliana* sunflower bean	circumnutations, tropisms, twining	[59] (2022)
Gibbs et al. system	*Triticum aestivum*	wind-induced plant movement in field-grown wheat	[60] (2019)

There are also studies of circumnutational movements of beans using stereovision [53]. The movement of the shoot tip was tracked in a 3D space, and reference was made to the support structures and their influence on the dynamics of the plant movement [53].

Studies of the rate of growth, development, and periodic endogenous movements were undertaken in the model plant *Arabidopsis thaliana*. To study the periodic movements of *Arabidopsis thaliana* leaves, the video system developed works on the basis of time-lapse photography and using fast Fourier transformation and nonlinear least squares fitting [54]. This system (PALMA, Plant Leaf Movement Analyzer) can effectively and automatically capture changes in the environment, such as iron deficiency, by detecting the extension of the period of leaf movement rhythm [54].

Also for *Arabidopsis thaliana* leaf movements, a new method for estimating the biological period was developed using a motion estimation algorithm that can be applied to whole-plant images (TRiP, Tracking Rhythms in Plants) [55]. The new system efficiently tracks the movement of cotyledons and leaves without having to select individual leaves for analysis [55]. The new method was also used to estimate the period of movement for five different plant species, underlining its broad applicability [55].

It should be emphasised that systems for video imaging of growth and plant organ movements are relatively low-cost in relation to the amount of information they can provide [61]. Such a system was proposed by Oskam, Snoek, et al. [56] to analyse leaf movements and elongation in relation to lighting conditions and especially the influence of infrared light. These movements, called nastic movements, are part of the shade avoidance response and are important in the processes of adaptation to changing light conditions and regulation of leaf temperature and hydration [56]. These factors are extremely important for plant cultivation, which is why this type of research is important for future crop cultivation.

The effects of external (e.g., abiotic stress) and/or internal (e.g., gene mutation) disturbances on plant growth can be approximated by analysing the movements of plant leaves [57]. A dense optical flow algorithm was used to measure movement directly rather than detecting the leaf tip or cotyledon in each image. The authors [57] tested wild-type and drought-tolerant *Arabidopsis thaliana* mutants and applied two water levels and two nitrogen levels. The video method effectively distinguished environmental and genotypic differences in the plant response. These studies attempt to compare video movement patterns with real environmental conditions and indicate that video monitoring is suitable for plant phenotyping [57].

Also, a system for studying *Arabidopsis* seedling circumnutation has been recently developed [58]. A deep learning-based model was proposed to track the circumnutation of the flowering shoot apex in *Arabidopsis thaliana* from time-lapse videos. U-Net was used for vertex segmentation and combined with a model updating mechanism and pre- and post-processing steps, which significantly improved the efficiency and correctness of the analysis system [58].

The rapidly developing procedures of convolutional neural networks used to estimate the position of a human or animal are being introduced to estimate the dynamics of plant growth and movement (SLEAP, Social LEAP Estimates Animal Poses) [59]. These procedures are particularly accurate in tracking lateral views of shoots and roots of plants. These algorithms open up new possibilities for fast research focusing on plant dynamics [59].

Recently, work has emerged demonstrating wind-induced plant movement as an important factor in most agricultural plant crops [60,62]. A strong method for characterising movement in field-grown wheat (*Triticum aestivum*) plants is presented based on time-ordered image sequences and training a convolutional neural network. The authors [60] present an automated data extraction that can be used to inform lodging models, breeding programs, and link movement characteristics to canopy light distribution and dynamic light fluctuations [60].

Currently, algorithms are also used to track the velocity of plant root growth using high-resolution microscopic image sequences [63]. An observation and analysis system called RTip is helpful in tracking the root tip with transient perturbations and is used for plant phenotyping [63]. The RTip tracker is an example of the possibility of video tracking of plant root growth in laboratory conditions. The topic of video tracking of root growth is not addressed in the present review.

6. Discussion

6.1. Electrical Signals–Signatures

6.1.1. Abiotic Factors

Studying the plant electrograms and thus electrical signature (Table 1) that accompanies water status changes [33–35] and nutrient deficiencies [28] may help in the future to save water resources and nutrient and mineral fertilisers that are necessary during plant growth. Research on electrical signals may also support the assessment of the perception of environmental stimuli such as wind, heat, light, and pollution [30–32]. Particularly promising is the role of whole plants as biosensors and, together with the electrical signatures they produce, whose specificity is classified using statistical and AI methods [31,35], they may support the maintenance of a sustainable environment in the future.

6.1.2. Biotic Factors

As shown above (Table 1), studying the electrical signature may contribute to detection of early-stage infections by fungi [36], viruses [37], and parasitic nematodes [38]. In the future, this may help to reduce the use of chemicals, such as fungicides and nematicides, for pest control. Therefore, understanding the plant electrome may aid in making adequate decisions about the management of chemicals and support the production of healthier food in future.

6.1.3. Statistic and AI Analysis for Studying Plant Electrograms

In this work, Table 1 presents the various methods currently used to analyse plant electrograms. Analysis of these methods allowed to distinguish the key steps currently used in research. They are summarized in Table 3 and can be taken into account in the design of further studies.

Table 3. Key steps in plant electrograms analysis (on the basis Table 1).

Key Steps	Statistics/AI Analysis
data preparation and preprocessing	average value sample space reduction data
feature extraction and signal processing	signal decomposition
	statistical features (arithmetic average, skewness, kurtosis, probability density function (PDF), autocorrelation, power spectral density (PSD))
	transforms and entropies (fast Fourier transform (FFT), approximate entropy (ApEn), multiscale approximate entropy (ApEn(s))
statistical analysis of variation and differences	descriptive statistics, median, one-way ANOVA, Duncan's multi-range test, Bayesian change point (BCP) analysis
multivariate analysis	principal component analysis (PCA)
machine learning methods (ML)	classification and regression (support vector machine (SVM), k-nearest neighbours (KNN), random forest (RF), discriminant analysis, different classification algorithms)
	predictive models (construction of regression and classification models using selected features: median, autoregressive coefficients, autocorrelation)
	other ML techniques (cluster analysis)
deep learning (DL)	lightweight convolutional neural network (lightweight CNN, e.g., PlantNet), deep-learning methods: neural networks with various architectures
automation and toolchains	automatic toolchain, integration of ML models within automated tools

The above analysis is based on a small number of studies conducted mainly in laboratory conditions. Terms of accuracy, computational cost, and practical applicability should be taken into account in future investigation.

6.2. Time-Lapse Video

As evidenced by the works presented above, considerable progress has been made in the methodology of studying movements and growth mainly in the model plant *Arabidopsis thaliana* [52,54,55,57,58]. Work is in progress on cultivated plants, such as sunflower [51], bean [53], wheat [60,64], and tomato [52]. Some video systems for studying the growth and movement of plant organs are universal for many species [51,52,55,59].

The current techniques used in modern basic science may find application in smart agriculture in the future. Non-invasive and relatively cheap imaging also deserves to be emphasised. There is increasing interest in vision systems involving time-lapse video imaging and deep learning algorithms that could be used in future agricultural activities [65,66]. The research presented above shows how the time-lapse video technique can support investigations of plant growth. The development of these methods in the future will provide better solutions to the problems of modern plant cultivation.

6.3. Advantages and Difficulties

It should also be added and emphasised that the presented methods of measuring electrical and motor activity are relatively cheap and not burdensome for the environment in relation to the increase in new information and knowledge about plant growth/development. Based on the analysed literature, it can be concluded that the research on electrical and motor activity of plants is associated with not only new application possibilities but also many difficulties to be solved (Figure 4).

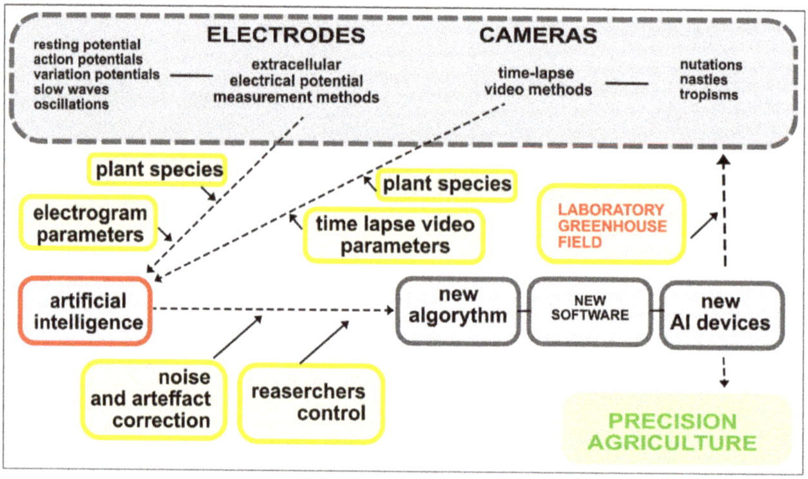

Figure 4. Key issues necessary to be solved (yellow colour) when introducing plant electrograms and time-lapse video into precision agriculture.

6.3.1. Advantages and Possibilities

- Application of measurement methods in laboratories, greenhouses, and field conditions;
- Successful attempts to record and analyse some electrograms and time-lapse videos in field conditions;
- Optimisation of universal measurement and analytical methods aimed at crop plants (not only the model plant *Arabidopsis*);

- Relatively cheap and either not at all or minimally invasive methods;
- Detection of stress and impact on the accuracy of biomass estimation;
- Real-time monitoring of environmental factors and stresses;
- Support for the decision-making process in agriculture in real time.

Difficulties and Problems

- Calibration of measurement and analysis to a specific plant species and environmental factors (currently a large variety of measurement and analytical methods that are incomparable (Tables 1 and 2));
- Determination of electrogram and time-lapse video parameters;
- Elimination of artifacts and noise;
- Researcher supervision during method introduction;
- Impact of field and atmospheric conditions;
- Electrodes resistant to field conditions;
- Wireless connections enabling remote sensing;
- Selection of appropriate and standardised AI analyses;
- Integrated simultaneous bioelectrical studies and organ movement estimation;
- Standardisation of recording methods, electrode placement, sampling frequency, and sampling time during the day, taking into account the plant growth phase.

7. Conclusions

Various types of methods for measuring changes in electrical voltage have potential in monitoring the physiological state of plants and diagnosing plant diseases. Specific electrical signatures occurring in association with a specific type of environmental stress can be used in the future in early detection of stress in plants even before visual symptoms appear. These studies are promising but require further work on the interpretation and standardisation of measurements. These relatively cheap environmental sensors—electrodes and digital video cameras—facilitate relatively cheap monitoring of the condition of a plant or a cultivated field. Their advantage is non-invasiveness or, in the case of electrodes, very low invasiveness. A huge advantage of such electrical and visual systems is the possibility of continuous real-time monitoring of the condition of plants or the environment. Such systems will facilitate decision-making related to irrigation or the use of nutrients or chemical plant protection substances. This application will be particularly precise, which will affect the health of the plants themselves and the soil environment. Equally important will be more economical calculations resulting from such precisely managed crops [67]. In the future, specific electrical signatures together video patterns of movement and growth associated with real environmental conditions could support decisions about plant cultivation.

8. Future Directions

Currently, precision agriculture is in its early stages of development. The ongoing cooperation with scientists dealing with plant biology will contribute to a proper transfer of modern solutions and technologies to farming practice in future. Most of the time-lapse analyses have been performed on *Arabidopsis thaliana*, which indicates the need to extend the research to other plants important in food production processes, as in the case of the recently undertaken research on wheat, sunflower, and bean. Similarly, the study of specific electrical signatures and the whole electrome in plants is promising but requires improvement of procedures and expansion of its scope. Threats to human health and the environment are related to the still insufficient knowledge of ecological, physiological, social, and economic processes, and therefore the use of inappropriate solutions in practice.

Observing natural processes by basic science and applying technical solutions that are in accordance with the laws of nature will help to improve human daily lives in the future. Attempts to apply both artificial intelligence and precision agriculture are bringing new solutions towards creating healthier food and environments, and therefore healthier lives in the future.

Funding: This research received no external funding.

Data Availability Statement: No new data were created or analyzed in this study. Data sharing is not applicable to this article.

Conflicts of Interest: The authors declare no conflict of interest.

Abbreviations

The following abbreviations are used in this manuscript:

AI	artificial intelligence
AP	action potential
SLEAP	Social LEAP Estimates Animal Poses
TRiP	Tracking Rhythms in Plants
PALMA	Plant Leaf Movement Analyzer
PCA	principal component analysis
PDF	probability density function
PSD	power spectral density
ApEn	approximate entropy
ApEn(s)	multiscale approximate entropy
FFT	fast Fourier transform
RF	random forest
ML	machine learning
DL	deep learning
CNN	convolutional neural network
SVM	support vector machine
KNN	k-nearest neighbours

References

1. Brzezicki, M. A Systematic Review of the Most Recent Concepts in Kinetic Shading Systems with a Focus on Biomimetics: A Motion/Deformation Analysis. *Sustainability* **2024**, *16*, 5697. [CrossRef]
2. Balyan, S.; Jangir, H.; Tripathi, S.N.; Tripathi, A.; Jhang, T.; Pandey, P. Seeding a Sustainable Future: Navigating the Digital Horizon of Smart Agriculture. *Sustainability* **2024**, *16*, 475. [CrossRef]
3. dos Reis, G.A.; Martínez-Burgos, W.J.; Pozzan, R.; Pastrana Puche, Y.; Ocán-Torres, D.; de Queiroz Fonseca Mota, P.; Rodrigues, C.; Lima Serra, J.; Scapini, T.; Karp, S.G.; et al. Comprehensive Review of Microbial Inoculants: Agricultural Applications, Technology Trends in Patents, and Regulatory Frameworks. *Sustainability* **2024**, *16*, 8720. [CrossRef]
4. Li, P.; Han, Z.; Chepkorir, D.; Fang, W.; Ma, Y. Effect of Exogenous Jasmonates on Plant Adaptation to Cold Stress: A Comprehensive Study Based on a Systematic Review with a Focus on Sustainability. *Sustainability* **2024**, *16*, 10654. [CrossRef]
5. Rogo, U.; Viviani, A.; Pugliesi, C.; Fambrini, M.; Usai, G.; Castellacci, M.; Simoni, S. Improving Crop Tolerance to Abiotic Stress for Sustainable Agriculture: Progress in Manipulating Ascorbic Acid Metabolism via Genome Editing. *Sustainability* **2025**, *17*, 719. [CrossRef]
6. Sanberg, P.R. "Neural Capacity" in *Mimosa pudica*: A Review. *Behav. Biol.* **1976**, *17*, 435–452. [CrossRef]
7. Applewhite, P.B. Plant and Animal Behavior: An Introductory Comparison. In *Aneural Organisms in Neurobiology*; Eisenstein, E.M., Ed.; Springer US: Boston, MA, USA, 1975; pp. 131–139.
8. Simons, P.J. The Role of Electricity in Plant Movements. *New Phytol.* **1981**, *87*, 11–37. [CrossRef]
9. Baluska, F. Recent Surprising Similarities between Plant Cells and Neurons. *Plant Signal. Behav.* **2010**, *5*, 87–89. [CrossRef]
10. Borges, R.M. Do Plants and Animals Differ in Phenotypic Plasticity? *J. Biosci.* **2005**, *30*, 41–50. [CrossRef]
11. Huey, R.B.; Carlson, M.; Crozier, L.; Frazier, M.; Hamilton, H.; Harley, C.; Hoang, A.; Kingsolver, J.G. Plants Versus Animals: Do They Deal with Stress in Different Ways? *Integr. Comp. Biol.* **2002**, *42*, 415–423. [CrossRef]

12. Jones, A.M.; Chory, J.; Dangl, J.L.; Estelle, M.; Jacobsen, S.E.; Meyerowitz, E.M.; Nordborg, M.; Weigel, D. The Impact of Arabidopsis on Human Health: Diversifying Our Portfolio. *Cell* **2008**, *133*, 939–943. [CrossRef] [PubMed]
13. Bhatla, S.C.; Lal, M.A. Plant Movements. In *Plant Physiology, Development and Metabolism*; Springer: Singapore, 2023; pp. 641–659.
14. Darwin, C.; Darwin, F. *The Power of Movement in Plants*; John Murray: London, UK, 1880.
15. Del Dottore, E.; Mondini, A.; Sadeghi, A.; Mattoli, V.; Mazzolai, B. An Efficient Soil Penetration Strategy for Explorative Robots Inspired by Plant Root Circumnutation Movements. *Bioinspir. Biomim.* **2018**, *13*, 015003. [CrossRef]
16. Qin, K.; Tang, W.; Zong, H.; Guo, X.; Xu, H.; Zhong, Y.; Wang, Y.; Sheng, Q.; Yang, H.; Zou, J. Parthenocissus Inspired Soft Climbing Robots. *Sci. Adv.* **2025**, *11*, eadt9284. [CrossRef]
17. Burdon-Sanderson, J.S. Note on the Electrical Phenomena Which Accompany Irritation of the Leaf of *Dionæa muscipula*. *Proc. Roy. Soc. Lond.* **1873**, *21*, 495–496.
18. Tills, O.; Ibbini, Z.; Spicer, J.I. Bioimaging and-the Future of Whole-Organismal Developmental Physiology. *Comp. Biochem. Physiol. Mol. Integr. Physiol.* **2024**, *300*, 111783. [CrossRef]
19. Wang, Q. Understanding Plant Movement: From Kinematics to Machine Learning. PhD Thesis, Università degli Studi di Padova, Padova, Italy, 2024.
20. de Toledo, G.R.; Parise, A.G.; Simmi, F.Z.; Costa, A.V.; Senko, L.G.; Debono, M.-W.; Souza, G.M. Plant Electrome: The Electrical Dimension of Plant Life. *Theor. Exp. Plant Physiol.* **2019**, *31*, 21–46. [CrossRef]
21. De Loof, A. The Cell's Self-Generated "Electrome": The Biophysical Essence of the Immaterial Dimension of Life? *Commun. Integr. Biol.* **2016**, *9*, e1197446. [CrossRef]
22. Reissig, G.N.; de Carvalho Oliveira, T.F.; Parise, A.G.; Posso, D.A.; Souza, G.M. An Electrifying View into the Life of Plants: The Plant Electrome. *Front. Young Minds* **2024**, *12*, 1400420. [CrossRef]
23. Zhao, D.-J.; Wang, Z.-Y.; Li, J.; Wen, X.; Liu, A.; Huang, L.; Wang, X.-D.; Hou, R.-F.; Wang, C. Recording Extracellular Signals in Plants: A Modeling and Experimental Study. *Math. Comput. Model.* **2013**, *58*, 556–563. [CrossRef]
24. Beilby, M.; Coster, H. The Action Potential in *Chara Corallina* III.* The Hodgkin-Huxley Parameters for the Plasmalemma. *Funct. Plant Biol.* **1979**, *6*, 337–353. [CrossRef]
25. Beilby, M.J.; Al Khazaaly, S. Re-Modeling Chara Action Potential: II. The Action Potential Form under Salinity Stress. *AIMS Biophys.* **2017**, *4*, 298–315.
26. Kisnieriene, V.; Lapeikaite, I.; Pupkis, V.; Beilby, M.J. Modeling the Action Potential in Characeae *Nitellopsis obtusa*: Effect of Saline Stress. *Front. Plant Sci.* **2019**, *10*, 82. [CrossRef]
27. Shu, L.; Hancke, G.P.; Abu-Mahfouz, A.M. Guest Editorial: Sustainable and Intelligent Precision Agriculture. *IEEE Trans. Ind. Inform.* **2021**, *17*, 4318–4321. [CrossRef]
28. Sai, K.; Sood, N.; Saini, I. Classification of Various Nutrient Deficiencies in Tomato Plants through Electrophysiological Signal Decomposition and Sample Space Reduction. *Plant Physiol. Biochem.* **2022**, *186*, 266–278. [CrossRef]
29. Kim, H.N.; Seok, Y.J.; Park, G.M.; Vyavahare, G.; Park, J.H. Monitoring of Plant-Induced Electrical Signal of Pepper Plants (*Capsicum annuum* L.) According to Urea Fertilizer Application. *Sci. Rep.* **2023**, *13*, 291. [CrossRef]
30. Bhadra, N.; Chatterjee, S.K.; Das, S. Multiclass Classification of Environmental Chemical Stimuli from Unbalanced Plant Electrophysiological Data. *PLoS ONE* **2023**, *18*, e0285321. [CrossRef]
31. Aust, T.; Buss, E.; Mohr, F.; Hamann, H. Automated Phytosensing: Ozone Exposure Classification Based on Plant Electrical Signals. In Proceedings of the IEEE Symposia on Computational Intelligence for Energy, Transport and Environmental Sustainability (CIETES), Trondheim, Norway, 17–20 March 2025; pp. 1–7.
32. Buss, E.; Aust, T.; Wahby, M.; Rabbel, T.-L.; Kernbach, S.; Hamann, H. Stimulus Classification with Electrical Potential and Impedance of Living Plants: Comparing Discriminant Analysis and Deep-Learning Methods. *Bioinspir. Biomim.* **2023**, *18*, 025003. [CrossRef]
33. Cattani, A.; De Riedmatten, L.; Roulet, J.; Smit-Sadki, T.; Alfonso, E.; Kurenda, A.; Graeff, M.; Remolif, E.; Rienth, M. Water Status Assessment in Grapevines Using Plant Electrophysiology: This Article Is Published in Cooperation with the XVth International Terroir Congress, 18–22 November 2024, Mendoza, Argentina. Guest Editors: Federico Berli, Jorge Prieto and Martín Fanzone. *OENO One* **2024**, *58*, 4.
34. de Toledo, G.R.; Reissig, G.N.; Senko, L.G.; Pereira, D.R.; da Silva, A.F.; Souza, G.M. Common Bean under Different Water Availability Reveals Classifiable Stimuli-Specific Signatures in Plant Electrome. *Plant Signal. Behav.* **2024**, *19*, 2333144. [CrossRef] [PubMed]
35. Qi, J.; Liu, C.; Wang, Q.; Shi, Y.; Xia, X.; Wang, H.; Sun, L.; Men, H. Clivia Biosensor: Soil Moisture Identification Based on Electrophysiology Signals with Deep Learning. *Biosens. Bioelectron.* **2024**, *262*, 116525. [CrossRef] [PubMed]
36. Simmi, F.Z.; Dallagnol, L.J.; Almeida, R.O.; da Rosa Dorneles, K.; Souza, G.M. Barley Systemic Bioelectrical Changes Detect Pathogenic Infection Days before the First Disease Symptoms. *Comput. Electron. Agric.* **2023**, *209*, 107832. [CrossRef]
37. Ghasemi, E.; Ebrahimie, E.; Niazi, A. Machine Learning for Early Detection of Plant Viruses: Analyzing Post-Infection Electrical Signal Patterns. *Smart Agric. Technol.* **2024**, *9*, 100668. [CrossRef]

38. Kurenda, A.; Jenni, D.; Lecci, S.; Buchholz, A. Bringing Light into the Dark—Plant Electrophysiological Monitoring of Root Knot Nematode Infestation and Real-Time Nematicide Efficacy. *J. Pest Sci.* **2024**, *98*, 493–507. [CrossRef]
39. Iosip, A.-L.; Scherzer, S.; Bauer, S.; Becker, D.; Krischke, M.; Al-Rasheid, K.A.; Schultz, J.; Kreuzer, I.; Hedrich, R. Dyscalculia, a Venus Flytrap Mutant without the Ability to Count Action Potentials. *Curr. Biol.* **2023**, *33*, 589–596.e5. [CrossRef]
40. Scherzer, S.; Böhm, J.; Huang, S.; Iosip, A.L.; Kreuzer, I.; Becker, D.; Heckmann, M.; Al-Rasheid, K.A.; Dreyer, I.; Hedrich, R. A Unique Inventory of Ion Transporters Poises the Venus Flytrap to Fast-Propagating Action Potentials and Calcium Waves. *Curr. Biol.* **2022**, *32*, 4255–4263.e5. [CrossRef]
41. Böhm, J.; Scherzer, S.; Krol, E.; Kreuzer, I.; von Meyer, K.; Lorey, C.; Mueller, T.D.; Shabala, L.; Monte, I.; Solano, R.; et al. The Venus Flytrap *Dionaea muscipula* Counts Prey-Induced Action Potentials to Induce Sodium Uptake. *Curr. Biol.* **2016**, *26*, 286–295. [CrossRef]
42. Roblin, G.; Moyen, C.; Fleurat-Lessard, P.; Dédaldéchamp, F. Rapid Osmocontractile Response of Motor Cells of *Mimosa pudica* Pulvini Induced by Short Light Signals. *Photochem. Photobiol.* **2025**, *101*, 728–745. [CrossRef]
43. Stolarz, M.; Trębacz, K. Spontaneous Rapid Leaf Movements and Action Potentials in *Mimosa pudica* L. *Physiol. Plant.* **2021**, *173*, 1882–1888. [CrossRef]
44. Mahta, J.; Spangaro, A.; Lenartowicz, M.; Mattiazzi Usaj, M. From Pixels to Insights: Machine Learning and Deep Learning for Bioimage Analysis. *BioEssays* **2024**, *46*, 2300114.
45. Boquet-Pujadas, A.; Olivo-Marin, J.-C.; Guillén, N. Bioimage Analysis and Cell Motility. *Patterns* **2021**, *2*, 100170. [CrossRef]
46. Bragantini, J.; Theodoro, I.; Zhao, X.; Huijben, T.A.; Hirata-Miyasaki, E.; VijayKumar, S.; Balasubramanian, A.; Lao, T.; Agrawal, R.; Xiao, S.; et al. Ultrack: Pushing the Limits of Cell Tracking across Biological Scales. *bioRxiv* **2024**. [CrossRef]
47. Ershov, D.; Phan, M.-S.; Pylvänäinen, J.W.; Rigaud, S.U.; Le Blanc, L.; Charles-Orszag, A.; Conway, J.R.; Laine, R.F.; Roy, N.H.; Bonazzi, D.; et al. Trackmate 7: Integrating State-of-the-Art Segmentation Algorithms into Tracking Pipelines. *Nat. Methods* **2022**, *19*, 829–832. [CrossRef] [PubMed]
48. Gallusser, B.; Weigert, M. Trackastra: Transformer-Based Cell Tracking for Live-Cell Microscopy. In Proceedings of the European Conference on Computer Vision, Paris, France, 26–27 March 2025; Springer Science & Business Media: Dordrecht, The Netherlands, 2025.
49. Stolarz, M.; Król, E.; Dziubińska, H. Lithium Distinguishes between Growth and Circumnutation and Augments Glutamate-Induced Excitation of *Helianthus annuus* Seedlings. *Acta Physiol. Plant.* **2015**, *37*, 69. [CrossRef]
50. Stolarz, M.; Krol, E.; Dziubinska, H.; Zawadzki, T. Complex Relationship between Growth and Circumnutations in *Helianthus annuus* Stem. *Plant Signal. Behav.* **2008**, *3*, 376–380. [CrossRef]
51. Stolarz, M.; Zuk, M.; Krol, E.; Dziubińska, H. Circumnutation Tracker: Novel Software for Investigation of Circumnutation. *Plant Methods* **2014**, *10*, 24. [CrossRef]
52. Bours, R.; Muthuraman, M.; Bouwmeester, H.; van der Krol, A. Oscillator: A System for Analysis of Diurnal Leaf Growth Using Infrared Photography Combined with Wavelet Transformation. *Plant Methods* **2012**, *8*, 29. [CrossRef] [PubMed]
53. Ruiz-Melero, D.R.; Ponkshe, A.; Calvo, P.; García-Mateos, G. The Development of a Stereo Vision System to Study the Nutation Movement of Climbing Plants. *Sensors* **2024**, *24*, 747. [CrossRef]
54. Wagner, L.; Schmal, C.; Staiger, D.; Danisman, S. The Plant Leaf Movement Analyzer (PALMA): A Simple Tool for the Analysis of Periodic Cotyledon and Leaf Movement in Arabidopsis Thaliana. *Plant Methods* **2017**, *13*, 2. [CrossRef]
55. Greenham, K.; Lou, P.; Remsen, S.E.; Farid, H.; McClung, C.R. Trip: Tracking Rhythms in Plants, an Automated Leaf Movement Analysis Program for Circadian Period Estimation. *Plant Methods* **2015**, *11*, 33. [CrossRef]
56. Oskam, L.; Snoek, B.L.; Pantazopoulou, C.K.; van Veen, H.; Matton, S.E.; Dijkhuizen, R.; Pierik, R. A Low-Cost Open-Source Imaging Platform Reveals Spatiotemporal Insight into Leaf Elongation and Movement. *Plant Physiol.* **2024**, *195*, 1866–1879. [CrossRef]
57. Rehman, T.U.; Zhang, L.; Wang, L.; Ma, D.; Maki, H.; Sánchez-Gallego, J.A.; Mickelbart, M.V.; Jin, J. Automated Leaf Movement Tracking in Time-Lapse Imaging for Plant Phenotyping. *Comput. Electron. Agric.* **2020**, *175*, 105623. [CrossRef]
58. Mao, Y.; Liu, H.; Wang, Y.; Brenner, E.D. A Deep Learning Approach to Track Arabidopsis Seedlings' Circumnutation from Time-Lapse Videos. *Plant Methods* **2023**, *19*, 18. [CrossRef]
59. Gall, G.E.C.; Pereira, T.D.; Jordan, A.; Meroz, Y. Fast Estimation of Plant Growth Dynamics Using Deep Neural Networks. *Plant Methods* **2022**, *18*, 21. [CrossRef]
60. Gibbs, J.A.; Burgess, A.J.; Pound, M.P.; Pridmore, T.P.; Murchie, E.H. Recovering Wind-Induced Plant Motion in Dense Field Environments Via Deep Learning and Multiple Object Tracking. *Plant Physiol.* **2019**, *181*, 28–42. [CrossRef]
61. Kunkowska, A.B. What They Do in the Shadows: A Low-Cost Imaging System for Recording Leaf Expansion and Movements. *Plant Physiol.* **2024**, *195*, 1745–1747. [CrossRef]
62. Burgess, A.J.; Gibbs, J.A.; Murchie, E.H. A Canopy Conundrum: Can Wind-Induced Movement Help to Increase Crop Productivity by Relieving Photosynthetic Limitations? *J. Exp. Bot.* **2019**, *70*, 2371–2380. [CrossRef]

63. Ufuktepe, D.K.; Palaniappan, K.; Elmali, M.; Baskin, T.I. Rtip: A Fully Automated Root Tip Tracker for Measuring Plant Growth with Intermittent Perturbations. In Proceedings of the 2020 IEEE International Conference on Image Processing (ICIP), Virtual Conference, 25–28 October 2020.
64. Golka, W.; Arseniuk, E.; Golka, A.; Góral, T. Sztuczne Sieci Neuronowe I Teledetekcja W Ocenie Porażenia Pszenicy Jarej Fuzariozą Kłosów. *Biuletyn Inst. Hod. Aklim. Rośl.* **2020**, *288*, 67–75. [CrossRef]
65. Fiaz, M.; Mahmood, A.; Javed, S.; Jung, S.K. Handcrafted and Deep Trackers: Recent Visual Object Tracking Approaches and Trends. *ACM Comput. Surv.* **2019**, *52*, 1–44. [CrossRef]
66. Farooq, M.A.; Gao, S.; Hassan, M.A.; Huang, Z.; Rasheed, A.; Hearne, S.; Prasanna, B.; Li, X.; Li, H. Artificial Intelligence in Plant Breeding. *Trends Genet.* **2024**, *40*, 891–908. [CrossRef]
67. Gupta, D.K.; Pagani, A.; Zamboni, P.; Singh, A.K. AI-Powered Revolution in Plant Sciences: Advancements, Applications, and Challenges for Sustainable Agriculture and Food Security. *Explor. Foods Foodomics* **2024**, *2*, 443–459. [CrossRef]

sustainability

MDPI

Review

Sustainable Approaches to Agricultural Greenhouse Gas Mitigation in the EU: Practices, Mechanisms, and Policy Integration

Roxana Maria Madjar [1], Gina Vasile Scăețeanu [1,*], Ana-Cornelia Butcaru [2] and Andrei Moț [1]

[1] Faculty of Agriculture, University of Agronomic Sciences and Veterinary Medicine of Bucharest, 59 Marasti Blvd., District 1, 011464 Bucharest, Romania; roxana.madjar@agro-bucuresti.ro (R.M.M.)

[2] Research Center for Studies of Food Quality and Agricultural Products, University of Agronomic Sciences and Veterinary Medicine Bucharest, 011464 Bucharest, Romania

* Correspondence: gina.scaeteanu@agro-bucuresti.ro

Abstract: The agricultural sector has a significant impact on the global carbon cycle, contributing substantially to greenhouse gas (GHG) emissions through various practices and processes. This review paper examines the significant role of the agricultural sector in the global carbon cycle, highlighting its substantial contribution to GHG emissions through diverse practices and processes. The study explores the trends and spatial distribution of agricultural GHG emissions at both the global level and within the European Union (EU). Emphasis is placed on the principal gases released by this sector—methane (CH_4), nitrous oxide (N_2O), and carbon dioxide (CO_2)—with detailed attention to their sources, levels, environmental impacts, and key strategies to mitigate and control their effects, based on the latest scientific data. The paper further investigates emissions originating from livestock production, along with mitigation approaches including feed additives, selective breeding, and improved manure management techniques. Soil-derived emissions, particularly N_2O and CO_2 resulting from fertilizer application and microbial activity, are thoroughly explored. Additionally, the influence of various agricultural practices such as tillage, crop rotation, and fertilization on emission levels is analyzed, supported by updated data from recent literature. Special focus is given to the underlying mechanisms that regulate these emissions and the effectiveness of management interventions in reducing their magnitude. The research also evaluates current European legislative measures aimed at lowering agricultural emissions and promoting climate-resilient, sustainable farming systems. Various mitigation strategies—ranging from optimized land and nutrient management to the application of nitrification inhibitors and soil amendments are assessed for both their practical feasibility and long-term impact.

Keywords: agriculture; carbon dioxide; greenhouse gas emissions; livestock; methane; nitrous oxide; mitigation strategies; soil carbon sequestration; sustainability

1. Introduction

Human activities, including modern agriculture, contribute significantly to the production of greenhouse gases (GHGs), which have been increasing since the beginning of the industrial era. Some greenhouse gases, such as hydrofluorocarbons (HFCs), perfluorocarbons (PFCs), and sulfur hexafluoride (SF_6), are emitted exclusively by human activities.

Agriculture is a major source of greenhouse gases, primarily carbon dioxide (CO_2), methane (CH_4), and nitrous oxide (N_2O). Each gas has a different global warming potential and atmospheric lifetime, contributing to climate change to varying degrees due to its distinct impacts and persistence in the atmosphere [1].

Beyond the overall contribution of agriculture, soils play a particularly important role, accounting for about 15% of anthropogenic climate warming, mainly through CO_2 (74%), N_2O (17%), and CH_4 (9%) emissions [2]. In agricultural systems, CH_4 originates largely from enteric fermentation in ruminants and manure management [3], while N_2O emissions result from soil microbial processes such as nitrification and denitrification, driven by nitrogen inputs from fertilizers and manure [4].

Overall, agrifood systems contribute approximately one-third of global GHG emissions [5], with agriculture alone responsible for about 49% of CH_4 and 63% of N_2O emissions within the European Union.

Beyond direct emissions, agriculture indirectly elevates GHG levels through land-use changes, most notably deforestation, releasing carbon stored in soils and vegetation [6].

Emission mitigation strategies in agriculture are based on a variety of strategies targeting both crop production and livestock systems, as follows: changes in agronomic practices [4,7], modifying animal diets [8,9], manure treatment [10], and biological innovations [11]. While numerous mitigation options have been identified, their implementation is influenced by the heterogeneity of farming systems and socio-economic conditions. Differences in farm structure, diversity of crops, resource availability, and policy support shape the feasibility and effectiveness of mitigation practices [12,13]. Moreover, trade-offs often exist between emission reduction, food security, and farmers' livelihoods, highlighting the need for context-specific and balanced approaches to sustainable agricultural management.

However, agriculture is not solely a source of GHG emissions; it can also serve as a carbon sink by storing carbon in soils and plant biomass through a process known as carbon sequestration. Soil carbon sequestration is a natural and effective way to lower atmospheric CO_2 levels, playing an important role in mitigating climate change. It works by capturing CO_2 from the atmosphere and storing it in the soil as organic carbon [14]. This not only reduces greenhouse gas concentrations but also improves soil health and fertility [15]. Agricultural practices such as reduced tillage, cover cropping, and improved soil management can help reduce emissions while increasing the soil's capacity to store carbon [16].

This review analyzes GHG emissions (CO_2, CH_4, N_2O) by examining their sources, mechanisms, and management practices at the global and regional levels. It highlights how agricultural activities such as fertilization, soil and residue management, and livestock production contribute to climate change. By linking emission sources with mitigation strategies and relevant policy frameworks, this review synthesizes current scientific knowledge and management approaches, offering an integrative reference for advancing sustainable agricultural practices.

The novelty of this work lies in its integrated approach, combining scientific evidence on emission sources with mitigation measures and policy analysis, offering a comprehensive perspective on sustainable agricultural development within the EU context.

This paper is structured into nine sections, which explore the temporal evolution of CO_2, CH_4, and N_2O emissions at both the European Union and global levels, identify the primary agricultural sources of these emissions, and discuss current strategies for their mitigation and control. For a better understanding of the overall content, the chapters and subchapters are illustrated in Figure 1.

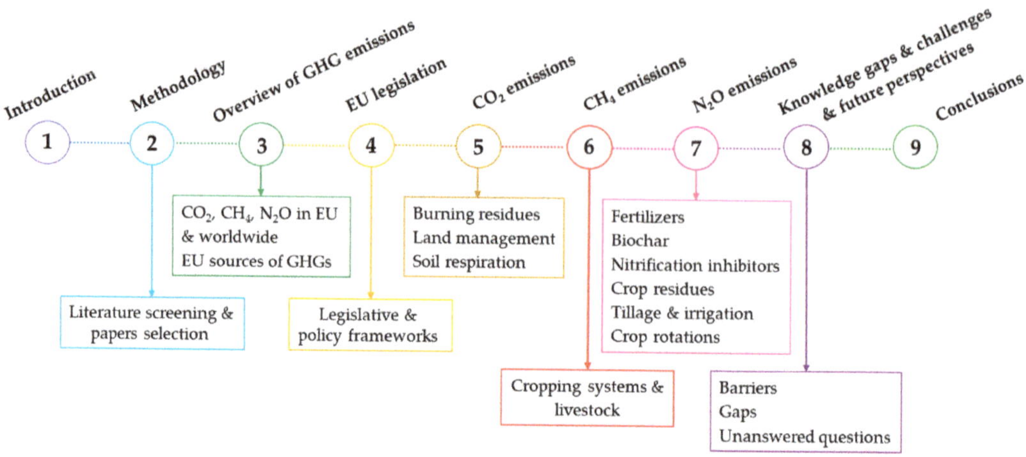

Figure 1. Research roadmap: agricultural GHG emissions.

2. Methodology

A literature search was conducted across commonly accessed scientific databases, including Web of Science, Scopus, PubMed, and Google Scholar, as well as in reports from international organizations such as FAO and EEA. The search focused on the role of the agricultural sector in GHG emissions and mitigation strategies. Various keywords and their combinations were used to ensure a comprehensive search that captured relevant papers across different agricultural practices, GHG types, and mitigation strategies. Few examples of used keywords include: "greenhouse gas emissions", "livestock", "manure management", "agriculture AND greenhouse gas emissions", "mitigation strategies AND agriculture", "methane emissions AND agriculture", "carbon dioxide emissions AND agriculture", "nitrous oxide AND agriculture".

Articles were included in the evaluation if they were published in peer-reviewed journals between 2000 and 2025, focused on GHG emissions from agriculture or livestock, addressed mitigation strategies or sustainable farming practices, and were available in English. Additionally, among the selected papers, some were used for backward and forward snowballing techniques to identify additional relevant studies, which were then included in our database for comprehensive data collection.

Articles were excluded if they did not provide data on GHG emissions, focused exclusively on forestry or non-agricultural sectors, or were published in journals with no impact factor. These criteria ensured that the selected literature was relevant and provided recent data on the subject.

Selected studies included research articles, review papers, and reports containing experimental data.

3. Overview of GHG Emissions in Agriculture

Agriculture contributes approximately 18% of total GHG emissions through activities involving soils, fertilizers, livestock, and residues [17]. Carbon dioxide, mainly produced from fossil fuel use and land-use changes, accounts for only 7% of agricultural emissions. Methane, primarily emitted from enteric fermentation in ruminants and manure management, is the largest contributor at nearly 50% of agricultural emissions. Nitrous oxide, released from agricultural soils and from manure application, accounts for about 30% of emissions and has a particularly strong impact on global warming due to its high Global Warming Potential (GWP), ranging from 265 to 310 [18,19].

Figure 2 illustrates the main sources of GHG emissions in agriculture and their links to specific agricultural activities and practices.

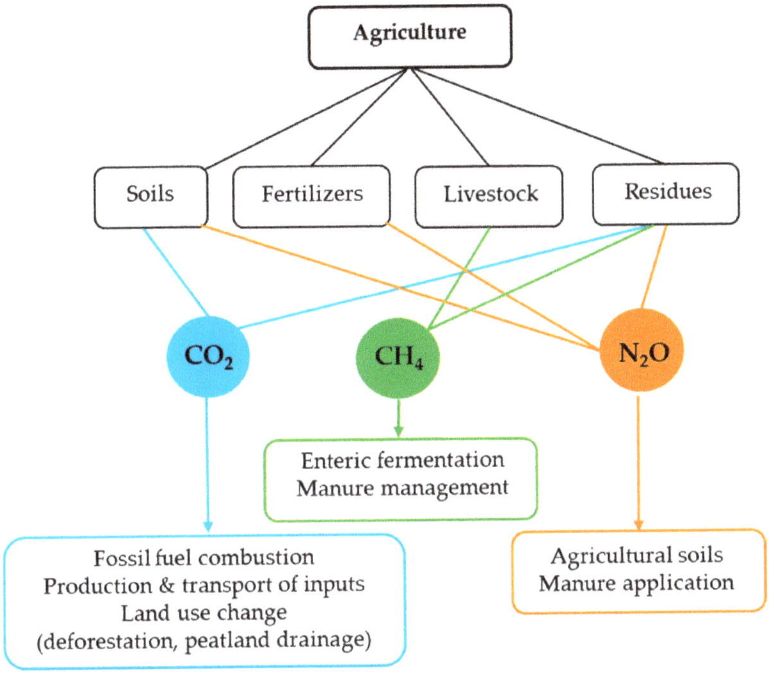

Figure 2. Hierarchical overview of agricultural GHG emissions, showing specific sources and their links to CO_2 (blue), CH_4 (green), and N_2O (orange).

In 2022, global agrifood systems released 16.2 billion tons of CO_2 equivalent (Gt CO_2e), a figure that remained almost unchanged from 2021 and represented a 10% increase compared to 2000 [20]. Fertilizer production and application (including manure and synthetic types) generate 2.6 Gt CO_2e per year [21].

Agricultural emissions vary regionally due to differences in farming practices, livestock density, and land-use patterns. These variations influence the types and magnitudes of emissions, with regions dominated by ruminant livestock showing higher CH_4 emissions [22], while areas with intensive cropping tend to have greater emissions from fertilizer use [23]. More than 60% of N_2O emissions are attributed to agriculture, largely due to the use of organic and inorganic fertilizers [23].

Moreover, in addition to its contributions to greenhouse gas emissions, agriculture can also serve as a potential carbon sink, removing carbon dioxide from the atmosphere through processes such as soil carbon sequestration and biomass accumulation. Maintaining and increasing soil organic carbon through practices like cover cropping, reduced tillage, agroforestry, and the addition of organic amendments can enhance this sink function while simultaneously improving soil health and fertility [24].

3.1. GHG Emissions from Agriculture in the EU and Worldwide

Between 1970 and 2023, agricultural GHG emissions showed contrasting trends in the European Union and globally (Figures 3–5).

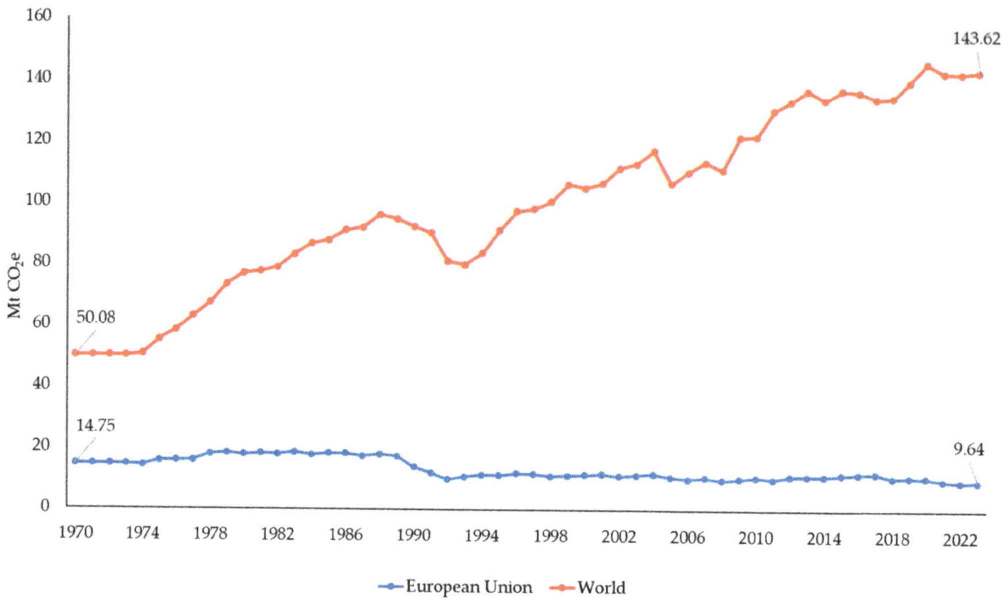

Figure 3. CO_2 emissions from agriculture: European Union vs. global trends (1970–2023).

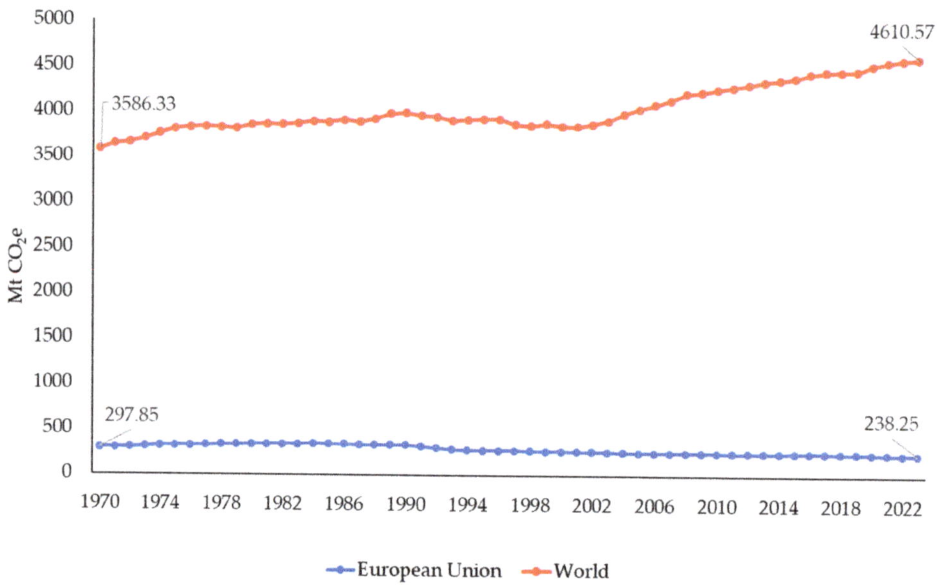

Figure 4. CH_4 emissions from agriculture: European Union vs. global trends (1970–2023).

In the EU, CO_2 emissions from agriculture peaked between 1983 and 1986 at 18.65 Mt CO_2e, then steadily declined to 9.64 Mt CO_2e by 2023, nearly halving compared to 1990 levels (Figure 3) [25]. This reduction reflects the impact of regulatory measures and agricultural policy reforms. Globally, agricultural CO_2 emissions rose significantly, accelerating around 1976 and almost doubling by 1988 to 96.23 Mt CO_2e. Emissions continued to increase, surpassing 140 Mt CO_2e by 2019, reaching 143.62 Mt CO_2e in 2023, almost three times the 1970 level.

EU CH_4 emissions peaked in 1984 at 343.10 Mt CO_2e and then declined by 30.55% to 238.25 Mt CO_2e in 2023 (Figure 4) [26]. This reduction began in the 1990s due to policy efforts, improved waste management, and structural shifts in agriculture and energy sectors. After 2010, emissions stabilized but maintained a downward trend. Globally, CH_4 emissions increased by 28.54% from 1970 to 2023, continuing a generally rising trend even during the COVID-19 pandemic, when EU emissions were declining. In 2023, the EU accounted for 5.16% of global CH_4 emissions.

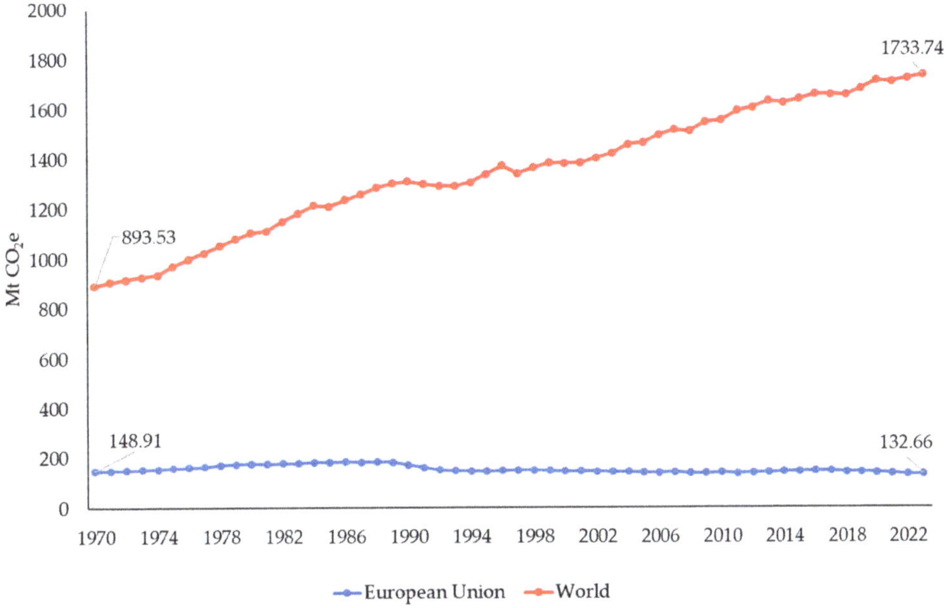

Figure 5. N_2O emissions from agriculture: European Union vs. global trends (1970–2023).

N_2O emissions in the EU rose until peaking at 184.85 Mt CO_2e in 1988, then steadily decreased by 28.23% to 132.66 Mt CO_2e by 2023 (Figure 5) [27]. This decline reflects environmental regulation and sustainable agricultural practices. Although there were fluctuations during the 2000s and 2010s, the overall downward trend persisted. Conversely, global N_2O emissions nearly doubled during the same period, continuing to increase even amid the COVID-19 pandemic.

Overall, the EU has successfully reduced agricultural emissions of CO_2, CH_4, and N_2O through comprehensive legislative frameworks, policy reforms, and the adoption of sustainable agricultural practices. This contrasts with the global trend of rising emissions, largely driven by expanding agricultural activities and weaker regulatory enforcement, highlighting the urgent need for international cooperation, especially in rapidly developing regions and areas with intensive agriculture.

The variation in the reduction in emissions in the EU compared to global trends can be explained by several factors, including regulatory and policy measures (further details are presented in Section 4), technological and management improvements such as crop rotation and cover cropping, and strong economic and political commitment [28]. In addition, global trends continue to show an increase in emissions, driven by expanding agricultural activities and less stringent regulations in developing countries, mainly [29].

3.2. Agricultural GHG Emissions in the EU by Source

According to data provided by the European Environment Agency [19] (Figure 6), CH_4 emissions from enteric fermentation have shown a gradual decline since 2005, decreasing from 188.83 Mt CO_2e to 176.76 Mt CO_2e in 2023. The decrease in CH_4 emissions from enteric fermentation likely reflects improvements in livestock management, such as better feeding strategies, dietary adjustments, or breeding programs aimed at reducing CH_4 production during digestion [30,31].

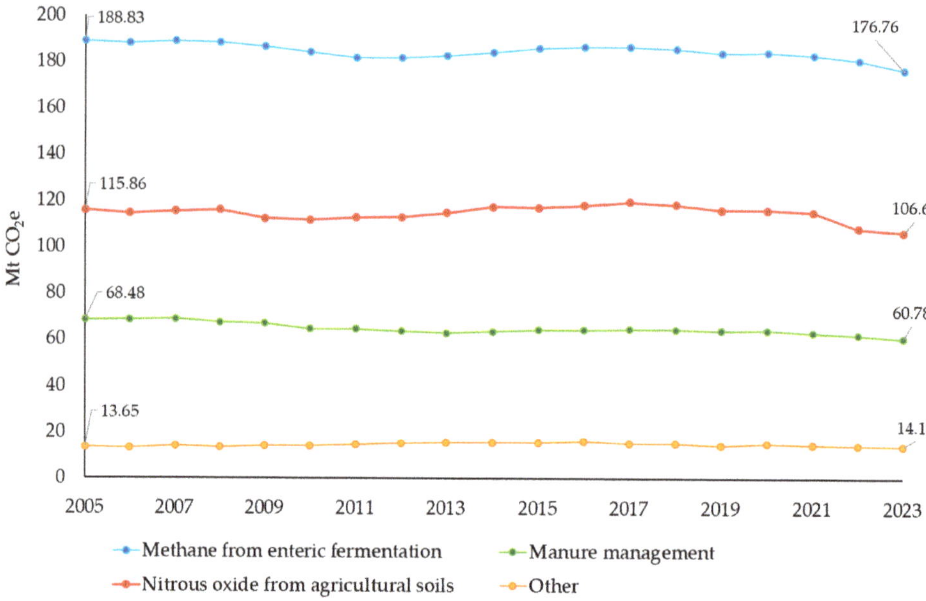

Figure 6. Breakdown of agricultural GHG emissions during 2005–2023 in the EU by source.

Similarly, N_2O emissions from agricultural soils declined overall from 115.86 Mt CO_2e in 2005 to 106.61 Mt CO_2e in 2023. However, within this period, emissions showed a pattern of fluctuations with both increases and decreases. The episodes of rising N_2O emissions can be attributed to factors such as increased nitrogen fertilizer application, shifts in agricultural practices, and intensified farming systems [32].

Further details regarding N_2O emissions from soil are discussed in Section 7. However, the reduction in recent years suggests that farmers may have adopted more precision agriculture techniques to manage fertilizer application more effectively, leading to reduced N_2O emissions [18].

Another possible factor is the promotion of sustainable agricultural practices [33], such as improved soil management or the use of slow-release or controlled-release fertilizers, nitrification inhibitors, as discussed in Section 7.2.

Emissions from manure management declined from 68.48 Mt CO_2e to 60.78 Mt CO_2e over the same period. This downward trend likely reflects improvements in manure-handling techniques, including better manure storage systems that reduce GHG emissions, as well as optimized land application practices [34]. Additional information regarding this issue is provided in Section 6.2.2.

In contrast, emissions from other agricultural sources remained relatively stable, increasing only slightly from 13.65 Mt CO_2e in 2005 to 14.13 Mt CO_2e in 2023. This stability suggests limited changes in land management and crop residue burning practices [17].

Concluding, significant, though insufficient, reductions in agricultural GHG emissions, particularly from enteric fermentation and manure management, have been achieved through improved livestock and manure practices. However, variability in soil N_2O emissions persists due to ongoing challenges in fertilizer use and soil management. To meet the EU's 2050 climate-neutrality goals [35], continued monitoring, coupled with broader adoption of precision agriculture and sustainable soil management techniques, is essential to ensure sustained progress in GHG emission mitigation.

4. EU Legislation Regulating GHG Emissions from Agriculture

In response to the need for controlling GHG emissions to more effectively manage climate change and mitigate its impacts, the European Union has developed a series of legislative and policy frameworks addressing GHG emissions across various sectors, including agriculture (Figure 7).

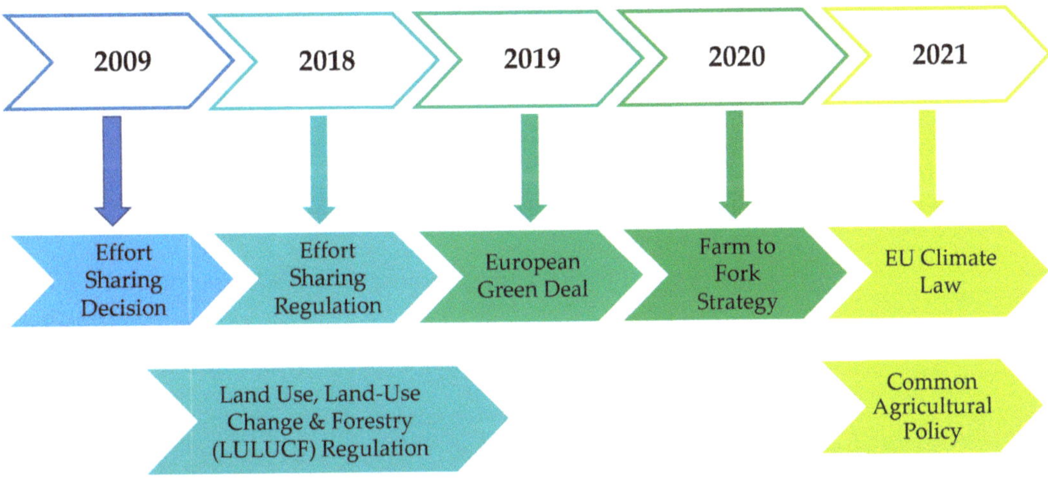

Figure 7. Timeline of EU legislation and policy frameworks that regulate GHG emissions resulting from agriculture.

These frameworks are designed to reduce emissions, promote sustainability, and align with the EU's broader climate goals, such as achieving climate neutrality by 2050 and reducing emissions by at least 55% by 2030 compared to 1990 levels.

Based on the knowledge provided by the documentation in this paper, the EU's climate objectives mentioned above are difficult to fulfill but are still achievable. There is progress that was observed during the time that suggests that ongoing policy frameworks and technological advances are effective. The objectives will be achieved through further innovation, widespread adoption of best practices, and compliance with all current regulations, as well as those that will complement the existing ones.

The development of these policies highlights the increasing focus on sustainable farming practices and carbon sequestration in agriculture, a sector considered central to the EU's climate strategy.

The European Union has implemented several policies and frameworks to guide the sector toward more sustainable practices (Table 1). Key among these is the Common Agricultural Policy (CAP), which has been consistently updated to encourage sustainable agricultural methods and contribute to the EU's climate objectives. Additionally, regulations like the Land Use, Land-Use Change, and Forestry (LULUCF) Regulation support practices that enhance carbon sequestration in soils and reduce emissions from land management. Policies such as the "Farm to Fork" strategy and the European Green Deal further integrate agriculture into the broader transition to a green and low-emission economy. These efforts collectively create a comprehensive policy framework that is essential for achieving the EU's emission-reduction targets and promoting sustainability in the agricultural sector.

The EU Emissions Trading System (EU ETS) is a cap-and-trade scheme that applies to sectors like power generation and heavy industry, setting a limit on emissions and allowing trading of allowances to reduce emissions cost-effectively [36]. Agricultural GHG emissions, which are not covered by the EU ETS, are managed under the Effort Sharing Regulation (ESR). The ESR sets binding national reduction targets for sectors like agriculture, transport, and buildings, ensuring that non-ETS emissions, such as CH_4 and N_2O from farming, contribute to the EU's overall climate goals [37]. Under the ESR, each Member State receives an annual emission allocation (AEA) for non-ETS sectors, including agriculture. States must implement measures like improving manure management, optimizing feed, and promoting sustainable farming practices. Flexibility mechanisms, such as banking, borrowing, and trading AEAs, allow for greater flexibility in meeting emission targets [37].

Table 1. EU legislation and policy frameworks for reducing agricultural GHG emissions.

Legislation/Policy Frameworks	Description	Refs.
Effort Sharing Decision (ESD) (adopted in 2009)	Member States were required: – to cap their GHG emissions in the Effort Sharing sectors annually from 2013 to 2020; – to achieve a collective reduction of approximately 10% in total EU emissions from the sectors covered by 2020, compared to 2005 levels. By 2020, the EU emissions within the scope of the Effort Sharing Decision were 16.3% lower than in 2005 (overachievement of the initial target). Agricultural emissions decreased by 3% compared to 2005.	[38]
Effort Sharing Regulation (ESR) (adopted in 2018, amended in 2023)	– Sets national targets for each EU Member State to reduce GHG emissions by 2030 in key sectors, including agriculture. – In 2023, the target was to reduce emissions by 30% compared to 2025 levels. – The amended form introduces new targets, with Member States collectively contributing to a 40% emissions reduction in the Effort Sharing sectors compared to 2005 levels. – The regulation also aims for a 55% reduction in total EU emissions by 2030 (compared to 1990 levels) and the achievement of climate neutrality by 2050.	[39]
Land Use, Land-Use Change and Forestry (LULUCF) Regulation (adopted in 2018, amended in 2023)	– Supports farming practices that store carbon and reduce land-use related CO_2 emissions. – The LULUCF Regulation outlines how the EU tracks and manages GHG emissions and carbon removals associated with land-use and forestry.	[40]

Table 1. *Cont.*

Legislation/Policy Frameworks	Description	Refs.
European Green Deal (EGD) (adopted in 2019)	The European Commission has introduced a series of proposals to align the EU's climate, energy, transport, and taxation policies with the goal of reducing net GHG emissions by at least 55% by 2030 compared to 1990 levels.	[41]
Farm to Fork Strategy (initiated in 2020)	It is a policy framework within the EGD aimed at creating a sustainable food system. It consists of actions designed to reduce GHG emissions by increasing organic farming and promoting sustainable agricultural practices (such as lower input usage).	[42]
EU Climate Law (ECL) (adopted in 2021)	The law also establishes an intermediate target of reducing net GHG emissions by at least 55% by 2030, compared to 1990 levels, and promotes sustainable land-use, emission-reduction strategies in agriculture.	[43]
Common Agricultural Policy (CAP) (launched in 1962; in 2021, the agreement on the reform of the CAP was formally adopted)	The CAP offers both legislative and financial instruments that encourage sustainable farming methods aimed at cutting GHG emissions, conserving natural resources, and thereby helping to implement the EU's climate goals within the agricultural sector.	[44]

Considering national targets under the Effort Sharing Regulation (ESR), GHG emissions varied significantly among EU Member States between 2005 and 2022. For example, Estonia, Latvia, and Bulgaria recorded substantial increases of 31.46%, 25.85%, and 19.72%, respectively (Figure 8) [19]. In contrast, Croatia and Slovakia achieved the largest reductions, with decreases of 26.59% and 23.68%. Under current measures, projections indicate further increases in emissions for some countries, such as Estonia and Bulgaria, while others, including Croatia and Denmark, are expected to see significant reductions.

Additionally, countries like Ireland, Austria, and Luxembourg, which currently show increasing trends, are projected to reverse these patterns and achieve notable emission reductions by 2030 if additional mitigation measures are implemented. These measures may include enhanced manure management techniques (such as anaerobic digestion, improved storage), adoption of precision agriculture to optimize fertilizer use, reduction in livestock numbers where feasible, and implementation of agroforestry or carbon sequestration practices.

When comparing the evolution of emissions among countries, Estonia and Bulgaria show the highest projected increases in GHG emissions by 2030 under both current and additional measures. In contrast, Croatia is projected to achieve the greatest overall reduction, with emissions expected to fall by more than 26% by 2030.

At the EU-27 level, emissions decreased by 5.45% between 2005 and 2022. With the implementation of additional measures, projections indicate a further reduction of 7.37% by 2030 [19].

Although the reductions in agricultural GHG emissions achieved by the EU need to be maintained, or even further decreased through the implementation of additional measures, it is important to highlight that these legislative regulations have been effective so far. This is especially evident when analyzing the trend of GHG emissions globally over time (Figures 3–5) [25–27,45]. This demonstrates the critical role of legislation in driving emissions reductions in one of the most challenging sectors for climate policy.

Country	Percentage change 2005–2022	Projected percentage change 2005–2030 with existing measures	Projected percentage change 2005–2030 with additional measures
Estonia	31.46	36.59	36.59
Latvia	25.85	22.52	22.52
Bulgaria	19.72	38.27	38.27
Ireland	11.31	9.89	-9.96
Poland	5.17	11.82	11.82
Luxembourg	4.17	6.84	-17.9
Hungary	3.81	19.14	19.14
Czechia	2.81	-5.35	-5.35
Cyprus	1.54	21.38	21.38
Austria	1.12	-8.95	-17.01
Portugal	0.2	-0.99	-3.54
Spain	-0.16	-8.85	-18.56
Lithuania	-0.38	-5.11	-9.18
The Netherlands	-1.79	-4.96	-4.96
Finland	-2.28	-10.89	-14.3
Sweden	-2.54	-10.54	-10.54
Slovenia	-3.55	6.64	0.52
EU-27	-5.45	-4.16	-7.37
Malta	-5.57	-9.18	-9.18
Denmark	-6.84	-19.07	-19.07
Belgium	-7.36	-8.11	-24.47
Germany	-8.5	-13.15	-14.34
Italy	-12.17	-11.66	-14.42
Romania	-13.52	5.72	3.41
France	-13.8	-6.65	-6.65
Greece	-14.53	0	0
Slovakia	-23.68	6.59	-3.82
Croatia	-26.59	-20.82	-26.37

Legend (percentage change in emissions)

40, 35, 30, 25, 20, 15, 10, 5, 0, -5, -10, -15, -20, -25, -30

Figure 8. Projected percentage change in emissions by country (2005–2030).

5. Carbon Dioxide Emissions from Agricultural Practices

Agricultural systems contribute significantly to CO_2 emissions through both natural processes and human activities. Major sources include deforestation and land-use change, which release large amounts of carbon stored in vegetation and soils [46].

Additional emissions arise from burning crop residues [47], fossil fuel use in machinery [48], and soil tillage [49], which accelerates the breakdown of organic matter (Figure 9).

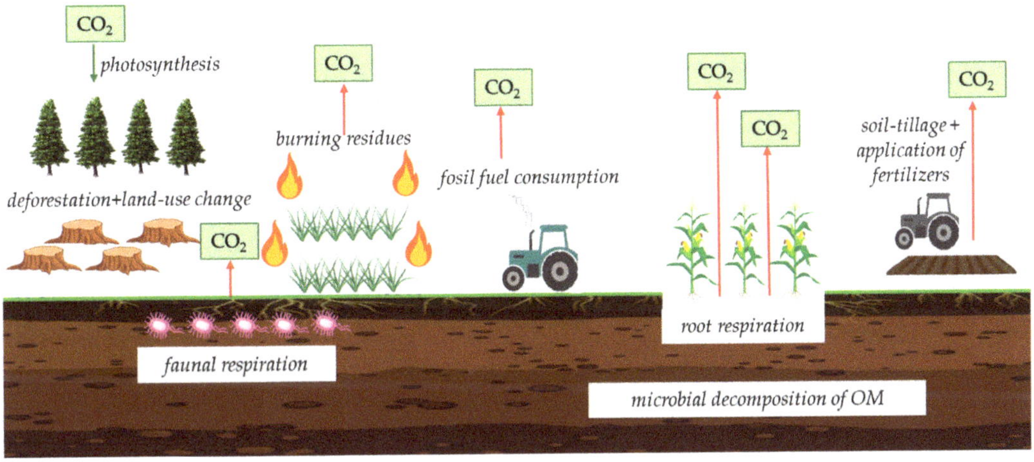

Figure 9. Key processes in agricultural soils responsible for CO_2 emissions.

Continuing with the sources of CO_2 in the agricultural area, it should also be mentioned that soil microbial respiration, root respiration, and organic matter decomposition significantly contribute to CO_2 emissions.

While photosynthesis in crops and forests can sequester carbon, unsustainable land management practices often result in agriculture being a net source of CO_2 emissions. Reducing land conversion, improving soil management, and minimizing fossil fuel use are key strategies for lowering the agricultural sector's carbon footprint.

5.1. Burning Crop Residues

Detecting areas with crop residue burning is performed using satellite imagery (Sentinel-2A and Sentinel-5P) and the Google Earth Engine platform, which allows for estimating the resulting CO_2 emissions [50,51]. In EU countries, the open burning of agricultural residues is prohibited under EU Regulation 1306/2013 [52]. Consequently, open burning practices have declined markedly, particularly in the Baltic states, where the area affected by burning has significantly decreased since their EU accession [53].

5.2. Land Management Practices

Considering the impact of tillage practices on CO_2 emissions, a long-term field study was conducted from 2017 to 2022 to evaluate their effects under varying weather conditions, with a focus on mitigation strategies [54]. The study compared conventional tillage (CT), reduced tillage (RT), and no-tillage (NT) within a crop rotation system. Results showed that CT produced the highest CO_2 emissions, particularly during the summer. In contrast, RT and NT, which retain mulch on the soil surface, significantly reduced CO_2 emissions by approximately 45% and 51%, respectively, compared to CT. The study highlights reduced and no-tillage as effective strategies to mitigate soil CO_2 emissions, supporting EU climate goals.

A similar study [55] conducted over four years investigated CO_2 emissions associated with tillage practices in a wheat–maize cropping system. Results showed that no-tillage (NT) reduced annual CO_2 emissions by 20.5% compared to conventional tillage (CT), with a 28.7% reduction during the maize season but an 8.99% increase during the wheat season. Maize accounted for 70–78% of the total soil CO_2 emissions, and its emission intensity was reduced by 35.9% under NT relative to CT. Furthermore, a study [56] revealed that zero tillage and chisel plow tillage significantly reduced CO_2 emissions compared to conventional plow tillage.

In Mediterranean maize cultivation under low water input management, a study [57] revealed that minimum tillage (MT) significantly reduced CO_2 emissions from soil bacterial activity compared to conventional tillage (CT) using moldboard plowing. In this context, cumulative CO_2 emissions were reduced by approximately 30% during the three months following autumn plowing and by about 28% during the spring–summer growing season. Notably, these reductions were achieved without compromising maize yields.

5.3. Microbial Soil Respiration

Heterotrophic respiration, the decomposition of soil organic matter by microorganisms, is a key process driving soil carbon emissions. Soil CO_2 emissions are influenced by both biological and environmental factors, with temperature and moisture playing particularly critical roles. Rising soil temperatures enhance enzymatic activity, thereby accelerating the decomposition of organic matter and increasing root respiration. This temperature—driven increase in soil respiration can intensify atmospheric CO_2 concentrations, reinforcing climate warming through a self-reinforcing process [58].

Another environmental factor that influences soil CO_2 emissions is soil organic matter. Elevated levels of soil organic matter provide a larger carbon substrate for microbial decomposition, potentially resulting in higher CO_2 emissions when environmental conditions promote respiration [59]. For example, the most significant source of soil-derived CO_2 emissions in Europe is the drainage of peatlands (organic soils). Nearly half of the total soil carbon stock in the EU-27 is concentrated in Sweden, Finland, and the United Kingdom, largely due to their extensive peatland areas. Ireland, Poland, Germany, Norway, and the Baltic states also have large areas of organic soils [60].

A recent model-based study [61], supported by empirical data, estimates that global heterotrophic respiration has risen by about 2% per decade since the 1980s. Under a worst-case emissions scenario, the model projects a 40% increase by the end of the 21st century, with the Arctic region showing the most pronounced rise.

Another study [62] investigated how different land uses (upland fields, rice paddies, and woodlands) affect the rate and carbon isotopic composition ($\delta^{13}CO_2$) of soil-respired CO_2. Through long-term soil incubation and aggregate fractionation, the authors observed that both CO_2 emissions and their $\delta^{13}C$ values declined exponentially over time. The highest respiration rates were recorded in woodland soils, likely due to a higher availability of carbon substrates for microbial activity. A positive correlation between the $\delta^{13}C$ of respired CO_2 and microbial biomass indicated a strong link between microbial activity and isotopic signatures. These results suggest that soil respiration and its isotopic characteristics respond in complex, land-use-dependent ways to warming, influenced by both soil type and microstructural properties.

The influence of nitrogen fertilization and native soil organic matter on CO_2 emissions under different climatic conditions was investigated using a two-year field experiment in contrasting upland cropping systems representing warm and cold climates [63]. To trace the sources of CO_2 emissions, ^{13}C-labeled urea was applied at 0%, 50%, 100%, and 200% of the Korean recommended nitrogen rates. Red pepper was grown during the warm season with 90 kg N ha^{-1}, and garlic during the cold season with 250 kg N ha^{-1}. In the warm season, CO_2 emissions and urea-derived CO_2 increased with higher nitrogen levels and warmer temperatures, likely due to enhanced microbial activity and root respiration. In contrast, both total and urea-derived CO_2 emissions were minimal and unaffected by nitrogen levels in the cold season.

The impact of three organic amendments (chicken manure, milorganite, and dairy manure) applied at four rates (0, 168, 336, and 672 kg N ha^{-1}) on soil CO_2 emissions was evaluated [64]. The results showed that soil CO_2 emissions were significantly influenced by both amendment type and application rate, with chicken manure and milorganite producing higher emissions than dairy manure. Furthermore, the type and rate of organic amendments had a greater effect on CO_2 emissions than environmental factors such as air/soil temperature and rainfall.

Concluding, sustainable land management practices (reduced tillage, elimination of residue burning, optimized nitrogen and organic amendment use) play an important role in mitigating soil-derived CO_2 emissions from agricultural systems. Climate factors like soil temperature and moisture strongly affect soil respiration rates, highlighting the need for context-specific mitigation. Peatland drainage remains a key CO_2 hotspot, requiring targeted management. Integrating these strategies is essential for reducing agriculture's carbon footprint and enhancing adaptation to climate change.

6. Methane Emissions from Cropping Systems and Livestock

6.1. *Emissions from Cropping Systems*

The main source of CH_4 in agriculture is paddy fields, with global emissions estimated at around 31–280 Tg yr^{-1}, accounting for approximately 10–20% of anthropogenic CH_4 emissions. The wide range reflects variability due to different estimation methods, geographical distributions, water management practices, and rice cultivars [65].

CH_4 is produced through several chemical processes that start with organic matter, mediated by methanogenic (anaerobic) bacteria. The resulting methane is then released via diffusion (the slow movement of molecules from soil pores to the surface), ebullition (the release of methane bubbles into the atmosphere), or plant-mediated transport (the movement of methane from soil through plant roots and stems) [66]. Alternatively, it can be oxidized to CO_2 by methanotrophic bacteria (Figure 10) [18,66]. Regarding environmental factors, daytime emissions tend to be higher than nighttime emissions and also increase with elevated CO_2 levels [67].

CH_4 emissions from rice fields depend on several factors, including soil characteristics, plant traits, environmental conditions, and agricultural practices [67]. A high level of organic matter (manure, crop residues) provides carbon substrates for methanogens, thereby generally increasing methane emissions. In contrast, the addition of biochar has been shown to reduce CH_4 emissions [68]. Methanogenic microbial activity is also strongly influenced by soil pH and redox potential [69]. Additionally, different rice cultivars can emit varying amounts of methane due to physiological differences.

Fertilizers can either stimulate or inhibit the activity of methanogens and methanotrophs. For example, the use of organic fertilizers and the incorporation of rice straw into the soil tend to increase CH_4 emissions, whereas chemical fertilizers may have a lower impact. A study [69] found that the application of phosphorus (P) and potassium (K) fertilizers significantly reduced methane emissions from rice paddy soils. Furthermore, the combined application of nitrogen, phosphorus, and potassium (N + P + K) fertilizers was shown to reduce CH_4 emissions per unit of grain yield in both dry and wet seasons (6.43 kg CH_4 per Mg of grain in the dry season and 83.57 kg CH_4 per Mg in the wet season).

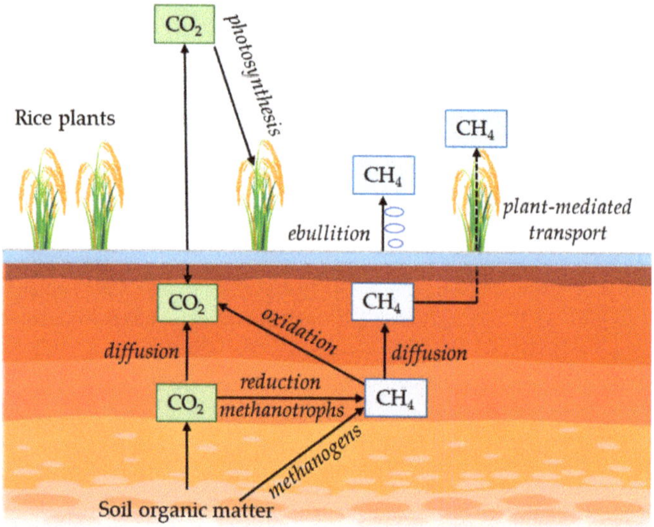

Figure 10. Key processes in paddy fields responsible for CH_4 emissions.

Considering the necessity to reduce CH_4 emissions from paddy rice cultivation, a study [67] evaluated the impact of various organic amendments on CH_4 release. The treatments were composed of control, blue-green algae (BGA), *Azolla*, farmyard manure (FYM), green leaf manure (GLM), BGA + *Azolla*, FYM + GLM, BGA + *Azolla* + FYM + GLM, vermicompost, and decomposed livestock manure. Among these, the combination of BGA and *Azolla* achieved the highest reduction in CH_4 emissions (37.9% compared to the control), followed by BGA alone. The findings demonstrate that biofertilizer application represents a viable and effective approach to mitigating CH_4 emissions in rice cultivation.

In addition, among biofertilizers, purple nonsulfur bacteria (PNSBs) are notable for their ability to reduce methane emissions in rice cultivation. For example, a study [70] demonstrated that *Rhodopseudomonas palustris* strains TN114, PP803, and TK103 reduced total methane emissions by 24–48% in organic and saline paddy soils, respectively, compared to those where chemical organic fertilizers were applied.

Given the increasing importance of reducing CH_4 emissions from paddy fields, agronomists should prioritize the evaluation of water management strategies, fertilization regimes, crop rotation systems, and the selection of low-emission rice cultivars as effective mitigation approaches [71].

Methane estimation using conventional methods has become laborious, time-consuming, and, to some extent, inaccurate. Recently, to overcome these limitations, modern technologies based on remote sensing have been adopted [72,73]. Furthermore, to facilitate the adoption of effective reduction measures, models for estimating emissions have been developed. In this context, empirical models have been developed based on variables such as soil organic carbon, pH, water management, application of organic amendments, and crop yield [74–76].

Furthermore, Nikolaisen and colleagues [77], relying on accepted methodologies, compared existing models across different global regions, assessing their performance and limitations. The study concludes that existing empirical models are adequate for predicting methane emission trends but are not reliable for estimating the magnitude of emissions. These models often either underestimate or overestimate emissions, with correlation values varying significantly. To improve emission predictions, the authors suggest that future models incorporate site-specific variables such as soil texture, planting methods, cultivar types, and growing seasons, which significantly influence emissions.

In conclusion, paddy fields are a major source of CH_4 emissions, primarily driven by methanogenic activity under anaerobic conditions. Emissions are amplified by high organic inputs but can be effectively mitigated through improved water and fertilizer management, the use of low-emission rice cultivars, and biofertilizers. Balanced NPK fertilization and alternative crop rotations further help reduce CH_4 intensity. Given the difficulty of direct measurements, empirical and remote-sensing models play a key role in estimating emissions and informing mitigation strategies in rice systems.

6.2. *Livestock Emissions*

Livestock emissions arise from feed production, enteric fermentation, manure management, and land-use change, contributing about 7.1 Gt of CO_2e annually, or 14.5% of global anthropogenic GHG emissions. Approximately 30% of global methane emissions come from enteric fermentation [78]. Cattle account for almost 77% of these emissions, while buffalo and small ruminants contribute significantly less, 13.55% and 9.52%, respectively [79].

6.2.1. Enteric Fermentation as Methane Source

Enteric fermentation is a natural digestive process in ruminant animals (cattle, goats, sheep, buffalos), during which the feed is broken down and fermented by microbes in their digestive system. In addition to producing energy and protein, this process also generates CH_4, which is primarily released through burping [79].

Methanogenesis in ruminants is a complex biochemical process by which methanogenic archaea produce CH_4 from substrates derived primarily from the fermentation of feed. This process has been comprehensively reviewed by Beauchemin and colleagues [80].

In summary, within the rumen, methanogens primarily convert H_2 and CO_2 into methane via hydrogenotrophic methanogenesis, alongside competing microbial processes, such as sulfate reduction, nitrate reduction, and fumarate reduction that also utilize hydrogen. These competing pathways help regulate hydrogen levels and shape the overall fermentation pattern [80,81]. A schematic representation of this process is presented in Figure 11.

Several factors affect CH_4 emissions from ruminants, such as feed intake levels [82,83], the type of carbohydrates in the diet [84], the method of feed processing, the inclusion of lipids or ionophores, and modifications to the rumen microbial community [85].

Furthermore, several studies [86–88] identify seaweeds as a potential dietary strategy for reducing CH_4 emissions from ruminants.

For example, red seaweeds such as *Asparagopsis taxiformis* and *A. armata* contain bioactive compounds capable of reducing enteric CH_4 emissions in ruminants by over 90%, and in some cases up to 99%. However, effectiveness depends on season, species, metabolite content, and animal type, while large-scale application is constrained by production, compound stability, and potential impacts on rumen health [87].

Adjusting these factors through targeted dietary and management interventions can help lower methane emissions and contribute to more sustainable livestock production (Table 2).

Furthermore, CH_4 emissions from ruminants vary a lot between species, breeds, and production conditions. A study [88] summarizing current knowledge on CH_4 production from ruminants found that dairy cows are among the highest emitters, with values ranging from 151 to 497 g day^{-1} (lactating cows: 354 g day^{-1}; dry cows: 269 g day^{-1}; heifers: 223 g day^{-1}). Mature beef cows follow, with emissions ranging from 240 to 396 g day^{-1}, while beef cattle overall emit 161–323 g day^{-1}. Sheep produce much lower

CH_4, Suffolk sheep producing 22–25 g day^{-1}. The study also highlighted breed differences within species and the influence of grazing management on emissions; for example, heifers grazing fertilized pasture emit 223 g day^{-1}, compared to 179 g day^{-1} on unfertilized pasture.

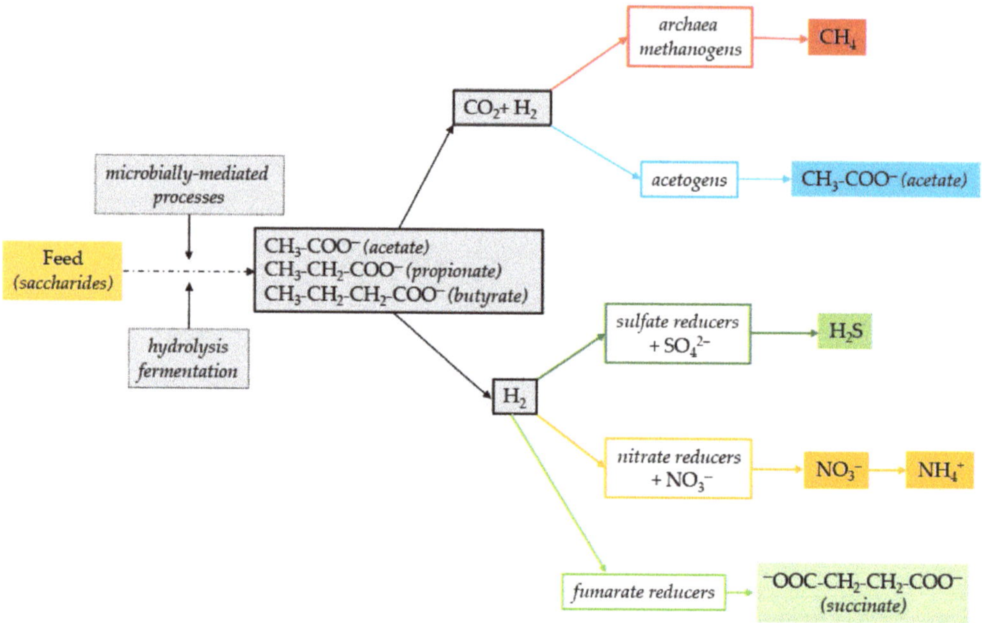

Figure 11. Microbially mediated hydrogen utilization and methanogenesis in ruminants.

Another study [89] found that CH_4 emissions in dairy cows are influenced by age, with levels peaking in mid-life. Similarly, in Brown Swiss cows, this trend cannot be directly linked to changes in intake, digestion, or chewing activity; however, the corresponding age-related pattern in fiber digestibility suggests it may contribute to the observed emissions pattern.

Also, dietary protein content plays a crucial role in shaping nitrogen excretion patterns and the resulting greenhouse gas (GHG) emissions during manure storage. In this regard, a study [90] demonstrated that lowering the dietary protein content of early-lactating cows reduced N_2O emission rates in most manure types, but increased CH_4 emissions from urine-rich slurry. Consequently, dietary manipulation emerges as an effective strategy for mitigating the environmental impact of livestock production, particularly in terms of reducing nitrogen losses.

In conclusion, enteric CH_4 emissions from ruminants are influenced by diet, feed processing, and animal type. Effectively reducing emissions requires an integrated approach combining strategic feeding interventions, additives (Table 2), genetic selection, and health management, enabling sustainable livestock production without compromising productivity or welfare.

Table 2. Strategies for mitigating CH_4 emissions in ruminant livestock.

	Improve Feeding Strategies	
Forage quality	Adjusting forage quality and concentrate-to-forage ratio can reduce CH_4 by up to 70%; younger/high-quality forage increases propionate and reduces H_2 for methanogenesis.	[91,92]
	Inclusion of legumes improves forage quality; tannin-containing legumes inhibit rumen fermentation. High alfalfa (78%) lowers CH_4 ~10%, moderate (40%) has no effect.	[93]
	Maize silage lowers CH_4 in dairy and cattle.	[94]
	Feed additives	
Seaweeds	*Asparagopsis taxiformis* reduces CH_4 up to 99% (bromoform inhibits methanogenesis); health concerns with heavy metals.	[95]
	Pelleted *Asparagopsis* in beef diets reduces CH_4 ~37.7%.	[96]
	Red, brown, and green seaweeds contain polysaccharides altering rumen microbes and CH_4 emissions.	[97]
Bee propolis extracts	Red propolis reduces CH_4 in ewes by modulating rumen fermentation.	[98]
Saponins (glycosides)	Decrease methanogenic archaea and protozoa, reducing CH_4.	[99]
Ionophores (antibiotics)	Monensin, lasalocid, narasin, and salinomycin reduce H_2 available for methanogens; commonly used in beef diets.	[100]
Yeast	Alternative to ionophores (monensin); modulate rumen microbiome and fermentation, lowering CH_4.	[101,102]
Organic acids	Formic, fumaric, malic acids and salts; alfalfa + fumaric acid enhances fermentation and reduces CH_4.	[102,103]
3-nitrooxypropanol (3-NOP)	Specific methanogenesis inhibitor; long-term reduction in dairy cows (~21% daily CH_4).	[104]
Nitrate (NO_3^-)	Alternative H_2 sink; reduces CH_4 ~13.9–30%; caution due to nitrite toxicity risk.	[105]
	Genetic breeding	
	Selective breeding for low-CH_4 cows; Nordic Methane Index targets ~20% reduction.	[106,107]
	Improve animal health	
	Enhances digestion efficiency, reduces CH_4; vaccination trials (sheep) inhibit methanogens; further research needed.	[108,109]

6.2.2. Manure as Methane Source

In addition to CH_4 emissions produced directly by ruminants, another significant source is manure. CH_4 from manure results from the anaerobic decomposition of organic materials in storage systems where oxygen availability is limited. The volume of CH_4 generated is influenced by factors such as the quantity and type of manure, the employed manure management practices, storage conditions, and climatic conditions [110].

For example, a study [111] demonstrated that the type of livestock manure significantly influences CH_4 emissions. In this study, six types of livestock manure (cow, buffalo, goat, rabbit, chicken, and duck) were analyzed over an 8-week period to determine CH_4 production. The results showed the following order of methane emissions (in mg g^{-1}): rabbit manure (2.70) < goat manure (6.01) < chicken manure (17.88) < cow manure (20.32) < buffalo manure (21.93) < duck manure (97.99). These differences were attributed to factors such as the animals' diet, physiological status, and manure management practices.

Another study [110] examined temperature effects on CH_4 emissions from liquid dairy manure. Under summer conditions, CH_4 production started after ~1 month, reaching 0.148 kg CH_4 kg^{-1} VS (volatile solid) over 40 weeks. In winter, emissions were much lower, peaking at 0.0011 kg CH_4 kg^{-1} VS over 20 weeks. The differences are due to temperature: higher summer temperatures enhance microbial activity and volatile solids decomposition, increasing CH_4, while lower winter temperatures suppress microbial metabolism and CH_4 production.

According to a study conducted in Romania [112], sheep manure emits significantly more greenhouse gases per unit area than cattle manure. Specifically, sheep manure produces approximately 0.83 tonnes of CH_4 and 61.3 tonnes of CO_2 per year, while cattle manure emits about 0.185 tonnes of CH_4 and 4.7 tonnes of CO_2 annually.

Table 3 summarizes key strategies used to mitigate CH_4 emissions from livestock manure, spanning biological, chemical, and physical approaches and reflecting the diversity of technologies applicable across different manure management systems.

Table 3. Strategies for mitigating CH_4 emissions from manure.

Control Measure	Effects	Ref.
Anaerobic digestion	Manure in sealed digesters produces CH_4 (biogas) for renewable energy; biochar addition can increase CH_4 yield up to 80%; ciprofloxacin can inhibit CH_4 production in hen manure.	[113–115]
Composting	Aerobic composting reduces CH_4 by altering nutrients and microbial activity; biochar (cornstalk, bamboo, woody, coir) further reduces CH_4; cornstalk biochar reduced CH_4 by 26.1%.	[116,117]
Manure covering	Reduces CH_4 emissions; polyethylene cover lowers pig manure CH_4 by 88%; straw reduces CH_4 by 26–50%.	[118]
Solid–Liquid separation	Mechanical separation divides manure into solid (organic matter) and liquid (nutrients); screen size affects CH_4 emissions.	[119,120]
Lower storage temperature	Higher temperatures increase CH_4 due to faster microbial activity and organic matter degradation.	[121]
Manure acidification	Adding acids (H_2SO_4, HCl, HNO_3, H_3PO_4) lowers pH, inhibiting methanogens; CH_4 reductions: 46–96% (pig slurry), 67% with 17.1 kg H_2SO_4/m^3, pH 5.5 $\rightarrow$ 95–99% (pig)/65–99% (cattle).	[122–125]
Additives	$CaSO_4$ reduces CH_4 by 63% (higher than H_3PO_4 54%, lower than H_2SO_4 91%); polyphenols + NaF reduce CH_4; commercial additives like SOP LAGOON also effective.	[126–128]

In conclusion, the effectiveness of CH_4 mitigation strategies is highly dependent on site-specific factors (manure composition, storage conditions), so tailored approaches that combine multiple methods could be adopted. Finding efficient additives may represent a new challenge, the resolution of which could contribute to CH_4 emission reduction. However, the identification of a method that is easy to implement yet effective largely depends on farm conditions and available resources.

In this regard, the most accessible and efficient methods are aerobic composting (combined with different types of biochar for increased effectiveness) and manure covering with various materials. Both methods are accessible, do not require large infrastructure investments, and can result in significant methane emission reductions, making them effective short- to medium-term solutions.

7. Nitrous Oxide Emissions from Agricultural Practices

Agricultural N_2O emissions primarily originate from the application of fertilizers and manure. Once added to the soil, these nitrogen sources undergo a series of microbially mediated processes, including the mineralization of organic matter, nitrification, and denitrification, which ultimately lead to the production and release of N_2O (Figure 12) [34,35,129,130].

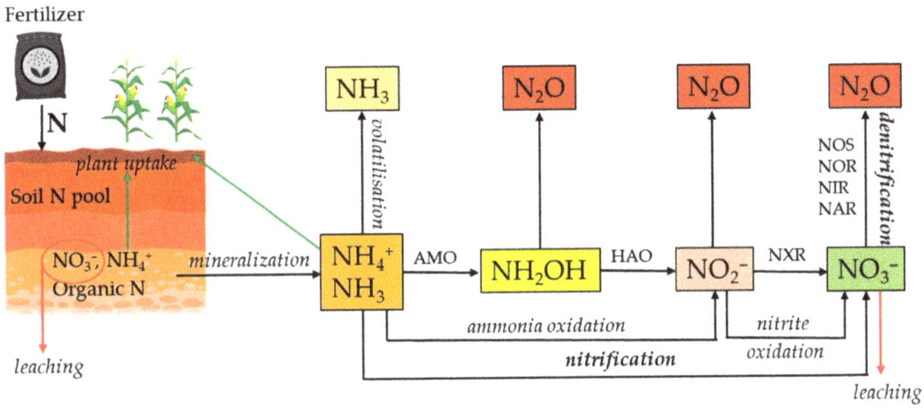

AMO = ammonia monooxygenase, **HAO** = hydroxylamine oxireductase, **NXR** = nitrous oxireductase, **NAR** = nitrate reductase, **NIR** = nitrite reductase, **NOR** = nitric oxide reductase, **NOS** = nitrous oxide reductase

Figure 12. Key processes in agricultural soils responsible for N_2O emissions.

N_2O emissions from soils are influenced by multiple factors related to soil properties and management practices (Figure 13) [32,131–140]. Soil properties such as moisture, texture, temperature, and pH, along with microbial activity, play critical roles in regulating N_2O emissions [32,131–135]. Management practices, including fertilizer application, tillage, residue handling, and irrigation, significantly influence these emissions by altering soil nitrogen availability and microbial processes [135–140].

Effective strategies for reducing agricultural N_2O emissions must integrate both soil properties and management practices, including their timing (Figure 14). These factors should be evaluated not only independently but also in combination, as their interactions significantly influence N_2O emission dynamics. Subsequent sections detail key mitigation strategies for N_2O emissions, as developed and applied by experts across various regions.

7.1. Fertilizers—Types, Application Rates, Timing

A recent study [136] emphasized the need for year-round monitoring and adaptive nitrogen management to mitigate N_2O emissions in semi-arid agroecosystems. Over a three-year canola–wheat rotation, applying the 4R nutrient strategy (Right source, Right rate, Right time, Right place) with reduced N rates and enhanced-efficiency fertilizers lowered cumulative N_2O emissions by 57% without affecting yield, mainly by improving nitrogen use efficiency and limiting emissions under high soil moisture conditions.

The type and rate of nitrogen fertilizer strongly affect N_2O emissions in sugarcane systems [141]. Both ammonium nitrate and urea caused emission pulses, but with different timing and intensity: ammonium nitrate produced rapid and high emissions within a day of application, while urea led to delayed peaks after several days. Emissions from ammonium nitrate increased linearly with N rate, whereas urea showed a non-linear response, peaking near 114 kg N ha^{-1}.

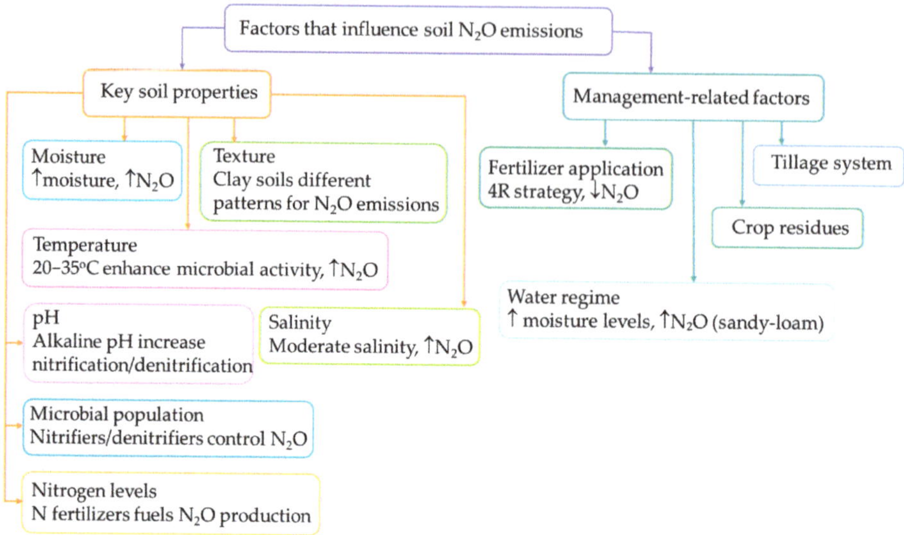

Figure 13. Factors that influence N_2O emissions.

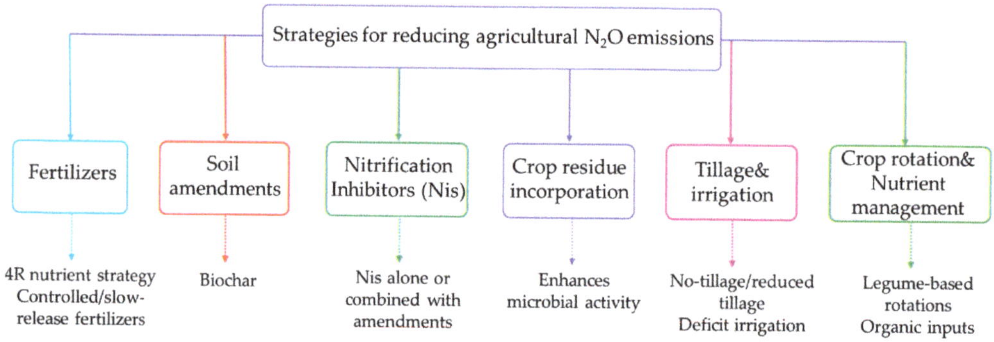

Figure 14. Effective strategies for N_2O mitigation.

Another study [142] indicated that calcium ammonium nitrate (CAN) tends to reduce both ammonia volatilization and N_2O emissions compared to urea, making it a preferred nitrogen source in some sugarcane systems. Furthermore, switching from calcium ammonium nitrate (CAN) to urea-based fertilizers, particularly those stabilized with urease and nitrification inhibitors, proved to be an effective mitigation strategy for reducing N_2O emissions in wet, temperate grassland systems, with emission factors as low as 0.25% [143].

N_2O emissions from sugarcane fields increased sharply with higher nitrogen application rates [144]. Peak emissions occurred within four weeks of fertilization, especially at 500 kg N ha^{-1} under warm and moist soil conditions. Reducing the rate to about 340 kg N ha^{-1} maintained yield while cutting N_2O emissions by over 65%, demonstrating the strong mitigation potential of optimized nitrogen management.

In addition, a study [145] presents findings on the impact of replacing mineral fertilizers with organic amendments on GHG emissions, showing that while this substitution significantly reduces N_2O emissions, it simultaneously increases CH_4 and CO_2 emissions, leading to a net increase in GWP.

A study in North-West Germany [146] compared N_2O emissions from winter wheat fields fertilized with urea (U), ammonium sulfate (AS), and calcium ammonium nitrate

(CAN) at 220 kg N ha^{-1}. All fertilizers significantly increased emissions compared with the unfertilized control, with cumulative losses following the order U > AS > CAN (0.11–0.28% of applied N). Among them, CAN appeared most effective in minimizing N_2O emissions.

Furthermore, slow-release fertilizers (Table 4) can reduce N_2O emissions by releasing nitrogen more gradually, thereby minimizing nitrogen losses and lowering peak emissions compared to conventional fertilizers.

Table 4. Effectiveness of controlled/slow-release fertilizers in reducing N_2O emissions across cropping systems.

Product	Crop/Details	Effects	Refs.
Slow-release fertilizer + cyhalofop-butyl	Rice paddies	Lowest cumulative N_2O emissions and highest yield among all treatments; inhibits nitrification in paddy soils.	[147]
Controlled-release coated urea (CRCU)/calcium nitrate (CRCN)	Soybean (after rice)	14–41% lower N_2O emissions vs. ammonium chloride; effective even at reduced N rates.	[148]
Polymer-coated urea (PCU)	Potato	Increases N availability but may raise N_2O emissions when nitrate leaching risk is low.	[149]
Controlled-release fertilizer (CRF) vs. ammonium sulfate (AS)	Temperate soils	CRF soils emitted less N_2O; controlled nutrient release reduced emission peaks.	[150]
Controlled-release fertilizer (CRF) vs. urea	Wheat	29–66% reduction in N_2O; inhibits nitrification processes; stable under high CO_2 conditions without yield loss.	[151]
Biochar-based CRF	Leaching columns	7–66% reduction in N_2O emissions and lower nitrate leaching.	[152]

7.2. Soil Amendments—Biochar

Biochar is a valuable soil amendment for reducing N_2O emissions through its effects on soil structure and microbial nitrogen cycling, especially denitrification [153].

Studies have demonstrated the potential of biochar to mitigate N_2O emissions from agricultural soils. In energy cane cultivation, biochar application (5 g kg^{-1} soil) reduced N_2O emissions by 56% with N fertilization alone and by 41% when combined with vinasse, without affecting yield [154]. Similarly, in wheat fields, biochar applied at 5–10 Mg ha^{-1} together with urea (150 kg N ha^{-1}) decreased N_2O emissions by 27–35%, mainly due to improved ammonium (NH_4^+) retention, which limited nitrification and N_2O production [155].

A controlled 40-day incubation study [156] examined how soil pH affects the ability of biochar to mitigate N_2O emissions. Biochar alone significantly reduced N_2O emissions in acidic vegetable soil by 47.6% and by 20.8% when combined with nitrogen fertilizer. Its application also raised soil pH across all soil types by 1.43–1.56 units in acidic, 0.57–0.70 in neutral, and 0.29–0.37 in alkaline soils, indicating that biochar's mitigation effect is strongest under acidic conditions.

A three-year study [157] assessed how biochar application rate and frequency affect soil N_2O emissions in a wheat–maize rotation. Biochar significantly reduced emissions, with annual applications of 8 and 12 Mg ha^{-1} decreasing N_2O by 31–56% compared with the control. In contrast, a single 12 Mg ha^{-1} application showed no lasting effect during 2020–2022, indicating that mitigation declines without reapplication. Modeling suggested that the inhibitory effect of one-time biochar application lasts about 1.77 years.

Overall, biochar demonstrates strong potential to reduce N_2O emissions by improving soil properties and microbial processes. However, studies indicate that regular reapplication is necessary to sustain these benefits over time.

7.3. Nitrification Inhibitors

Nitrification inhibitors (NIs) have been recommended by the IPCC [158] as an effective strategy to mitigate nitrous oxide (N_2O) emissions. A variety of nitrification inhibitors (NIs) and their specific applications have been reviewed by Malyan and co-workers [159].

Among these, 3,4-dimethylpyrazole phosphate (DMPP) is the most widely used in Europe due to its high efficiency, low application rate, and minimal ecotoxicity [160].

A study [161] evaluated the effectiveness of a DMPP nitrification inhibitor in reducing N_2O emissions when combined with various alkaline amendments: quicklime (CaO), chicken manure (CM), cow dung (CD), and biochar (BC) on acid soils. It was found that DMPP significantly reduced N_2O emissions when combined with various alkaline amendments, with reductions ranging from 3.3% to 60.2%. The highest mitigation effect occurred with the CaO + DMPP treatment, due to reduced activity of ammonia-oxidizing bacteria (AOB).

The NIs, 3,4-dimethylpyrazole phosphate (DMPP) and 3-methylpyrazole 1,2,4-triazole (Piadin), were investigated for sweet corn, and it was found to reduce N_2O emissions. Additionally, applying NI-coated urea at 20% reduced nitrogen rates decreases cumulative N_2O emissions by 51% without reducing the yield [162].

Corn field experiments demonstrated that applying urea combined with the NI dicyandiamide (DCD) at 170 kg urea ha^{-1} + 20 kg DCD ha^{-1} reduced N_2O emissions to approximately 3% of those from standard urea application, with emissions comparable to the unfertilized control and no impact on grain yield; in contrast, controlled-release fertilizer (LP-30) reduced emissions relative to urea but was less effective than the urea + DCD treatment, making the latter the most efficient mitigation strategy tested [163].

These studies sustain that NIs are highly effective in mitigating N_2O emissions, especially when combined with soil amendments or reduced nitrogen rates, without affecting crop yields.

7.4. Crop Residue Incorporation

Incorporating crop residues into the soil typically increases N_2O emissions by an average of 43% compared to residue removal. This is because residues provide nitrogen and readily decomposable carbon, which enhance microbial activity, particularly the nitrification and denitrification processes responsible for N_2O production [139].

Evaluation of N_2O emissions from crop residues is very challenging because it depends on various factors, including soil properties (clay content, aeration, pH), climatic conditions [139].

A study [164] describes that incorporation of mature crop residues (soybean: C/N = 75; corn: C/N = 130) led to rapid nitrogen immobilization, reducing N_2O emissions in high-carbon (silty clay) soils. In contrast, in low-carbon (sandy loam) soils, residue addition also caused nitrogen immobilization, but under moderate freeze–thaw (FT) conditions, N_2O emissions increased, likely due to early stimulation of microbial activity by the added carbon. These findings highlight that residue quality and soil carbon content strongly influence the impact of FT cycles on N_2O emissions. Therefore, incorporating mature crop residues can be an effective strategy to reduce FT-induced N_2O emissions, particularly in carbon-rich soils, by promoting mineral nitrogen immobilization.

7.5. Tillage and Irrigation Management

Tillage influences N_2O emissions primarily by altering soil physical conditions, microbial activity, and nitrogen cycling processes [32].

No-tillage (NT) and reduced tillage (RT) practices generally lower N_2O emissions compared with conventional tillage (CT). For example, NT reduced emissions by 51.1% in maize [165], while RT lowered emissions by 15–40%, likely due to reduced easily mineralizable soil components [166]. Combining NT with organic amendments, such as liquid pig manure, further decreased N_2O emissions compared to CT and chemical fertilizers in soybean systems, with reductions ranging from 20% (no fertilizer) to 35.7% (hairy vetch) [167].

However, NT does not always reduce N_2O emissions, as they depend on soil properties, moisture, temperature, residue management, and fertilization. A study in Ontario (2003–2005) showed variable responses: no-till corn had higher emissions in wet years, while tilled soils emitted more in 2005 due to enhanced aeration; soybean fields exhibited consistently low emissions due to limited biological N_2 fixation [168].

Deficit irrigation is another effective mitigation strategy, maintaining optimal soil moisture and reducing prolonged saturation that favors N_2O production [169]. Deficit irrigation reduced N_2O by 17.4% in wheat and 15.5% in maize [170], while moderate and severe deficit irrigation under drip-fertigation in maize decreased emissions by 15% and 40%, respectively, compared to full irrigation [171].

7.6. Crop Rotations and Integrated Nutrient Management

A field experiment evaluated the effects of continuous mineral and organic N fertilization on N_2O emissions in a maize–wheat rotation [172]. Mineral N at 200 kg ha^{-1} produced 2.71 kg N ha^{-1} of N_2O annually. Replacing 50% of mineral N with cattle, chicken, or pig manure did not significantly change emissions, whereas a 25% replacement reduced N_2O by 21–38%, mainly during the maize season. These reductions occurred without compromising crop yield, indicating that partial substitution of mineral N with organic manure can effectively mitigate N_2O emissions.

A study [173] evaluated the effect of reduced inorganic N input combined with organic amendments on N_2O emissions in a winter wheat–summer maize rotation. Three treatments were tested: RN (100% recommended N), RN40% + HOM (40% N + homemade organic matter), and HAN (zinc and humic acid-coated urea at full N rate). RN40% + HOM reduced cumulative N_2O emissions by 41% compared to RN and by 20.9% compared to HAN, while maintaining high yields, making it the preferred option.

A two-year on-farm study in northwest Germany [174] assessed the impact of converting from conventional to organic arable farming on N_2O emissions. Two organic rotations were tested: (1) with 25% legumes (grass + clover–winter wheat–winter rye–oats) and (2) with 40% legumes (grass + clover–winter wheat–winter rye–spring field peas–winter rye), compared to (3) a conventional rotation (winter oilseed rape–winter wheat–winter wheat–sugar beet–winter wheat). Two non-cropped systems served as references: (4) extensive grassland and (5) beech forest. The lowest N_2O emissions occurred in grassland (0.3 kg N $ha^{-1}yr^{-1}$) and forest (0.4 kg N ha^{-1} yr^{-1}), while organic systems emitted about 0.7 kg N ha^{-1} yr^{-1}, 66% less than the conventional system (2.1 kg N ha^{-1} yr^{-1}).

These studies demonstrate that integrating organic inputs or legume-based rotations can significantly reduce N_2O emissions across various cropping systems, without compromising crop productivity.

Overall conclusions that result from the above studies show that various agricultural practices hold significant potential to mitigate N_2O emissions. The most effective strategies include optimized fertilizer management, the use of nitrification inhibitors, biochar application, and organic amendments. Furthermore, integrating diverse crop rotations, adopting conservation or reduced tillage systems, and implementing efficient irrigation management can further contribute to emission reductions. However, the effectiveness of these mitigation approaches depends strongly on local soil characteristics, climatic conditions, and specific cropping systems. Therefore, continuous research, long-term field monitoring, and the adaptation of these practices to regional contexts are essential to achieve sustainable and long-term reductions in greenhouse gas emissions from agriculture, particularly in the context of ongoing climate change.

8. Gaps, Challenges, and Future Perspectives for Reducing Agricultural GHG Emissions

This section presents a comprehensive overview of the challenges associated with greenhouse gas (GHG) emissions from agriculture, highlighting key barriers such as economic constraints and policy issues, existing gaps in scientific knowledge, like measurement limitations and regional variability, and important unanswered questions related to the effectiveness and net impact of mitigation technologies (Figure 15). Understanding these interconnected factors is essential for developing effective strategies to reduce agricultural GHG emissions.

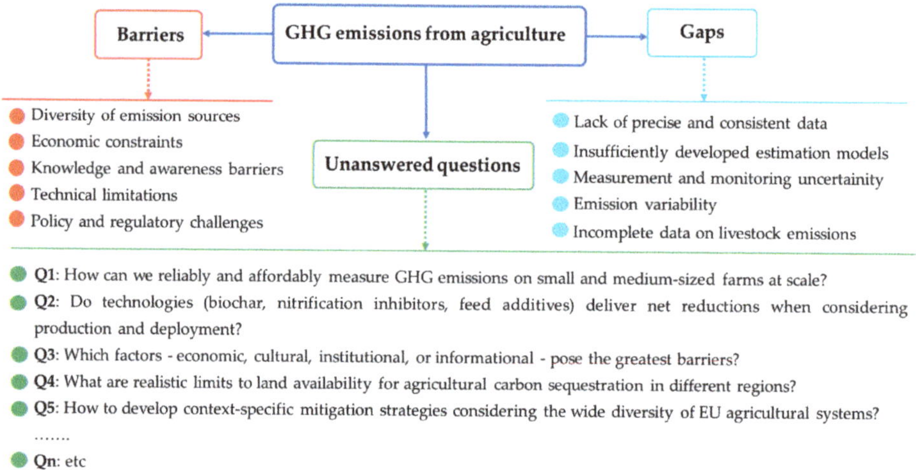

Figure 15. Conceptual framework of challenges and research needs for agricultural GHG emissions.

Although mitigation in the agriculture sector is both urgent and important, passing and implementing measures to reduce emissions can be challenging. Thus, a range of barriers exists at various levels—from the farm to the national and even global scale.

8.1. Barriers

Diversity of emission sources is a key challenge in addressing GHG emissions from agriculture. The sector produces GHGs from multiple sources, including enteric fermentation, manure management, soil microbial activity, and land-use changes. This diversity reflects the wide range of biological and management processes involved in agricultural production. Moreover, the variability introduced by different animal species, manure types, farming practices, and climate conditions adds further complexity to the task of managing and mitigating these emissions effectively.

Considering **economic constraints**, many farmers, particularly those operating small-scale farms, may lack the financial capacity to invest in the infrastructure and technologies required to reduce GHG emissions. The adoption of effective mitigation strategies may be severely constrained in regions where financial resources, institutional support, and technical capacity are insufficient.

Furthermore, **knowledge and awareness barriers** also pose a challenge to reducing agricultural GHG emissions, showing that insufficient information continues to hinder the adoption of sustainable practices. A significant challenge is the resistance from farmers to adopting new technologies or practices that could reduce GHG emissions. This resistance can stem from several factors, including a lack of awareness about the environmental and economic benefits of such practices, perceived risks, and financial constraints. In some cases, farmers may view new technologies as too costly or too complicated to implement. For instance, a study [175] found that while 30% of farmers were open to adopting smart farming technologies (SFT), others preferred production (21%) or organic farming (49%) approaches to mitigate emissions.

Technical limitations in measuring agricultural emissions create significant uncertainty in evaluating both the effectiveness and long-term permanence of mitigation efforts. Although various mitigation measures are available, their implementation is often constrained by technological and practical challenges. For instance, strategies aimed at reducing manure-related emissions typically require substantial investments in infrastructure, technical expertise, and resources—factors often lacking in many agricultural regions, particularly in developing countries or less technologically advanced areas.

Moreover, commonly used measurement methods are frequently labor-intensive, costly, and prone to inaccuracies due to their sensitivity to variables such as soil conditions, climate variability, and farm management practices [176]. Environmental fluctuations and the limited capacity to capture emissions accurately over time and across spatial scales further reduce the overall precision of these methods.

To address these challenges, researchers have developed various **process-based models** that simulate soil–plant–atmosphere interactions, enabling the estimation of greenhouse gas emissions under diverse environmental and management conditions. Several such models have been applied in different contexts [177–179], each offering specific strengths and limitations. Their accuracy depends heavily on factors such as climate, management practices, and the availability of reliable input data. While some models are better suited to certain crops or greenhouse gases, no single model has proven universally accurate.

To improve model performance and reliability, future research should focus on expanding model applicability, ensuring the use of consistent and high-quality validation data, and refining algorithms to better incorporate key biological and environmental processes.

Policy and regulatory barriers to reducing agricultural GHG emissions include opposition to carbon pricing and the sector's strong dependence on subsidies, which often have limited mitigation potential. Additional challenges stem from the difficulty of accurately quantifying emissions in a highly dispersed and variable sector, as well as from economic and social conditions that may reduce the effectiveness of policy interventions [180,181].

Furthermore, policy incentives, such as financial rewards or tax reductions, can act as effective motivators for farmers to implement GHG mitigation practices [182]. However, although many countries have introduced regulations targeting agricultural emissions, policy frameworks are often fragmented, and enforcement tends to be inconsistent. Many policies also fail to provide sufficient incentives or support mechanisms for the adoption of mitigation technologies.

Without robust, coherent policies and well-designed incentive structures, agricultural emissions are likely to continue rising, even in the presence of regulatory measures.

8.2. Gaps

Considering the existing gaps related to agricultural GHG emissions (Figure 15), it is important to highlight the **lack of precise and consistent data**, as well as the difficulties in quantifying indirect emissions, particularly those arising from fertilizer production and land-use change [183].

In addition, current **estimation models** remain **insufficiently developed**, with various limitations in their ability to simulate the complex interactions between biological, chemical, and environmental processes involved in GHG emissions [179].

Furthermore, **measurement and monitoring uncertainties** complicate mitigation efforts, as **emissions** are **highly variable** and influenced by factors such as soil and crop types, farming practices [184].

Lastly, **incomplete** and **inconsistent data on livestock emissions**, especially from enteric fermentation and manure management, represent another critical challenge in building accurate inventories and informing effective mitigation strategies [185].

Compared to other major agricultural regions, the European Union continues to face technical gaps in reducing agricultural GHG emissions, primarily due to the slow adoption of innovative mitigation technologies. Although strong policy frameworks exist, the adoption of precision agriculture and digital monitoring systems remains slower in many EU member states [186]. For comparison, the United States demonstrates a significantly higher adoption rate of precision agriculture technologies (approximately 70%) largely due to the prevalence of large-scale farms, stronger private investment, and better access to digital and technical infrastructure [187].

8.3. Unanswered Questions

Despite major scientific and technological advances, significant uncertainties and unresolved questions remain regarding agricultural GHG emissions (Figure 15). Critical issues, such as how to measure emissions reliably and affordably on small and medium-sized farms, or whether mitigation technologies (biochar, nitrification inhibitors, feed additives) deliver net reductions, remain unanswered.

The diversity of agricultural systems across the EU further complicates the development of universally effective mitigation strategies. Economic, cultural, institutional, and informational barriers vary widely by context, making it difficult to identify the most limiting factors. Similarly, while agricultural carbon sequestration is often promoted as a key solution, land availability and the long-term sustainability of such practices differ greatly between regions.

These knowledge gaps arise not only from a lack of research but also from the complexity of agricultural systems, variability of local conditions, and the intricate relationships between environmental, economic, and social factors.

8.4. Further Perspectives

The difficulty in finding clear answers lies not in the absence of research but in the challenge of integrating fragmented data, context-specific realities, and evolving scientific understanding into coherent, scalable, and actionable strategies.

To overcome the challenges of reducing GHG emissions from agriculture, future efforts should focus on the following strategies:

(a) Technological advances: this includes the development of methane inhibitors, biochar amendments, advanced anaerobic digestion, and precision nutrient management to increase mitigation efficiency.

(b) Integrated mitigation strategies: practices such as conservation agriculture, cover cropping, optimized fertilizer use, and improved manure management can reduce GHGs while enhancing soil health.
(c) Enhancing soil carbon sequestration: this has the potential to offset a significant portion of CO_2 emissions from agriculture. Practices like agroecology, organic farming, and agroforestry can improve soil carbon stocks, turning farms into carbon sinks.
(d) Integrated policy frameworks: these should promote sustainable farming practices, incentivize farmers to adopt low-carbon technologies, and provide clear frameworks for measuring and reporting emissions.
(e) Monitoring technologies: remote sensing, sensors, and modeling tools can provide real-time data on emissions.
(f) Education and continuous research: these are essential for developing new technologies and practices.

Considering the environmental impacts of GHGs, it is crucial to mitigate emissions from agriculture through innovative practices such as precision nutrient management, improved manure handling, soil carbon sequestration, and the implementation of stringent regulations.

9. Conclusions

The research highlights the significant role of agriculture in GHG emissions both within the European Union (EU) and globally. The primary GHGs from agricultural activities—carbon dioxide (CO_2), methane (CH_4), and nitrous oxide (N_2O)—contribute to climate change, emphasizing the need to address agricultural practices in mitigation strategies. EU-level emissions remain high, although existing reduction measures have improved control compared to the global level.

Key EU regulations aim to support emission reductions and broader climate goals, such as climate neutrality by 2050 and a 55% reduction by 2030 relative to 1990 levels. Regulations targeting fertilizers, livestock, and land-use are crucial, but their effectiveness depends on enforcement, innovation in farming practices, and farmer engagement.

GHG emissions can be mitigated through practices such as nitrification inhibitors, balanced fertilization, crop rotations, integrated nutrient management, and soil amendments like biochar. Soil carbon sequestration via sustainable farming can also significantly reduce emissions by capturing carbon in soils [188].

Livestock emissions, particularly CH_4 from enteric fermentation in ruminants, are major contributors. Feed type, forage quality, and additives such as seaweeds, saponins, ionophores, and 3-nitrooxypropanol can reduce methane. Complementary strategies include targeted diets, genetic breeding, and improving animal health and digestive efficiency.

Manure-related CH_4 emissions can be mitigated through anaerobic digestion (potentially with biochar), composting, covering, acidification, and temperature management during storage.

In addition to conventional mitigation practices, such as those presented in detail in the present study, the agricultural sector can also benefit from technological innovations (precision farming tools, emission-monitoring sensors), advisory services that guide farmers in adopting best management practices, and market-based mechanisms, including carbon credits, carbon farming schemes, and payments for ecosystem services, all of which, when combined with proper monitoring, can enhance the effectiveness and scalability of GHG emissions reductions across diverse farming systems.

Despite advancements, many questions remain unresolved, and emission control is not yet fully achieved. Nevertheless, ongoing research contributes to understanding and improving mitigation strategies, supporting the objectives of EU regulations.

Author Contributions: Conceptualization, R.M.M. and G.V.S.; methodology, R.M.M.; software, A.M.; validation, R.M.M., G.V.S. and A.-C.B.; formal analysis, G.V.S.; investigation, R.M.M. and G.V.S.; resources, A.M.; data curation, G.V.S.; writing—original draft preparation, G.V.S.; writing—review and editing, G.V.S. and R.M.M.; visualization, A.-C.B.; supervision, R.M.M.; project administration, A.-C.B. All authors have read and agreed to the published version of the manuscript.

Funding: This research received no external funding.

Institutional Review Board Statement: Not applicable.

Informed Consent Statement: Not applicable.

Data Availability Statement: No new data were created or analyzed in this study. Data sharing is not applicable to this article.

Acknowledgments: This paper is based on research concepts developed within the framework of the "A Comprehensive Digital Platform for Land Use Planning, Carbon Footprinting, and Decision Making in European Agriculture" (acronym HOLOSEU) and aligns closely with its scientific objectives.

Conflicts of Interest: The authors declare no conflicts of interest.

References

1. Johnson, J.M.-F.; Franzluebbers, A.; Weyers, S.L.; Reicosky, D. Agricultural opportunities to mitigate greenhouse emissions. *Environ. Pollut.* **2007**, *150*, 107–124. [CrossRef]
2. Kopittke, P.; Dalal, R.; McKenna, B.; Smith, P.; Wang, P.; Weng, Z.; Van der Bom, F.; Menzies, N. Soil is a major contributor to global greenhouse gas emissions and climate change. *Soil* **2024**, *10*, 873–885. [CrossRef]
3. Gerber, P.J.; Steinfeld, H.; Henderson, B.; Mottet, A.; Opio, C.; Dijkman, J.; Falcucci, A.; Tempio, G. Tackling Climate Change Through Livestock—A Global Assessment of Emissions and Mitigation Opportunities. Food and Agriculture Organization of the United Nations (FAO), Rome. 2013. Available online: https://www.fao.org/3/i3437e/i3437e.pdf (accessed on 25 July 2025).
4. Abalos, D.; Recous, S.; Butterbach-Bahl, K.; De Notaris, C.; Rittl, T.; Topp, C.; Petersen, S.; Hansen, S.; Bleken, M.; Rees, R.; et al. A review and meta-analysis of mitigation measures for nitrous oxide emissions from crop residues. *Sci. Total Environ.* **2022**, *828*, 154388. [CrossRef]
5. Crippa, M.; Solazzo, E.; Guizzardi, D.; Monforti-Ferrario, F.; Tubiello, F.N.; Leip, A. Food systems are responsible for a third of global anthropogenic GHG emissions. *Nat. Food* **2021**, *2*, 198–209. [CrossRef] [PubMed]
6. Foley, J.; Ramankutty, N.; Brauman, K.; Cassidy, E.; Gerber, J.; Johnson, M.; Mueller, N.; O'Connell, C.; Ray, D.; West, P.; et al. Solutions for a cultivated planet. *Nature* **2011**, *478*, 337–342. [CrossRef] [PubMed]
7. Karki, S.; Lal, R.; Lorenz, K. Greenhouse gas emissions under conservation agriculture: A synthesis of field observations on integrating conservation tillage and cover crops. *Acta Agric. Scand. B Soil Plant Sci.* **2025**, *75*, 2515024. [CrossRef]
8. Hristov, A.; Oh, J.; Giallongo, F.; Frederick, T.; Harper, M.; Weeks, H.; Branco, A.; Moate, P.; Deighton, M.; Williams, R.; et al. An inhibitor persistently decreased enteric methane emission from dairy cows with no negative effect on milk production. *Proc. Natl. Acad. Sci. USA* **2015**, *112*, 10663–10668. [CrossRef]
9. Llonch, P.; Haskell, M.J.; Dewhurst, R.J.; Turner, S.P. Current available strategies to mitigate greenhouse gas emissions in livestock systems: An animal welfare perspective. *Animal* **2017**, *11*, 274–284. [CrossRef]
10. Petersen, S.O.; Blanchard, M.; Chadwick, D.; Del Prado, A.; Edouard, N.; Mosquera, J.; Sommern, S.G. Manure management for greenhouse gas mitigation. *Animal* **2013**, *7*, 266–282. [CrossRef]
11. Xie, X.; Cao, Y.; Li, Q.; Yang, X.; Wang, R.; Zhang, X.; Tan, Z.; Lin, B.; Wang, M. Mitigating enteric methane emissions: An overview of methanogenesis, inhibitors and future prospects. *Anim. Nutr.* **2025**, *21*, 84–96. [CrossRef]
12. Rosenzweig, C.; Tubiello, F.N. Adaptation and mitigation strategies in agriculture: An analysis of potential synergies. *Mitig. Adapt. Strat. Glob. Change* **2007**, *12*, 855–873. [CrossRef]
13. Wang, Z.; Xuan, H.; Liu, B.; Zhang, H.; Zheng, T.; Liu, Y.; Dai, L.; Xie, Y.; Shang, X.; Zhang, L.; et al. Diversified Cropping Modulates Microbial Communities and Greenhouse Gas Emissions by Enhancing Soil Nutrients. *Agronomy* **2025**, *15*, 1472. [CrossRef]

14. Manono, B.O.; Gichana, Z. Agriculture-Livestock-Forestry Nexus: Pathways to Enhanced Incomes, Soil Health, Food Security and Climate Change Mitigation in Sub-Saharan Africa. *Earth* **2025**, *6*, 74. [CrossRef]
15. Das, S.; Beegum, S.; Acharya, B.S.; Panday, D. Soil Carbon Sequestration: A mechanistic perspective on limitations and future possibilities. *Sustainability* **2025**, *17*, 6015. [CrossRef]
16. Nazir, M.J.; Li, G.; Nazir, M.M.; Zulfiqar, F.; Siddique, K.; Iqbal, B.; Du, D. Harnessing soil carbon sequestration to address climate change challenges in agriculture. *Soil Tillage Res.* **2024**, *237*, 105959. [CrossRef]
17. Kabange, N.R.; Kwon, Y.; Lee, S.-M.; Kang, J.-W.; Cha, J.-K.; Park, H.; Dzorkpe, G.D.; Shin, D.; Oh, K.-W.; Lee, J.-H. Mitigating Greenhouse Gas Emissions from Crop Production and Management Practices, and Livestock: A Review. *Sustainability* **2023**, *15*, 15889. [CrossRef]
18. Basheer, S.; Wang, X.; Farooque, A.A.; Nawaz, R.A.; Pang, T.; Neokye, E.O. A Review of Greenhouse Gas Emissions from Agricultural Soil. *Sustainability* **2024**, *16*, 4789. [CrossRef]
19. European Environment Agency (EEA). Greenhouse Gas Emissions from Agriculture. Available online: https://www.eea.europa.eu/en/analysis/indicators/greenhouse-gas-emissions-from-agriculture (accessed on 31 July 2025).
20. FAO. Greenhouse Gas Emissions from Agrifood Systems—Global, Regional and Country Trends, 2000–2022. FAOSTAT Analytical Brief Series; No. 94; Food and Agriculture Organization of the United Nations: Rome, Italy, 2024. Available online: https://openknowledge.fao.org/handle/20.500.14283/cd3167en (accessed on 26 July 2025).
21. Carbon Chain. Available online: https://www.carbonchain.com/blog/understand-your-synthetic-fertilizer-emissions (accessed on 26 July 2025).
22. Evangelista, C.; Milanesi, M.; Pietrucci, D.; Chillemi, G.; Bernabucci, U. Enteric Methane Emission in Livestock Sector: Bibliometric Research from 1986 to 2024 with Text Mining and Topic Analysis Approach by Machine Learning Algorithms. *Animals* **2024**, *14*, 3158. [CrossRef]
23. Saeed, Q.; Zhang, A.; Mustafa, A.; Sun, B.; Zhang, S.; Yang, X. Effect of long-term fertilization on greenhouse gas emissions and carbon footprints in northwest China: A field scale investigation using wheat-maize-fallow rotation cycles. *J. Clean. Prod.* **2022**, *332*, 130075. [CrossRef]
24. Silva, G.R.d.; Liska, A.J.; Bayer, C. Cover Crops Can Reduce Greenhouse Gas Emissions from No-Till Maize in Southern Brazil: Insights from a Long-Term Field Experiment. *Agronomy* **2025**, *15*, 267. [CrossRef]
25. World Bank Group Data. Carbon Dioxide (CO_2) Emissions from Agriculture (Mt CO_2e). Available online: https://data.worldbank.org/indicator/EN.GHG.CO2.AG.MT.CE.AR5?end=2023&name_desc=false&start=1970&view=chart&year=2023; (accessed on 2 August 2025).
26. World Bank Group Data. Methane (CH_4) Emissions from Agriculture (Mt CO_2e). Available online: https://data.worldbank.org/indicator/EN.GHG.CH4.AG.MT.CE.AR5?end=2023&name_desc=false&start=1970&view=chart&year=2023; (accessed on 2 August 2025).
27. World Bank Group Data. Nitrous Oxide (N_2O) Emissions from Agriculture (Mt CO_2e). Available online: https://data.worldbank.org/indicator/EN.GHG.N2O.AG.MT.CE.AR5?end=2023&name_desc=false&start=1970&view=chart&year=2023 (accessed on 2 August 2025).
28. Verschuuren, J. Achieving agricultural greenhouse gas emission reductions in the EU post-2030: What options do we have? *RECIEL* **2022**, *31*, 246–257. [CrossRef]
29. Mohammed, S.; Raihan, A.; Arshad, S.; Ata, B.; Ocwa, A.; Al-Dalahmeh, M.; Harsanyi, E. European Union agro-climate policies toward sustainability: Analyzing emission trends and land use dynamics (1990–2021). *Resour. Environ. Sustain.* **2025**, *21*, 100239. [CrossRef]
30. Bilotto, F.; Christie-Whitehead, K.M.; Malcolm, B.; Barnes, N.; Cullen, B.; Ayre, M.; Harrison, M.T. Costs of transitioning the livestock sector to net-zero emissions under future climates. *Nat. Commun.* **2025**, *16*, 3810. [CrossRef]
31. European Environment Agency. Methane, Climate Change and Air Quality in Europe: Exploring the Connections. Available online: https://www.eea.europa.eu/en/analysis/publications/methane-climate-change-and-air-quality-in-europe-exploring-the-connections?activeTab=6397c084-2e5f-4545-a873-f99323d40846 (accessed on 1 August 2025).
32. Wang, C.; Amon, B.; Schulz, K.; Mehdi, B. Factors That Influence Nitrous Oxide Emissions from Agricultural Soils as Well as Their Representation in Simulation Models: A Review. *Agronomy* **2021**, *11*, 770. [CrossRef]
33. Vasile Scăețeanu, G.; Madjar, R.M. The Control of Nitrogen in Farmlands for Sustainability in Agriculture. *Sustainability* **2025**, *17*, 5619. [CrossRef]
34. Symeon, G.K.; Akamati, K.; Dotas, V.; Karatosidi, D.; Bizelis, I.; Laliotis, G.P. Manure Management as a Potential Mitigation Tool to Eliminate Greenhouse Gas Emissions in Livestock Systems. *Sustainability* **2025**, *17*, 586. [CrossRef]
35. European Commission. Energy, Climate Change, Environment. 2050 Long-Term Strategy. Available online: https://climate.ec.europa.eu/eu-action/climate-strategies-targets/2050-long-term-strategy_en (accessed on 1 August 2025).
36. European Commission. EU Emissions Trading System (EU ETS). 2023. Available online: https://climate.ec.europa.eu/eu-action/eu-emissions-trading-system-eu-ets_en (accessed on 28 October 2025).

37. European Union. Regulation (EU) 2018/842 of the European Parliament and of the Council of 30 May 2018 on binding annual greenhouse gas emission reductions by Member States from 2021 to 2030 contributing to climate action to meet commitments under the Paris Agreement and amending Regulation (EU) No 525/2013. *Off. J. Eur. Union* **2018**, *156*, 26–42.
38. European Union. Decision No 406/2009/EC of the European Parliament and of the Council of 23 April 2009 on the effort of Member States to reduce their greenhouse gas emissions to meet the Community's greenhouse gas emission reduction commitments up to 2020. *Off. J. Eur. Union* **2009**, *140*, 136–148.
39. European Union. Regulation (EU) 2023/857 of the European Parliament and of the Council of 19 April 2023 amending Regulation (EU) 2018/842 on binding annual greenhouse gas emission reductions by Member States from 2021 to 2030 contributing to climate action to meet commitments under the Paris Agreement, and Regulation (EU) 2018/1999 (Text with EEA relevance). *Off. J. Eur. Union* **2023**, *111*, 1–14.
40. European Union. Regulation (EU) 2023/839 of the European Parliament and of the Council of 19 April 2023 amending Regulation (EU) 2018/841 as regards the scope, simplifying the reporting and compliance rules, and setting out the targets of the Member States for 2030, and Regulation (EU) 2018/1999 as regards improvement in monitoring, reporting, tracking of progress and review. *Off. J. Eur. Union* **2023**, *107*, 1–28.
41. European Council. European Green Deal. Available online: https://www.consilium.europa.eu/en/policies/european-green-deal/ (accessed on 1 August 2025).
42. European Commission. Farm to Fork Strategy. Available online: https://food.ec.europa.eu/horizontal-topics/farm-fork-strategy_en (accessed on 1 August 2025).
43. European Union. Regulation (EU) 2021/1119 of the European Parliament and of the Council of 30 June 2021 establishing the framework for achieving climate neutrality and amending Regulations (EC) No 401/2009 and (EU) 2018/1999 ('European Climate Law'). *Off. J. Eur. Union* **2021**, *243*, 1–17.
44. European Commission. Agriculture and Rural Development. Available online: https://agriculture.ec.europa.eu/common-agricultural-policy/cap-overview/cap-glance_en (accessed on 1 August 2025).
45. World Bank Group Data. Available online: https://data.worldbank.org/indicator/EN.GHG.ALL.LU.MT.CE.AR5 (accessed on 2 August 2025).
46. Psistaki, K.; Tsantopoulos, G.; Paschalidou, A.K. An Overview of the Role of Forests in Climate Change Mitigation. *Sustainability* **2024**, *16*, 6089. [CrossRef]
47. Deshpande, M.V.; Kumar, N.; Pillai, D.; Krishna, V.; Jain, M. Greenhouse gas emissions from agricultural residue burning have increased by 75% since 2011 across India. *Sci. Total Environ.* **2023**, *904*, 166944. [CrossRef] [PubMed]
48. Flammini, A.; Pan, X.; Tubiello, F.N.; Qiu, S.Y.; Souza, L.R.; Quadrelli, R.; Bracco, S.; Benoit, P.; Sims, R. Emissions of greenhouse gases from energy use in agriculture, forestry and fisheries: 1970–2019. *Earth Syst. Sci. Data* **2022**, *14*, 811–821. [CrossRef]
49. Mühlbachová, G.; Růžek, P.; Kusá, H.; Vavera, R. CO_2 Emissions from Soils under Different Tillage Practices and Weather Conditions. *Agronomy* **2023**, *13*, 3084. [CrossRef]
50. Bahsi, K.; Ustaoglu, B.; Aksoy, S.; Sertel, E. Estimation of emissions from crop residue burning in Türkiye using remotely sensed data and the Google Earth Engine platform. *Geocarto Int.* **2023**, *38*, 2157052. [CrossRef]
51. Çınar, T.; Cakır, M.F.; Aydın, A. Assessment of environmental and atmospheric impacts of stubble burning in Mardin-Diyarbakır (Southeastern of Türkiye): A remote sensing approach. *Nat. Hazards* **2025**, *121*, 17895–17912. [CrossRef]
52. European Union. Regulation (EU) No 1306/2013 of the European parliament and of the council of 17 December 2013 on the financing, management and monitoring of the common agricultural policy and repealing Council Regulations (EEC) No 352/78, (EC) No 165/94, (EC) No 2799/98, (EC) No 814/2000, (EC) No 1290/2005 and (EC) No 485/2008. *Off. J. Eur. Union* **2013**, *347*, 549–607.
53. Hall, J.; Zibtsev, S.; Giglio, L.; Skakun, S.; Myroniuk, V.; Zhuravek, O.; Goldammer, J.G.; Kussul, N. Environmental and political implications of underestimated cropland burning in Ukraine. *Environ. Res. Lett.* **2021**, *16*, 064019. [CrossRef]
54. Li, Z.; Zhang, Q.; Li, Z.; Qiao, Y.; Du, Y.; Yue, Z.; Tian, C.; Leng, P.; Cheng, H.; Chen, G.; et al. Responses of soil CO_2 emissions to tillage practices in a wheat-maize cropping system a 4-year field study. *Field Crop Res.* **2023**, *294*, 108832. [CrossRef]
55. Wang, W.; Zhang, H.; Mo, F.; Liao, Y.; Wen, X. Reducing greenhouse gas emissions and improving net ecosystem economic benefit through long-term conservation tillage in a wheat-maize multiple cropping system in the Loess Plateau, China. *Eur. J. Agron.* **2022**, *141*, 126619. [CrossRef]
56. Forte, A.; Fiorentino, N.; Fagnano, M.; Fierro, A. Mitigation impact of minimum tillage on CO_2 and N_2O emissions from a Mediterranean maize cropped soil under low-water input management. *Soil Tillage Res.* **2017**, *166*, 167–178. [CrossRef]
57. Carey, J.C.; Tang, J.W.; Templer, P.H.; Kroeger, K.D.; Crowther, T.W.; Burton, A.J.; Dukes, J.S.; Emmett, B.; Frey, S.D.; Heskel, M.A.; et al. Temperature response of soil respiration largely unaltered with experimental warming. *Proc. Natl. Acad. Sci. USA* **2016**, *113*, 13797–13802. [CrossRef]

58. Gougoulias, C.; Clark, J.M.; Shaw, L.J. The role of soil microbes in the global carbon cycle: Tracking the below-ground microbial processing of plant-derived carbon for manipulating carbon dynamics in agricultural systems. *J. Sci. Food Agric.* **2014**, *94*, 2362–2371. [CrossRef]
59. European Commission. *Soil: The Hidden Part of the Climate Cycle*; Publications Office of the European Union: Luxembourg, 2011; pp. 8–10.
60. Nissan, A.; Alcolombri, U.; Peleg, N.; Galili, N.; Jimnez-Martinez, J.-J.; Molnar, P.; Holzner, M. Global warming accelerates soil heterotrophic respiration. *Nat. Commun.* **2023**, *14*, 3452. [CrossRef]
61. Chen, P.; Yuan, X.-L.; Li, L.-Y.; Li, J.-Y.; Zhang, R.-Q.; Li, Z.-G.; Liu, Y. Aggregational differentiation of soil-respired CO_2 and its $\delta^{13}C$ variation across land-use types. *Geoderma* **2023**, *432*, 116384. [CrossRef]
62. Khan, M.I.; Sarfraz, R.; Kim, T.; Park, H.-J.; Kim, P.J.; Kim, G.W. Partitioning carbon dioxide emissions from soil organic matter and urea in warm and cold cropping seasons. *Atmos. Pollut. Res.* **2024**, *15*, 101995. [CrossRef]
63. Ray, R.L.; Griffin, R.W.; Fares, A.; Elhassan, A.; Awal, R.; Woldesenbet, S. Soil CO_2 emission in response to organic amendments, temperature, and rainfall. *Sci. Rep.* **2020**, *10*, 5849. [CrossRef] [PubMed]
64. Gu, X.; Weng, S.; Li, Y.; Zhou, X. Effects of Water and Fertilizer Management Practices on Methane Emissions from Paddy Soils: Synthesis and Perspective. *Int. J. Environ. Res. Public Health* **2022**, *19*, 7324. [CrossRef]
65. Rajendran, S.; Park, H.; Kim, J.; Park, S.J.; Shin, D.; Lee, J.-H.; Song, Y.H.; Paek, N.-C.; Kim, C.M. Methane emission from rice fields: Necessity for molecular approach for mitigation. *Rice Sci.* **2024**, *31*, 159–178. [CrossRef]
66. Senthilraja, K.; Venkatesan, S.; Udhaya Nandhini, D.; Dhasarathan, M.; Prabha, B.; Boomiraj, K.; Mohan Kumar, S.; Bhuvaneswari, K.; Raveendran, M.; Geethalakshmi, V. Mitigating Methane Emission from the Rice Ecosystem through Organic Amendments. *Agriculture* **2023**, *13*, 1037. [CrossRef]
67. Middelanis, T.; Looschelders, D.; Mueller, P.; Knorr, K.-H. Potential of biochar to mitigate methane production in paddy soils—Application of a new incubation and modelling approach. *Biogeochemistry* **2025**, *168*, 6. [CrossRef]
68. Qiu, S.; Zhang, X.; Xia, W.; Li, Z.; Wang, L.; Chen, Z.; Ge, S. Effect of extreme pH conditions on methanogenesis: Methanogen metabolism and community structure. *Sci. Total Environ.* **2023**, *877*, 162702. [CrossRef]
69. Datta, A.; Santra, S.C.; Adhya, T.K. Effect of inorganic fertilizers (N,P,K) on methane emission from tropical rice field of India. *Atmos. Environ.* **2013**, *66*, 123–130. [CrossRef]
70. Kantachote, D.; Nunkaew, T.; Kantha, T.; Chaiprapat, S. Biofertilizers from *Rhodopseudomonas palustris* strains to enhance rice yields and reduce methane emissions. *Appl. Soil Ecol.* **2016**, *100*, 154–161. [CrossRef]
71. Wang, S.; Liu, Y.; Zhang, F.; Jin, K.; Liu, H.; Zhai, L. Methane emissions sources and impact mechanisms altered by the shift from rice-wheat to rice-crayfish rotation. *J. Clean. Prod.* **2024**, *434*, 139968. [CrossRef]
72. Pazhanivelan, S.; Sudarmanian, N.S.; Geethalakshmi, V.; Deiveegan, M.; Ragunath, K.; Sivamurugan, A.P.; Shanmugapriya, P. Assessing Methane Emissions from Rice Fields in Large Irrigation Projects Using Satellite-Derived Land Surface Temperature and Agronomic Flooding: A Spatial Analysis. *Agriculture* **2024**, *14*, 496. [CrossRef]
73. Song, Y.; Song, C.; Choi, S.-E.; Kim, J.; Kim, M.; Hwang, W.; Roh, M.; Lee, S.; Lee, W.-K. Estimating Methane Emissions in Rice Paddies at the Parcel Level Using Drone-Based Time Series Vegetation Indices. *Drones* **2024**, *8*, 459. [CrossRef]
74. Scarff, H.; Jacobs, J. Applying guidance for methane emission estimation for landfills. *Waste Manag.* **2006**, *26*, 417–429. [CrossRef]
75. Delgado, M.; Lopez, A.; Esteban-Garcia, A.L.; Lobo, A. The importance of particularizing the model to estimate landfill GHG emissions. *J. Environ. Manag.* **2023**, *325*, 116600. [CrossRef] [PubMed]
76. Shi, Y.; Lou, Y.; Zhang, Z.; Ma, L.; Ojara, M.A. Estimation of methane emissions based on crop yield and remote sensing data in a paddy field. Greenh. *Gases Sci. Technol.* **2020**, *10*, 196–207. [CrossRef]
77. Nikolaisen, M.; Hillier, J.; Smith, P.; Nayak, D. Modelling CH_4 emission from rice ecosystem: A comparison between existing empirical models. *Front. Agron.* **2023**, *4*, 1058649. [CrossRef]
78. Food and Agriculture Organization of the United Nations. Livestock Solutions for Climate Change. Available online: https://www.fao.org/family-farming/detail/en/c/1634679/ (accessed on 7 August 2025).
79. Food and Agriculture Organization of the United Nations. Livestock and Enteric Methane. Available online: https://www.fao.org/in-action/enteric-methane/background/en (accessed on 14 August 2025).
80. Beauchemin, K.A.; Ungerfeld, E.M.; Eckard, R.J.; Wang, M. Fifty years of research on rumen methanogenesis: Lessons learned and future challenges for mitigation. *Animal* **2020**, *14*, s2–s16. [CrossRef]
81. Morgavi, D.P.; Forano, E.; Martin, C.; Newbold, C.J. Microbial ecosystem and methanogenesis in ruminants. *Animal* **2010**, *4*, 1024–1036. [CrossRef]
82. Pas, M.A.W.; Uddin, M.E.; Letelier, P.; Jackson, R.D.; Larson, R.A. Emission and mitigation of greenhouse gases from dairy farms: The cow, the manure and the field. *Appl. Animal Sci.* **2019**, *35*, 238–254. [CrossRef]
83. Santander, D.; Clariget, J.; Banchero, G.; Alecrim, F.; Simon Zinno, C.; Mariotta, J.; Gere, J.; Ciganda, V.S. Beef Steers and Enteric Methane: Reducing Emissions by Managing Forage Diet Fiber Content. *Animals* **2023**, *13*, 1177. [CrossRef]

84. Sun, X.; Cheng, L.; Jonker, A.; Munidasa, S.; Pacheco, D. A Review: Plant Carbohydrate Types-The Potential Impact on Ruminant Methane Emissions. *Front. Vet. Sci.* **2022**, *9*, 880115. [CrossRef] [PubMed]
85. Johnson, K.A.; Johnson, D.E. Methane emissions from cattle. *J. Anim. Sci.* **1995**, *73*, 2483–2492. [CrossRef] [PubMed]
86. Min, B.; Parker, D.; Brauer, D.; Waldrip, H.; Lockard, C.; Hales, K.; Akbay, A.; Augyte, S. The role of seaweed as a potential dietary supplementation for nteric methane mitigation in ruminants: Challenges and opportunities. *Animal Nutr.* **2021**, *7*, 1371–1387. [CrossRef] [PubMed]
87. Wanapat, M.; Prachumchai, R.; Dagaew, G.; Matra, M.; Phupaboon, S.; Sommai, S.; Suriyapa, C. Potential use of seaweed as a dietary supplement to m itigate enteric methane emission in ruminants. *Sci. Total Environ.* **2024**, *931*, 173015. [CrossRef]
88. Broucek, J. Production of Methane Emissions from Ruminant Husbandry: A Review. *J. Environ. Prot.* **2014**, *5*, 1482–1493. [CrossRef]
89. Grandl, F.; Amelchanka, S.L.; Furger, M.; Clauss, M.; Zeitz, J.O.; Kreuzer, M.; Schwarm, A. Biological implications of longevity in dairy cows: 2. Changes in methane emissions and efficieny with age. *J. Dairy Sci.* **2016**, *99*, 3472–3485. [CrossRef]
90. Külling, D.R.; Menzi, H.; Kröber, T.F.; Neftel, A.; Sutter, F.; Lischer, P.; Kreuzer, M. Emissions of ammonia, nitrous oxide and methane from different types of dairy manure during storage as affected by dietary protein content. *J. Agric. Sci.* **2001**, *137*, 235–250. [CrossRef]
91. Tseten, T.; Sanjorjo, R.A.; Kwon, M.; Kim, S.-W. Strategies to m itigate enteric methane emissions from ruminant animals. *J. Microbiol. Biotechnol.* **2022**, *32*, 269–277. [CrossRef]
92. Haque, M. Dietary manipulation: A sustainable way to mitigate methane emissions from ruminants. *J. Anim. Sci. Technol.* **2018**, *60*, 15. [CrossRef]
93. Vargas, J.; Ungerfeld, E.; Muñoz, C.; DiLorenzo, N. Feeding Strategies to Mitigate Enteric Methane Emission from Ruminants in Grassland Systems. *Animals* **2022**, *12*, 1132. [CrossRef]
94. Eugene, M.; Klumpp, K.; Sauvant, D. Methane mitigating options with forages fed ruminants. *Grass Forage Sci.* **2021**, *76*, 196–204. [CrossRef]
95. Liu, Y.; Zhou, M.; Diao, Q.; Ma, T.; Tu, Y. Seaweed as a feed additive to mitigate enteric methane emissions in ruminants: Opportunities and challenges. *J. Integr. Agric.* **2025**, *24*, 1327–1341. [CrossRef]
96. Meo-Filho, P.; Ramirez-Aguelo, J.; Kebreab, E. Mitigating methane emissions in grazing beef cattle with a seaweed-based feed additive: Implications for climate-smart agriculture. *Agric. Sci.* **2024**, *121*, e2410863121. [CrossRef]
97. Cheong, K.-L.; Zhang, Y.; Li, Z.; Li, T.; Ou, Y.; Shen, J.; Zhong, S.; Tan, K. Role of Polysaccharides from Marine Seaweed as Feed Additives for Methane Mitigation in Ruminants: A Critical Review. *Polymers* **2023**, *15*, 3153. [CrossRef]
98. Morsy, A.S.; Soltan, Y.A.; El-Zaiat, H.M.; Alencar, S.M.; Abdalla, A.L. Bee propolis extract as a phytogenic feed additive diet digestibility, rumen microbial biosynthesis, mitigating methane formation and health status of late pregnant ewes. *Animal Feed. Sci. Technol.* **2021**, *273*, 114834. [CrossRef]
99. Ridla, M.; Laconi, E.B.; Jayanegara, A. Effects of saponin on enteric methane emission and nutrient digestibility of ruminants: An in vivo meta-analysis. *IOP Conf. Ser. Earth Environ. Sci.* **2021**, *788*, 012028. [CrossRef]
100. Marques, R.d.S.; Cooke, R.F. Effects of Ionophores on Ruminal Function of Beef Cattle. *Animals* **2021**, *11*, 2871. [CrossRef] [PubMed]
101. Cagliari, A.R.; Magnani, E.; Rigon, F.; Loregian, K.E.; Casagrande, A.C.; Amancio, B.R.; Bueno da Silva, J.; Lisboa Santos, V.; Marcondes, M.I.; Paula, E.M.; et al. Evaluation of yeast-based additives, as an alternative to ionophores, on rumen fermentation of ruminant diets using an in vitro gas production system. *Front. Anim. Sci.* **2023**, *4*, 1233273. [CrossRef]
102. Palangi, V.; Lackner, M. Management of Enteric Methane Emissions in Ruminants Using Feed Additives: A Review. *Animals* **2022**, *12*, 3452. [CrossRef] [PubMed]
103. Palangi, V.; Macit, M. Indictable Mitigation of Methane Emission Using Some Organic Acids as Additives Towards a Cleaner Ecosystem. *Waste Biomass. Valor.* **2021**, *12*, 4825–4834. [CrossRef]
104. Van Gastelen, S.; Burgers, E.; Dijkstra, J.; De Mol, R.; Muizelaar, W.; Walker, N.; Bannink, A. Long-term effects of 3-nitrooxypropanol on methane emission and milk production characteristics in Holstein-Friesian dairy cows. *J. Dairy Sci.* **2024**, *107*, 5556–5573. [CrossRef]
105. Feng, X.Y.; Dijkstra, J.; Bannink, A.; Van Gastelen, S.; France, J.; Kebreab, E. Antimethanogenic effects of nitrate supplementation in cattle: A meta-analysis. *J. Dairy Sci.* **2020**, *103*, 11375–11385. [CrossRef]
106. De Haas, Y.; Veerkamp, R.F.; De Jong, G.; Aldridge, M.N. Selective breeding as a mitigation tool for methane emissions from dairy cattle. *Animal* **2021**, *15*, 100294. [CrossRef]
107. Viking Genetics. New Nordic Methane Index to Reduce Cow Emissions by 20%. Available online: https://www.vikinggenetics.com/news/nordic-methane-index (accessed on 18 August 2025).
108. Króliczewska, B.; Pecka-Kiełb, E.; Bujok, J. Strategies Used to Reduce Methane Emissions from Ruminants: Controversies and Issues. *Agriculture* **2023**, *13*, 602. [CrossRef]

109. Baca-González, V.; Asensio-Calavia, P.; González-Acosta, S.; Pérez de la Lastra, J.M.; Morales de la Nuez, A. Are Vaccines the Solution for Methane Emissions from Ruminants? A Systematic Review. *Vaccines* **2020**, *8*, 460. [CrossRef]
110. Cardenas, A.; Ammon, C.; Schumacher, B.; Stinner, W.; Herrmann, C.; Schneider, M.; Weinrich, S.; Fischer, P.; Amon, T.; Amon, B. Methane emissions from storage of liquid dairy manure: Influences of season, temperature and storage duration. *Waste Manag.* **2021**, *121*, 393–402. [CrossRef]
111. Hidayat, C.; Widiawati, Y.; Tiesnamurti, B.; Pramono, A.; Krisnan, R.; Shiddieqy, M.I. Comparisom of methane production from cattle, buffalo, goat, rabbit, chicken, and duck manure. *IOP Conf. Ser. Earth Environ. Sci.* **2021**, *648*, 012112. [CrossRef]
112. Vac, S.C.; Popita, G.-E.; Frunzeti, N.; Popovici, A. Evaluation of greenhouse gas emission from animal manure using the closed chamber method for gas fluxes. *Not.Bot.Agrobo.* **2013**, *41*, 576–581. [CrossRef]
113. Ahlberg-Eliasson, K.; Westerholm, M.; Isaksson, S.; Schnürer, A. Anaerobic Digestion of Animal Manure and Influence of Organic Loading Rate and Temperature on Process Performance, Microbiology, and Methane Emission from Digestates. *Front. Energy Res.* **2021**, *9*, 740314. [CrossRef]
114. Hu, J.; Stenchly, K.; Gwenzi, W.; Wachendorf, M.; Kaetzl, K. Critical evaluation of biochar effects on methane production and process stability in anaerobic digestion. *Front. Energy Res.* **2023**, *11*, 1205818. [CrossRef]
115. Wang, Y.; Liu, Q.; Ran, H.; Peng, P.; Wang, Y.; Peng, G.; Wu, Y.; Wen, X. Residual ciprofloxacin in chicken manure inhibits methane production in an anaerobic digestion system. *Poultry Sci.* **2025**, *104*, 104539. [CrossRef] [PubMed]
116. Kim, S.Y.; Jeng, S.T.; Ho, A.; Hong, C.O.; Lee, C.H.; Kim, P.J. Cattle manure composting: Shifts in the methanogenic community structure, chemical composition, and consequences on methane production potential in a rice paddy. *Appl. Soil Ecol.* **2018**, *124*, 344–350. [CrossRef]
117. Chen, W.; Liao, X.; Wu, Y.; Liang, J.B.; Mi, J.; Huang, J.; Zhang, H.; Wu, Y.; Qiao, Z.; Li, X.; et al. Effects of different types of biochar on methane and ammonia mitigation during layer manure composting. *Waste Manag.* **2017**, *61*, 506–515. [CrossRef]
118. Hansen, R.R.; Nielsen, D.A.; Schramm, A.; Nielsen, L.P.; Revsbech, N.P.; Hansen, M.N. Greenhouse gas microbiology in wet and dry straw crust covering pig slurry. *J. Environ Qual.* **2009**, *38*, 1311–1319. [CrossRef]
119. Zhang, Y.; Bo, Q.; Ma, X.; Du, Y.; Du, X.; Xu, L.; Yang, Y. Solid–Liquid Separation and Its Environmental Impact on Manure Treatment in Scaled Pig Farms—Evidence Based on Life Cycle Assessment. *Agriculture* **2023**, *13*, 2284. [CrossRef]
120. Zhang, Y.; Bai, W.; Xu, J.; Wang, W.; Wu, G.; Zhan, X.; Hu, Z.H. Evaluation of solid-liquid separation of dairy manure with different separator screen sizes on the resource recovery and greenhouse gas emissions reduction. *J. Clean. Prod.* **2024**, *448*, 141680. [CrossRef]
121. Feng, L.; Ward, A.J.; Moset, V.; Møller, H.B. Methane emission during on-site pre-storage of animal manure prior to anaerobic digestion at biogas plant: Effect of storage temperature and addition of food waste. *J. Environ. Manag.* **2018**, *225*, 272–279. [CrossRef]
122. Fabgueiro, D.; Hjorth, M.; Gioelli, F. Acidification of animal slurry—A review. *J. Environ. Manag.* **2015**, *149*, 46–56. [CrossRef]
123. Ma, C.; Dalbym, F.; Feilberg, A.; Jacobsen, B.; Petersen, S. Low-dose acidification as a methane mitigation strategy for manure management. *ACS Agric. Sci. Technol.* **2022**, *2*, 437–442. [CrossRef]
124. Overmeyer, V.; Trimborn, M.; Clemens, J.; Hölscher, R.; Büscher, W. Acidification of slurry to reduce ammonia and methane emissions: Deployment of a retrofittable system in fattening pig barns. *J. Environ. Manag.* **2023**, *331*, 117263. [CrossRef] [PubMed]
125. Ambrose, H.W.; Dalby, F.; Feilberg, A.; Kofoed, M. Additives and methods for the mitigation of methane emission from stored liquid manure. *Biosyst. Eng.* **2023**, *229*, 209–245. [CrossRef]
126. Lima, F.M.P.; Laniel, M.; Balde, H.; Gordon, R.; Vander Zaag, A. Methane emission reduction by adding sulfate to liquid dairy manure. *J. Environ. Qual.* **2025**, *54*, 349–358. [CrossRef] [PubMed]
127. Svane, S.; Karring, H. Combining fluoride with either phenolic compounds or plant extracts offers potential mitigation strategy for ammonia and methane emissions from livestock manure. *Environ. Technol. Innov.* **2022**, *28*, 102830. [CrossRef]
128. Chiodini, M.E.; Costantini, M.; Zoli, M.; Bacenetti, J.; Aspesi, D.; Poggianella, L.; Acutis, M. Real-Scale Study on Methane and Carbon Dioxide Emission Reduction from Dairy Liquid Manure with the Commercial Additive SOP LAGOON. *Sustainability* **2023**, *15*, 1803. [CrossRef]
129. Boku University. Metabolomics for the Investigation of Biological Nitrification Inhibitors in the Rhizosphere. Available online: https://boku.ac.at/en/docservice/doctoral-studies/doktoratsschulen/agrigenomics/research/projects/metabolomics-for-the-investigation-of-biological-nitrification-inhibitors-in-the-rhizosphere (accessed on 1 August 2025).
130. Voigt, C.; Marushchak, M.; Abbott, B.; Biasi, C.; Elberling, B.; Siciliano, S.; Sonnentag, O.; Stewart, K.; Yang, Y.; Martikainen, P. Nitrous oxide emissions from permafrost-affected soils. *Nat. Rev. Earth Environ.* **2020**, *1*, 420–434. [CrossRef]
131. Liu, H.; Zheng, X.; Li, Y.; Ding, H.; Sveen, T.R.; Zhang, Y. Soil moisture determines nitrous oxide emission and uptake. *Sci. Total Environ.* **2022**, *822*, 153566. [CrossRef]
132. Cui, P.; Chen, Z.; Fan, F.; Yin, C.; Song, A.; Li, T.; Zhang, H.; Liang, Y. Soil texture is an easily overlooked factor affecting temperature sensitivity of N_2O emissions. *Sci. Total Environ.* **2023**, *862*, 160648. [CrossRef]

133. Hassan, M.U.; Aamer, M.; Mahmood, A.; Awan, M.I.; Barbanti, L.; Seleiman, M.F.; Bakhsh, G.; Alkharabsheh, H.M.; Babur, E.; Shao, J.; et al. Management Strategies to Mitigate N_2O Emissions in Agriculture. *Life* **2022**, *12*, 439. [CrossRef]
134. Li, Y.; Xu, J.; Liu, B.; Wang, H.; Qi, Z.; Wei, Q.; Liao, L.; Liu, S. Enhanced N_2O Production Induced by Soil Salinity at a Specific Range. *Int. J. Environ. Res. Public Health* **2020**, *17*, 5169. [CrossRef]
135. Xu, C.; Zhu, H.; Wang, J.; Liu, Y.; Chen, D.; Zhang, H.; Wang, J.; Zhang, Y. Fertilizer N triggers native soil N-derived N_2O emissions by priming gross N mineralization. *Soil Biol. Biochem.* **2023**, *178*, 108961. [CrossRef]
136. Ferland, D.; Wagner-Riddle, C.; Brown, S.E.; Bourgault, M.; Helgason, W.; Farrell, R.E.; Congreves, K.A. Improved nitrogen fertilizer management reduces nitrous oxide emissions in a northern Prairie cropland. *Sci. Total Environ.* **2024**, *956*, 177211. [CrossRef]
137. Pelster, D.; Chantigny, M.; Royer, I.; Angers, D.; Vanasse, A. Reduced tillage increased growing season N_2O emissions from a fine but not a coarse textured soil under the cool, humid climate of eastern Canada. *Soil Tillage Res.* **2021**, *206*, 104833. [CrossRef]
138. Hu, N.; Chen, Q.; Zhu, L. The Responses of Soil N_2O Emissions to Residue Returning Systems: A Meta-Analysis. *Sustainability* **2019**, *11*, 748. [CrossRef]
139. Abalos, D.; Rittl, T.; Recous, S.; Thiebeau, P.; Topp, C.; Van Groenigen, K.J.; Butterbach-Bahl, K.; Thorman, R.; Smith, K.; Ahuja, I.; et al. Predicting field N_2O emissions from crop residues based on their biochemical composition: A meta-analysis approach. *Sci. Total Environ.* **2022**, *812*, 152532. [CrossRef] [PubMed]
140. Ottaiano, L.; Di Mola, I.; Di Tommasi, P.; Mori, M.; Magliulo, V.; Vitale, L. Effects of Irrigation on N_2O Emissions in a Maize Crop Grown on Different Soil Types in Two Contrasting Seasons. *Agriculture* **2020**, *10*, 623. [CrossRef]
141. Signor, D.; Cerri, C.E.P.; Conant, R. N_2O emissions due to nitrogen fertilizer applications in two regions of sugarcane cultivation in Brazil. *Environ. Res. Lett.* **2013**, *8*, 015013. [CrossRef]
142. Degaspari, I.A.M.; Soares, J.R.; Montezano, Z.F.; Del Grosso, S.J.; Vitti, A.C.; Rossetto, R.; Cantarella, H. Nitrogen sources and application rates affect emissions of N_2O and NH_3 in sugarcane. *Nutr. Cycl. Agroecosyst.* **2020**, *116*, 329–344. [CrossRef]
143. Harty, M.A.; Forrestal, P.J.; Watson, C.J.; McGeough, K.L.; Carolan, R.; Krol, D.; Laughlin, R.J.; Richards, K.G.; Lanigan, G.J. Reducing nitrous oxide emissions by changing N fertilizer used from calcium ammonium nitrate (CAN) to urea based formulations. *Sci. Total Environ.* **2016**, *563–564*, 576–586. [CrossRef]
144. Li, Z.; Xu, Q.; Lu, Y.; Ning, W.; Wu, R.; Li, T.; Mao, B.; Yang, Y.; Su, H.; Yang, Y.; et al. Reducing nitrogen fertilizer applications mitigates N_2O emissions and maintains sugarcane yields in South China. *Agric. Ecosyst. Environ.* **2025**, *377*, 109250. [CrossRef]
145. He, Z.; Ding, B.; Pei, S.; Cao, H.; Liang, J.; Li, Z. The impact of organic fertilizer replacement on greenhouse gas emissions and its influencing factors. *Sci. Total Environ.* **2023**, *905*, 166917. [CrossRef] [PubMed]
146. Lebender, U.; Senbayram, M.; Lammel, J.; Kuhlman, H. Effect of mineral nitrogen fertilizer forms on N_2O emissions from arable soils in winter wheat production. *J. Plant Nutr. Soil Sci.* **2014**, *177*, 722–732. [CrossRef]
147. Zhang, Z.; Chen, G.; Yu, X.; Liang, D.; Xu, C.; Ji, C.; Wang, L.; Ma, H.; Wang, J. A slow-release fertilizer containing cyhalofop-butyl reduces N_2O emissions by slowly releasing nitrogen and down-regulating the relative abundance of *nirK*. *Sci. Total Environ.* **2024**, *906*, 167493. [CrossRef] [PubMed]
148. Hasukawa, H.; Inoda, Y.; Takayama, T.; Takemisa, K.; Sudo, S.; Akiyama, H.; Yanai, J. Effects of controlled release N fertilizers and reduced application rate on nitrous oxide emissions from soybean fields converted from rice paddies. *Soil Sci. Plant Nutr.* **2021**, *67*, 716–726. [CrossRef]
149. Zebarth, B.; Snowdon, E.; Burton, D.; Goyer, C.; Dowbenko, R. Controlled release fertilizer product effects on potato crop response and nitrous oxide emissions under rain-fed production on a medium-textured soil. *Can. J. Soil Sci.* **2012**, *92*, 759–769. [CrossRef]
150. Bandara, W.B.M.A.C.; Sakai, K.; Anan, M.; Nakamura, S.; Setouchi, H.; Noborio, K.; Kai, T.; Rathnappriya, R.H.K. N_2O emissions from controlled-release and conventional N-fertilizers applied to red-yellow soil in Okinawa, Japan. *Soil Tillage Res.* **2025**, *248*, 106376. [CrossRef]
151. Liu, Q.; Liu, Y.; Hao, X.; Song, C.; Zong, Y.; Zhang, D.; Shi, X.; Li, P. Effects of controlled-release fertilizer on N_2O emissions in wheat under elevated CO_2 concentration and temperature. *Plant Soil* **2023**, *488*, 343–361. [CrossRef]
152. Pereira, V.V.; Morales, M.M.; Pereira, D.H.; de Rezende, F.A.; de Souza Magalhães, C.A.; de Lima, L.B.; Marimon-Junior, B.H.; Petter, F.A. Activated Biochar-Based Organomineral Fertilizer Delays Nitrogen Release and Reduces N_2O Emission. *Sustainability* **2022**, *14*, 12388. [CrossRef]
153. Zhang, Y.; Yan, C.; Wang, T.; Zhang, G.; Bahn, M.; Mo, F.; Han, J. Biochar strategy for long-term N_2O emission reduction: Insights into soil physical structure and microbial interaction. *Soil Biol. Biochem.* **2025**, *202*, 109685. [CrossRef]
154. Moitinho, M.R.; Philippot, L.; Gongaza, L.C.; Bru, D.; Carvalho, J.L.N. Biochar addition reduces N_2O emissions in fertilized soils under energy cane cultivation. *Agric. Ecosyst. Environ.* **2024**, *359*, 108744. [CrossRef]
155. Dawar, K.; Saif-ur-Rahman; Fahad, S.; Alam, S.S.; Khan, S.A.; Dawar, A.; Younis, U.; Danish, S.; Datta, R.; Dick, R. Influence of variable biochar concentration on yield-scaled nitrous oxide emissions, Wheat yield and nitrogen use efficiency. *Sci. Rep.* **2021**, *11*, 16774. [CrossRef]

156. Lin, F.; Wang, H.; Shaghaleh, H.; Ali Adam Hamad, A.; Zhang, Y.; Yang, B.; Alhaj Hamoud, Y. Effects of Biochar Amendment on N_2O Emissions from Soils with Different pH Levels. *Atmosphere* **2024**, *15*, 68. [CrossRef]
157. Liu, H.; Wang, N.; Wang, Y.; Li, Y.; Zhang, Y.; Qi, G.; Dong, H.; Wang, H.; Zhang, X.; Li, X. Inhibitory Effects of Biochar on N_2O Emissions through Soil Denitrification in Huanghuaihai Plain of China and Estimation of Influence Time. *Sustainability* **2024**, *16*, 5813. [CrossRef]
158. Edenhofer, O.; Pichs-Madruga, R.; Sokona, Y.; Farahani, E.; Kadner, S.; Seyboth, K.; Adler, A.; Baum, I.; Brunner, S.; Eickemeier, P.; et al. (Eds.) *Climate Change 2014: Mitigation of Climate Change. Contribution of Working Group III to the Fifth Assessment Report of the Intergovernmental Panel on Climate Change*; Cambridge University Press: Cambridge, UK; New York, NY, USA, 2014.
159. Malyan, S.K.; Maithani, D.; Kumar, V. Nitrous Oxide Production and Mitigation Through Nitrification Inhibitors in Agricultural Soils: A Mechanistic Understanding and Comprehensive Evaluation of Influencing Factors. *Nitrogen* **2025**, *6*, 14. [CrossRef]
160. Tariq, A.; Larsem, K.S.; Hansen, L.V.; Jensen, L.S.; Bruun, S. Effect of nitrification inhibitor (DMPP) on nitrous oxide emissions from agricultural fields; automated and manual measurements. *Sci. Total Environ.* **2022**, *847*, 157650. [CrossRef] [PubMed]
161. Wang, J.; Huang, Q.; Yu, D.; Zhang, Y.; Uwiragiye, Y.; Fallah, N.; Chen, M.; Cheng, Y. Effectiveness of Nitrification Inhibitor in Reducing N_2O Emissions Depends on Soil Acidification Mitigation in Acid Soils. *Agronomy* **2025**, *15*, 1536. [CrossRef]
162. Muller, J.; De Rosa, D.; Friedl, J.; De Antoni Migliorati, M.; Rowlings, D.; Grace, P.; Scheer, C. Combining nitrification inhibitors with a reduced N rate maintains yield and reduces N_2O emissions in sweet corn. *Nutr. Cycl. Agroecosyst.* **2023**, *125*, 107–121. [CrossRef]
163. Hadi, A.; Jumadi, O.; Inubushi, K.; Yagi, K. Mitigation options for N_2O emissions from a corn field in Kalimantan, Indonesia. *Soil Sci. Plant Nutr.* **2008**, *54*, 644–649. [CrossRef]
164. Pelster, D.E.; Chantigny, M.H.; Rochette, P.; Angers, D.A.; Laganière, J.; Zebarth, B.; Goyer, C. Crop residue incorporation alters soil nitrous oxide emissions during freeze–thaw cycles. *Can. J. Soil Sci.* **2013**, *93*, 415–425. [CrossRef]
165. Li, Z.; Zhang, Q.; Li, Z.; Qiao, Y.; Du, K.; Tian, C.; Zhu, N.; Leng, P.; Yue, Z.; Cheng, H.; et al. Effects of no-tillage on greenhouse has emissions in maize fields in a semi-humid temperate climate region. *Environ. Pollut.* **2022**, *309*, 119747. [CrossRef]
166. Rutkowska, B.; Szulc, W.; Szara, E.; Skowronska, M.; Jadczyszyn, T. Soil N_2O emissions under conventional and reduced tillage methods and maize cultivation. *Plant Soil Environ.* **2017**, *63*, 342–347. [CrossRef]
167. Yoo, J.; Woo, S.H.; Park, K.D.; Chung, K.-Y. Effect of no-tillage and conventional tillage practices on the nitrous oxide (N_2O) emissions in an upland soil: Soil N_2O emission as affected by the fertilizer applications. *Appl. Biol. Chem.* **2016**, *59*, 787–797. [CrossRef]
168. Gregorich, E.G.; Rochette, P.; St-Georges, P.; McKim, U.F.; Chan, C. Tillage effects on N_2O emission from soils under corn and soybeans in Eastern Canada. *Can. J. Soil Sci.* **2008**, *88*, 153–161. [CrossRef]
169. Aguilera, E.; Lassaletta, L.; Sanz-Cobena, A.; Garnier, J.; Vallejo, A. The potential of organic fertilizers and water management to reduce N_2O emissions in Mediterranean climate cropping systems. A review. *Agric. Ecosyst. Environ.* **2013**, *164*, 32052. [CrossRef]
170. Zhang, P.; Liu, J.; Zhang, H.; Wang, M.; Xu, J.; Yu, L.; Cai, H. Deficit irrigation interacting with biochar mitigates N_2O emissions from farmland in a wheat-maize rotation system. *Agric. Water Manag.* **2024**, *297*, 108843. [CrossRef]
171. Flynn, N.; Stewart, C.; Comas, L.; Del Grosso, S.; Schnarr, C.; Schipanski, M.; Von Fisher, J.; Stuchiner, E.; Fonte, S. Deficit irrigation impacts on greenhouse gas emissions under drip-fertigated maize in the Great Plains of Colorado. *J. Environ.Qual.* **2022**, *51*, 877–889. [CrossRef] [PubMed]
172. Liu, D.; Sun, H.; Liao, X.; Luo, J.; Lindsey, S.; Yuan, J.; He, T.; Zaman, M.; Ding, W. N_2O and NO Emissions as Affected by the Continuous Combined Application of Organic and Mineral N Fertilizer to a Soil on the North China Plain. *Agronomy* **2020**, *10*, 1965. [CrossRef]
173. Shu, X.; Wang, Y.; Wang, Y.; Ma, Y.; Men, M.; Zheng, Y.; Xue, C.; Peng, Z.; Noulas, C. Response of soil N_2O emission and nitrogen utilization to organic matter in the wheat and maize rotation system. *Sci. Rep.* **2021**, *11*, 4396. [CrossRef]
174. Biernat, L.; Taube, F.; Loges, R.; Kluß, C.; Reinsch, T. Nitrous Oxide Emissions and Methane Uptake from Organic and Conventionally Managed Arable Crop Rotations on Farms in Northwest Germany. *Sustainability* **2020**, *12*, 3240. [CrossRef]
175. Otieno, F.; Nalakurthi, S.-R.; Raji, M.; Tiwari, A.; Anton, I.; Gharbia, S. Farmer's attitudes towards GHG emissions and adoption to low-cost sensor-driven smart farming for mitigation: The case of Ireland tillage and horticultural farmers. *Smart Agric. Technol.* **2024**, *9*, 100622. [CrossRef]
176. Zhang, H.; Batchelor, W.D.; Hu, K.; Liang, H.; Han, H.; Li, J. Simulation of N_2O emissions from greenhouse vegetable production under different management systems in North China. *Ecol. Modell.* **2022**, *470*, 110019. [CrossRef]
177. Begum, K.; Kuhnert, M.; Yeluripati, J.; Ogle, S.; Parton, W.; Kader, M.A.; Smith, P. Model Based Regional Estimates of Soil Organic Carbon Sequestration and Greenhouse Gas Mitigation Potentials from Rice Croplands in Bangladesh. *Land* **2018**, *7*, 82. [CrossRef]
178. Chen, X.; Tao, T.; Zhou, J.; Yu, H.; Guo, H.; Chen, H. Simulation and Prediction of Greenhouse Gas Emissions from Beef Cattle. *Sustainability* **2023**, *15*, 11994. [CrossRef]
179. Tang, Y.; Qiao, Y.; Ma, Y.; Huang, W.; Komal, K.; Miao, S. Quantifying greenhouse gas emissions in agricultural systems: A comparative analysis of process models. *Ecol. Model.* **2024**, *490*, 110646. [CrossRef]

180. Van Hoof, S. Climate Change Mitigation in Agriculture: Barriers to the Adoption of Carbon Farming Policies in the EU. *Sustainability* **2023**, *15*, 10452. [CrossRef]
181. Spiegel, A.; Heidecke, C.; Rommel, J.; Matthews, A. Unravelling complexity of policies for climate change mitigation in agriculture. *Q Open* **2025**, *5*, qoaf007. [CrossRef]
182. Xing, Y.; Wang, X. Impact of Agricultural Activities on Climate Change: A Review of Greenhouse Gas Emission Patterns in Field Crop Systems. *Plants* **2024**, *13*, 2285. [CrossRef] [PubMed]
183. Menegat, S.; Ledo, A.; Tirado, R. Greenhouse gas emissions from global production and use of nitrogen synthetic fertilisers in agriculture. *Sci. Rep.* **2022**, *12*, 14490. [CrossRef] [PubMed]
184. Virkkunen, H.; Riihimaki, J.; Salminen, J.; Savolainen, H. Framework for quantification of land use-based greenhouse gas emissions according to crop and farm type. *J. Clean. Prod.* **2025**, *498*, 145111. [CrossRef]
185. Kimei, E.H.; Nyambo, D.G.; Mduma, N.; Kaijage, S. Review of Sources of Uncertainty and Techniques Used in Uncertainty Quantification and Sensitivity Analysis to Estimate Greenhouse Gas Emissions from Ruminants. *Sustainability* **2024**, *16*, 2219. [CrossRef]
186. Masi, M.; De Rosa, M.; Vecchio, Y.; Bartoli, L.; Adinolfi, F. The long way to innovation adoption: Insights from precision agriculture. *Agric. Econ.* **2022**, *10*, 27. [CrossRef]
187. Research and Markets. Global Precision Agriculture Market Analysis, 2023–2033: Efficiency, Yields, Cost Reduction, Sustainability All to Benefit with Growing Adoption. Available online: https://www.globenewswire.com/news-release/2024/4/3/2856713/28124/en/Global-Precision-Agriculture-Market-Analysis-2023-2033-Efficiency-Yields-Cost-Reduction-Sustainability-All-to-Benefit-with-Growing-Adoption.html (accessed on 6 November 2025).
188. Oroian, C.; Odagiu, A.; Vizitiu, O. What carbon storage farming activities adopted by farmers? A survey from the North-Western region of Romania. *Agrolife Sci. J.* **2024**, *13*, 192–199. [CrossRef]

sustainability

MDPI

Article

Pathways Through Which Digital Technology Use Facilitates Farmers' Adoption of Green Agricultural Technologies: A Comprehensive Study Based on Grounded Theory and Empirical Testing

Xiyang Yin [1], Wanyi Li [1], Shuyu Tang [1], Yanjiao Li [2], Jianhua Zhao [3] and Pengpeng Tian [1,*]

[1] College of Public Administration, Sichuan Agricultural University, Yaan 625014, China; 14984@sicau.edu.cn (X.Y.); 17380670775@163.com (W.L.); tangshuyu@stu.sicau.edu.cn (S.T.)
[2] College of Economics and Management, Northwest A&F University, Yangling 712100, China; yanjiaoli123@nwafu.edu.cn
[3] Business and Communication, INTI International University, Nilai 71800, Malaysia; zhaojianhua@tfswufe.edu.cn
* Correspondence: sicagpp@163.com

Abstract: The use of digital technologies can break down information barriers in rural areas, thereby creating crucial conditions for the widespread adoption of green agricultural technologies (GATs) among farmers. To explore the relationship between digital technology use (DTU) and farmers' adoption of GATs, this study draws on 18 in-depth interviews and 608 survey responses collected from rice farmers in Sichuan Province, China. By adopting a mixed-methods design, it offers a comprehensive examination of the mechanisms through which digital technology use (DTU) promotes the adoption of green agricultural technologies (GATs) among farmers. Grounded theory analysis reveals that the DTU–GATs adoption pathway can be conceptualized within a "condition–process–outcome" framework. Specifically, digital infrastructure, farmers' capital endowment, and practical needs constitute the foundational conditions, while technology perception and the regional soft environment act as key mediating processes. The ultimate outcomes include improvements in economic performance, social well-being, and ecological sustainability. Empirical evidence confirms that DTU significantly promotes the adoption of GATs, primarily by enhancing farmers' perceptions of technology and improving the agricultural soft environment at the regional level. Moreover, the effects of DTU display substantial heterogeneity across different types of green technologies and among various farmer groups. These findings highlight the importance of strengthening digital infrastructure in rural areas, enhancing farmers' digital literacy and capacity, and leveraging digital tools to tailor the dissemination and guidance of GATs. Such efforts are essential to raise farmers' awareness, foster a supportive soft environment for sustainable agriculture, and ultimately advance the adoption of GATs.

Keywords: digital technology; technology perception; agricultural regional soft environment; green agricultural technologies; grounded theory

1. Introduction

Green agricultural development is a crucial approach to addressing the increasing tension in global agricultural resources and the environment [1,2]. Currently, the insufficient total supply of agricultural products remains a challenge for many developing countries. To increase agricultural yields, chemical inputs such as pesticides, fertilizers, and plastic films are being excessively used in many developing countries, leading to severe agricultural non-point source pollution [3,4]. As the largest developing country in the world, China's agricultural development has long been focused on "pursuing increased agricultural product quantity," resulting in severe misuse of agricultural chemicals [5]. Taking fertilizers as an example, compared to 1996, the amount of fertilizer applied in China in 2015 increased by 21.95 million tons, with a fertilizer application intensity of 446 kg/ha, approximately twice the international safety threshold [6]. In fact, during this period, the overall fertilizer utilization rate in China was less than 30%, with a large amount of unused fertilizer entering water bodies or soils through surface runoff, severely polluting the ecological environment [7,8]. In this context, the Chinese government has gradually recognized the urgency of green agricultural development and introduced a series of policies, such as the "Zero Growth Action Plan for Fertilizer Use by 2020" and the "Zero Growth Action Plan for Pesticide Use by 2020." These policies aim to guide farmers in adopting GATs through a combination of incentives and penalties (e.g., organic fertilizer subsidy policies, straw burning prohibition policies, etc.). Unfortunately, the implementation of these policies has not yielded ideal results, and the adoption of GATs by Chinese farmers remains relatively low. As of 2022, China's fertilizer usage intensity remained well above the global average, while its pesticide consumption continued to rank first worldwide [9].

The lack of information is often considered the primary reason for the difficulty in promoting green agricultural production technologies [10,11]. The Household Contract Responsibility System implemented in China has made small-scale farming households the main agricultural operators. These small farmers have long been embedded in traditional rural social networks based on kinship and locality, where agricultural information is primarily concentrated within a small elite group (e.g., village leaders) [5,12,13]. The majority of small farmers, however, are distant from information hubs, making it difficult for them to access timely information on GATs, thus hindering the promotion and widespread adoption of such technologies. Therefore, breaking down the barriers to information dissemination in rural areas is key to promoting the widespread adoption of GATs in China and other developing countries.

The rapid advancement of digital technologies has fundamentally reshaped traditional modes of economic development, giving rise to new forms of industry and driving China's economic expansion. Official statistics indicate that in 2023, the scale of China's digital economy reached 53.9 trillion yuan, with its growth rate outpacing nominal GDP growth by 2.76 percentage points and contributing 66.45% to overall GDP growth [14]. Despite this remarkable progress, the development of the digital economy remains uneven. It is predominantly concentrated in urban areas, where digital technologies are more extensively applied in manufacturing and service sectors, while rural regions continue to lag far behind [15]. This urban–rural divide is largely attributable to the underdeveloped digital infrastructure in rural areas, where issues such as poor network quality and low internet speeds persist, thereby constraining farmers' incentives to participate in digital economic activities [16].

In response, the Chinese government has increasingly prioritized the promotion of digital technologies in agriculture and rural development. Policy initiatives such as the Digital Agriculture and Rural Development Plan (2019–2025) and the Digital Countryside Development Strategy Outline have provided important institutional support for advancing rural digitalization. As a result, more than 99% of Chinese villages are now covered by 4G networks, the rural Internet user base has expanded to 300 million, and the informatization rate of agricultural extension services has reached 61.3% [17]. The diffusion of digital technologies in the agricultural sector has accelerated the digital transformation of production, processing, distribution, and management, thereby laying a foundation for meeting increasingly diverse and higher-quality demands for agricultural products [18].

Against this backdrop, some scholars have begun to focus on the impact of DTU on the promotion of GATs in China. They generally believe that DTU expands farmers' information channels, helping them gain a deeper understanding of the advantages and effectiveness of GATs, thereby enhancing their awareness and willingness to adopt such technologies [19]. At the same time, digital technologies can provide farmers with efficient, cost-effective, and convenient technical services such as online training, remote education, and expert consultations, reducing the difficulty of adopting GATs and facilitating their adoption [20]. However, the aforementioned studies primarily conduct empirical analysis through econometric models, and the measurement of relevant variables is not comprehensive (e.g., measuring DTU based on whether farmers use the Internet), which may affect the reliability of the research findings. In addition, the promotion and application of digital technologies in rural China is relatively recent, and theoretical research combining digital technology with farmer behavior is still in its early stages. Relying solely on quantitative research makes it difficult to deeply analyze the impact mechanisms of DTU on farmers' adoption of GATs. In fact, quantitative and qualitative research are not mutually exclusive; their combined use can provide a more comprehensive understanding of the research findings [21,22]. Against the backdrop of accelerating agricultural digitalization and growing demand for green agricultural products in China, further investigation into the relationship between DTU and farmers' adoption of green technologies is of great significance.

Therefore, to systematically and comprehensively explore the pathways and mechanisms through which DTU promotes farmers' adoption of green technologies, this study applies grounded theory coding techniques to analyze the interview data from 18 rice farmers, extracting and synthesizing the pathways through which DTU facilitates the adoption of GATs by farmers, based on their practical cases. At the same time, this paper uses 608 farmer questionnaire survey data, adopts the Ordered Probit model and Control Function Method (CFM) for empirical testing, and further verifies the results of the case analysis, thereby forming a mutual confirmation between quantitative and qualitative research and ensuring the reliability of the research conclusions of this paper. This study makes three main contributions: First, based on grounded theory, it employs a mixed-methods approach combining in-depth interviews and surveys to analyze the pathways through which DTU promotes the adoption of GATs by farmers from both qualitative and quantitative perspectives. This addresses the gap in the existing literature regarding qualitative analysis and enriches the research on DTU and farmer behavior. Second, this study comprehensively measures farmers' use of digital technologies based on their application in areas such as agricultural input purchasing, agricultural product sales, technical learning, information access, and social interactions. It also assesses farmers' adoption of GATs through a composite measure of five key technologies: green cultivation, pest and disease control, fertilization, irrigation, and waste management. Therefore, this study provides a more comprehensive variable measurement, enhancing the reliability of the empirical analysis. Third, this study explores the mechanisms through which DTU influences farmers' adoption of GATs from the perspectives of technology perception and regional soft environment, providing targeted recommendations for further promoting GATs among Chinese farmers.

2. Literature Review

With the development of China's socio-economic landscape, the demand for agricultural products among Chinese consumers has shifted from meeting basic needs to a focus on green, healthy, and safe products [5]. The traditional agricultural supply model, which relied on increased pesticide and fertilizer use, has become outdated in response to the upgrading of consumer demand [7]. As the main participants in China's agricultural production, how to promote farmers' adoption of GATs and advancing green agricultural development has attracted significant scholarly attention. Currently, scholars primarily focus on researching the factors influencing farmers' adoption of GATs and their effects [3,23–26]. Behavioral economics theory suggests that human behavior is characterized by bounded rationality, where, under certain environmental influences, individuals make relatively satisfactory behavioral decisions based on their limited cognitive abilities [27]. Therefore, existing research primarily analyzes the factors influencing farmers' adoption of GATs from two perspectives: farmers' internal characteristics and the external environment. Relevant studies indicate that individual and household attributes are the fundamental internal factors influencing farmers' adoption of technologies, affecting their adoption of green technologies to varying degrees. Relevant studies indicate that individual and household attributes are the fundamental internal factors influencing farmers' technology adoption, with a significant impact on the adoption of GATs [28–31]. Among these factors, farmers' perceptions of technology play a crucial role in shaping their adoption of GATs. Generally, the easier farmers perceive the technology to use, the higher the perceived benefits, and the lower the perceived risks, the stronger their willingness to adopt GATs [32–35]. Some studies categorize the external environmental factors influencing farmers' adoption of GATs into the regional soft environment and the regional hard environment. Research indicates that regional soft environment factors, such as government actions [36,37], market conditions [38], and social capital [39,40], have a significant impact on farmers' adoption of GATs. At the same time, regional hard environment factors such as geographical location, climate, terrain, and infrastructure form the foundation of agricultural development and also have a significant impact on farmers' GATs [41]. Additionally, some studies have found that farmers' adoption of GATs not only has economic effects but also generates social and ecological impacts [42–44].

Digital technology refers to modern technologies that enable the digitalization of production and management through hardware devices such as computers, mobile phones, and sensors, supported by digital software [45,46]. The widespread application of digital technologies has brought extensive, continuous, and profound transformations to traditional economic systems, fundamentally reshaping people's modes of production and daily life [47,48]. With the ongoing development of the global digital economy, the information dividends generated by digital technologies have gradually diffused into the agricultural and rural sectors. Consequently, an increasing number of scholars have begun to examine the effects of DTU on farmers' production behaviors [49,50]. Among these behaviors, farmers' adoption of GATs is directly linked to the sustainability of agricultural production, and the relationship between DTU and such adoption has received growing academic attention [51,52]. Existing research indicates that digital technologies can effectively reduce the costs, risks, and complexity associated with adopting GATs [53–56] while increasing farmers' expected returns from GATs adoption [57], thereby motivating farmers to adopt GATs. Specifically, digital technologies can provide farmers with more transparent, convenient, and low-cost information and technical support, which not only significantly lowers the barriers to adopting GATs but also enhances farmers' awareness of these technologies, thereby promoting their adoption [19,58–60]. Although some research has attempted to explore new pathways for promoting farmers' adoption of GATs from the perspective of DTU, most of these studies rely primarily on micro-level data for empirical testing and lack in-depth analysis based on typical case studies. Research combining both qualitative and quantitative approaches is even rarer.

Overall, current academic research on farmers' adoption of GATs is relatively extensive, with scholars exploring various pathways to promote this adoption from different perspectives. Some researchers have also attempted to analyze farmers' adoption of GATs is relatively extensive, with scholars exploring various pathways to promote this adoption from the perspective of DTU, but these studies are still in the early stages, and the related theories and analytical methods need further development and refinement. To explore in greater depth the pathways through which DTU facilitates farmers' adoption of GATs, this study combines qualitative and quantitative analysis methods, employing a grounded theory framework. It derives a theoretical model of the relationship between DTU and farmers' adoption of GATs from real-world case studies, and empirically tests this model using econometric analysis. This approach combines case study analysis with empirical testing to provide a detailed examination of the important role digital technology plays in the adoption process, offering valuable insights for further promoting the adoption of GATs among Chinese farmers.

3. Exploratory Research Based on Grounded Theory

3.1. Research Design

3.1.1. Research Methods

Grounded theory, proposed by Glaser and Strauss [61], is a qualitative research method that generates abstract theories from empirical data and real-world observations. Subsequently, through the contributions of Strauss [62] and Charmaz [63], grounded theory evolved into three major schools: classic grounded theory, Straussian (or procedural) grounded theory, and constructivist grounded theory. Among them, Straussian grounded theory places greater emphasis on human agency, arguing that data contain implicit causal relationships that require researchers' informed interpretation and systematic organization. Its analytical process consists of three stages: open coding, axial coding, and selective coding [64]. Therefore, this study follows the data analysis procedures of Straussian grounded theory—open coding, axial coding, and selective coding—to systematically analyze the interview data. Through this step-by-step process, labels, concepts, and categories are extracted, causal relationships among variables at different levels are identified, and an abstract theoretical framework is ultimately constructed.

3.1.2. Data Collection

Compared to quantitative research, which relies on large-scale random sampling, case studies place greater emphasis on purposive sampling—that is, selecting cases that are most likely to provide rich and relevant information for addressing the research question [65]. The selection of case farmers in this study was closely aligned with the research objectives, while also considering their use of digital technologies and green production technologies. The interview materials were collected through field interviews conducted by the research team in major rice-producing areas of Sichuan Province, China, from July to September 2022. Prior to the interviews, an interview outline was developed based on the research needs, and the interview team received training to ensure familiarity with the questions and proficiency in interview techniques, thereby enhancing the quality of the interviews. During the fieldwork, 20 farmers who met the selection criteria were chosen for one-on-one in-depth interviews, each lasting approximately 30 min and recorded with consent. A total of 18 valid interview transcripts were obtained. After transcription, 15 interviews were randomly selected for coding and model construction, while the remaining 3 were used for theoretical saturation testing.

3.1.3. Research Process

To enhance the robustness of the research, extensive efforts were made to collect and supplement the data. Following the coding procedures of Straussian grounded theory, the data were analyzed step by step, with continuous refinement and abstraction of concepts and their relationships, ultimately leading to the construction of a substantive theory. The specific coding strategy is illustrated in Figure 1.

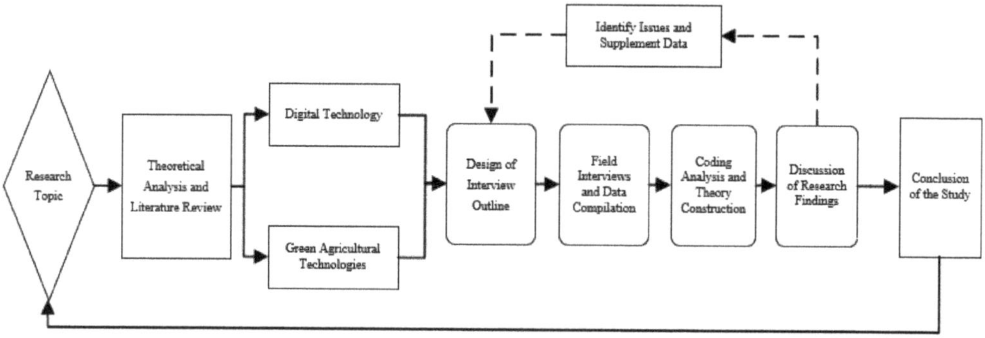

Figure 1. Research Process.

To minimize analytical bias caused by individual subjectivity, the case coding in this study was conducted collaboratively by three researchers experienced in grounded theory coding. Prior to the coding process, all team members received standardized technical training. After a thorough discussion to ensure the operability of the coding procedure, each member independently coded five cases. Final labels were determined through group discussion based on the individual coding results. During the processes of labeling, conceptualization, categorization, and theorization, the coding team conducted follow-up phone interviews with respondents as needed to gather additional information and continuously refine the emerging labels, concepts, categories, and relationships. Finally, the three reserved cases were re-coded and analyzed to test the theoretical saturation of the coding process.

3.2. *Data Coding*

3.2.1. Open Coding

This study adheres to the fundamental principle of remaining maximally open to all potential theoretical insights and conducts open coding around the core theme of "the pathways through which DTU facilitates farmers' adoption of green technologies." First, the coding team extracted original statements closely related to the research theme from the interview transcripts and conducted initial conceptualization, identifying 70 substantively meaningful initial concepts (prefixed with "A"). These concepts were then grouped and abstracted into categories based on similarity in meaning or attributes, resulting in 30 initial categories. The correspondence between concepts and categories is presented in Table 1.

Table 1. Open coding categorization.

Initial Categories	Initial Concepts
Social Capital Endowment	A01 Cadre Experience A02 Political Identity

Table 1. *Cont.*

Initial Categories	Initial Concepts
Digitalization of Information Sharing	A03 Online Message Transmission A30 Expansion of Online Information Channels A37 Online information search A38 Online Information Acquisition
Human Capital Endowment	A04 Age A09 Familiarity with GATs A56 Education level
Digitalization of Technology Sharing	A05 Online Technical Training A27 Online Technical Learning A31 Online Technical Consultation
Digitization of Knowledge Sharing	A06 Online Video-Based Learning A54 Online Experience-Based Learning
Digitalization of Agricultural Input Sharing	A07 Online Land Leasing
Digitalization of Product Sharing	A36 Online Sales of Green Agricultural Products
Economic Benefits	A08 Reduce Land Leasing Costs A21 Improve Agricultural Product Quality A24 Reduce Agricultural Input Costs A39 Reduce Labor Costs A41 Stable and Increased Agricultural Production A47 Increase in Agricultural Product Prices
Digitalization of Agricultural Machinery Sharing	A10 Online Rental of Agricultural Machinery
Green Pest Management Technology	A11 Physical Pest Control Technology A62 Biological Pest Control Technology
Green Fertilization Technology	A13 Application of Organic (Farmyard) Manure A67 Application of Commercial Organic Fertilizers
Improvement of the policy environment	A17 Online Policy Promotion A28 Online Access to Policy Information
Green Waste Treatment Technologies	A15 Straw Crushing and Return to Field A18 Agricultural film recycling and processing
Perception of Economic Benefits	A12 Reduction in pesticide use A16 Reduction in fertilizer use A23 Green agricultural products command higher prices A64Increase in agricultural output
Perception of Environmental Benefits	A14 Organic fertilizers cause less environmental pollution A19 Chemical pesticides and fertilizers pollute the environment A48 GATs are beneficial for soil protection
Environmental benefits	A25 GATs are beneficial for environmental protection A53 GATs are beneficial for ecological conservation
Social benefits	A55 Beneficial for promoting green production among other farmers A68 Beneficial for meeting consumer demand for green agricultural products
Demand for improving agricultural product quality	A20 The quality of agricultural products in the market remains low A46 Family members and friends have a demand for green agricultural products
Demand for product sales	A22 Green agricultural products are of high quality but are underval-ued in the market A34 Green agricultural products suffer from insufficient consumer trust A35 Green consumption remains weak in the local market

Table 1. *Cont.*

Initial Categories	Initial Concepts
Improvement of the market environment	A26 Online promotion of agricultural products A29 Engage with consumers online
Demand for cost reduction and income enhancement	A32 The expectation of income improvement among farmers is relatively high A40 High labor costs
Improvement of the social environment	A33 Online communication among farmers A43 Learning about peers' farming models via digital channels
Demand for safety and health	A44 Farmers place a high priority on food safety A50 Farmers care about the village environment
Perception of social benefits	A51 Green production contributes to human health A52 Green production contributes to ensuring food security
digital devices	A57 Feature phones used by elderly users often lack Internet access A60 smartphones and computers
Perceived ease of use	A58 Green production technologies are easy to use
Digital networks	A59 Rural areas have an extensive network coverage
Perception of technological risks	A61 The adoption of GATs entails a risk of failure
Perception of market risk	A66 Green agricultural products may involve sales risks
Green cultivation technologies	A69 Cultivating during appropriate seasons helps reduce pest infestations A70 Deep tillage with machinery

3.2.2. Axial Coding

Axial coding involves reorganizing and integrating the preliminary categories generated during open coding to identify interconnections among them and develop higher-level core categories [66]. To further explore the relationships among the initial categories, this study reorganized and clustered 30 initial categories, resulting in 8 higher-level axial categories. The relationships among the axial categories can be interpreted using the "conditions–process–outcome" model. The detailed results of axial coding are presented in Table 2.

Table 2. The main category of axial coding forms.

Correspondence	Main Categories	Initial Categories
Conditions	Digital Infrastructure	Digital networks; digital devices
	Capital Endowment	Social Capital Endowment; Human Capital Endowment
	Practical Needs	Demand for improving agricultural product quality; Demand for product sales; Demand for cost reduction and income enhancement; Demand for safety and health
Process	Digital Technology Use	Digitalization of Information Sharing; Digitalization of Technology Sharing; Digitization of Knowledge Sharing; Digitalization of Agricultural Input Sharing; Digitalization of Agricultural Machinery Sharing; Digitalization of Product Sharing
	Regional Soft Environment	Improvement of the policy environment; Improvement of the market environment; Improvement of the social environment
	perception of technology	Perception of Economic Benefits; Perception of Environmental Benefits; Perception of social benefits; Perception of technological risks; Perception of market risk; Perceived ease of use

Table 2. *Cont.*

Correspondence	Main Categories	Initial Categories
Process	Adoption of GATs	Green Pest Management Technology; Green Fertilization Technology; Green Waste Treatment Technologies; Green cultivation technologies
Outcome	Benefits of Adopting GATs	Economic Benefits; Environmental benefits; Social benefits

3.2.3. Selective Coding

Selective coding abstracts the core category from all identified categories and links it with the main categories to form a coherent "storyline," thereby constructing a theoretical model that reflects the full narrative. Based on the analysis and synthesis of original case data, key concepts, and hierarchical categories, this study develops a theoretical model illustrating how the use of digital technologies facilitates farmers' adoption of GATs, as shown in Figure 2.

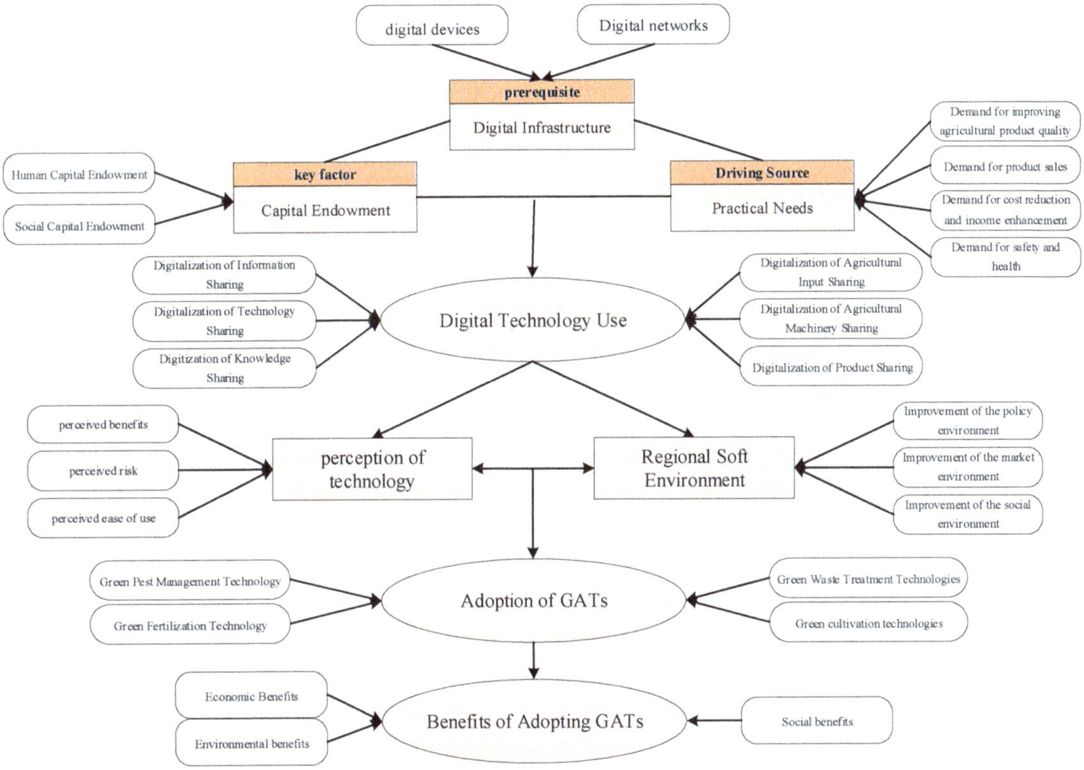

Figure 2. Theoretical Model of How DTU Facilitates Farmers' Adoption of GATs.

3.2.4. Theoretical Saturation Test

To further verify theoretical saturation and ensure that no new concepts emerge, this study conducted additional coding analysis using three reserved interview transcripts. The results indicate that no new concepts were identified in the remaining data. Therefore, the categories and relationships presented in the theoretical model (Figure 2) are considered theoretically saturated and constitute a well-grounded framework based on empirical data and contextual analysis.

4. Empirical Examination Based on Questionnaire Survey

4.1. Research Hypotheses

The theoretical framework of this study is illustrated in Figure 3. The diffusion of digital technologies in rural China has profoundly transformed farmers' cognitive and behavioral patterns, thereby providing new momentum for advancing the adoption of GATs. First, DTU can reduce the costs associated with adopting GATs. By virtualizing these technologies, digital platforms enable farmers to engage in learning, training, and information exchange related to green agriculture at a lower cost, thereby reducing the expenses involved in acquiring such technologies [58]. Moreover, digital trading platforms help farmers strengthen their connections with input suppliers, effectively streamlining the supply chain and lowering the cost of purchasing green pro-duction inputs [67]. Second, DTU can effectively reduce the risks associated with adopting GATs. Existing research indicates that green technologies often require targeted application and involve complex operations; improper use by farmers can lead to yield reductions and technical risks [68]. In addition, the market for green agricultural products is characterized by information asymmetry, which may result in the phenomenon of "high quality but low price," posing significant market risks [69]. With the information-sharing capabilities of digital technology, farmers can access detailed guidance on the proper use of green technologies anytime and anywhere, thereby minimizing technical risks. Furthermore, digital platforms enable direct interaction with consumers, allowing farmers to align production with green consumer demand, reduce information asymmetry, and effectively mitigate market risks [70]. Finally, digital technologies can reduce the perceived difficulty of adopting GATs. By breaking the spatial and temporal constraints of technology dissemination, digital tools expand farmers' access to green agricultural knowledge. Through mobile Internet platforms, farmers can search for, learn about, exchange, and consult on GATs anytime and anywhere, significantly lowering the barriers to accessing and using such technologies [54,55]. Therefore, this study proposes the following hypotheses:

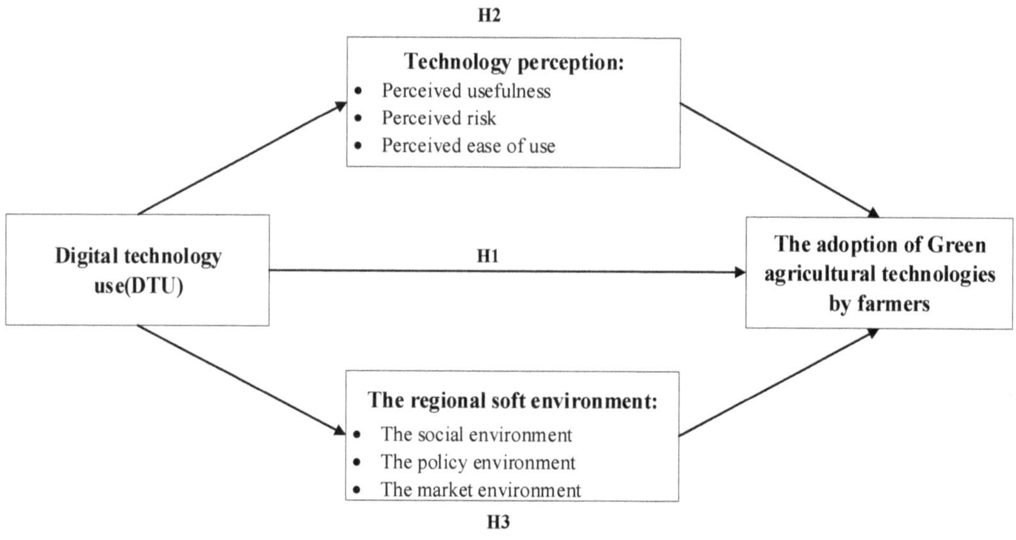

Figure 3. Structural model.

H1. *DTU has a significant positive effect on farmers' adoption of GATs.*

Farmers' intrinsic motivation to adopt GATs originates from their internal perceptions of such technologies, which primarily includes perceptions of benefits, risks, and ease of use [69,71]. The fundamental logic behind farmers' production decisions is to maximize benefits and minimize risks. While the pursuit of benefit maximization serves as the core motivation for adopting GATs, the presence of risk introduces uncertainty into the potential gains, making risk minimization a crucial precondition for adoption [72]. In addition, adoption is influenced by the perceived ease of use: the learning and application of GATs require investments of time, money, and effort. If farmers perceive these technologies as difficult to adopt, they are generally less willing to try them [73]. Based on the technical information they possess, farmers make comprehensive evaluations regarding the benefits, risks, and ease of adoption of GATs. These evaluations shape their overall technological cognition, which ultimately influences their decision on whether to adopt such technologies. On the one hand, digital technologies enhance farmers' ability to access information, enabling them to acquire more comprehensive and systematic knowledge about GATs [74]. This deepens their cognition regarding the benefits, risks, and ease of adoption of such technologies, and helps them develop rational strategies for adopting green practices. On the other hand, by breaking down information barriers, improving the efficiency of information dissemination, and reducing the cost of information acquisition, digital technologies help lower farmers' perceived risks, while enhancing their expected benefits and perceived ease of use, thereby encouraging the proactive adoption of GATs [20]. Therefore, this study proposes the following hypotheses:

H2. *Technological cognition mediates the relationship between DTU and farmers' adoption of GATs.*

Behavioral economics posits that human rationality is bounded, resulting from the joint influence of internal characteristics and external environmental factors. Under such bounded rationality, individuals tend to adopt satisficing strategies rather than pursuing absolute optimality [75]. In this context, farmers' adoption of GATs is influenced not only by internal factors such as cognitive perceptions and personal abilities, but also by external conditions, including regional hard environments—such as geography, climate, and infrastructure—and soft environments, such as policies, markets, and social norms [76]. Among these, the hard environment serves as the foundational constraint for agricultural development, typically stable and difficult to change in the short term. In contrast, the soft environment is more flexible and diverse, playing a guiding role in shaping farmers' production behaviors and technological choices [77]. Therefore, elements of the soft environment—such as regional green development policies, green technology extension efforts, and the dissemination of green concepts—can significantly influence farmers' decisions regarding the adoption of GATs. Rural China is characterized by a social network rooted in kinship and geographic proximity, within which information dissemination follows a hierarchical and segmented pattern [12]. Information related to green agricultural development—such as policy measures, promotional activities, and advocacy of green concepts—is typically first accessed by a small group of rural elites and then gradually diffused to the broader farming population. However, farmers located on the periphery of information networks often experience restricted access to digital channels and tend to have relatively low levels of information literacy. As a result, they may lack comprehensive and detailed knowledge of local green development initiatives. Consequently, the influence of the regional soft environment on these farmers' adoption of GATs tends to be limited. The advent of modern digital technologies has fundamentally transformed the modes of information dissemination in rural areas. These technologies have enabled the decentralization of rural information flows, breaking through the traditional spatial and temporal constraints of information transmission. As a result, information channels have been significantly expanded and the efficiency of communication has improved. This transformation facilitates farmers' access to timely, diversified, and comprehensive in-formation regarding the social, policy, and market aspects of green agricultural development, thereby increasing the likelihood that they will adopt GATs. Therefore, this study proposes the following hypotheses:

H3. *The regional soft environment mediates the relationship between DTU and farmers' adoption of GATs.*

4.2. Data Source

The research team conducted field interviews with rice farmers in Sichuan Province, China, from July to September 2022. Sichuan Province is one of China's major rice-producing regions, with rice cultivation primarily concentrated in eight cities located in the central, northeastern, and southeastern parts of the province. This study first employed a random sampling method to select five sample cities: Chengdu, Mianyang, Guang'an, Luzhou, and Yibin. Following the administrative hierarchy and using stratified random sampling, one to three sample counties were randomly selected within each city, followed by one to three townships within each county, and then one to three villages within each township. Finally, 15 to 25 rice-farming households were randomly selected in each village to complete the questionnaire survey. After removing invalid responses, a total of 608 valid questionnaires were obtained. The sample distribution is illustrated in Figure 4.

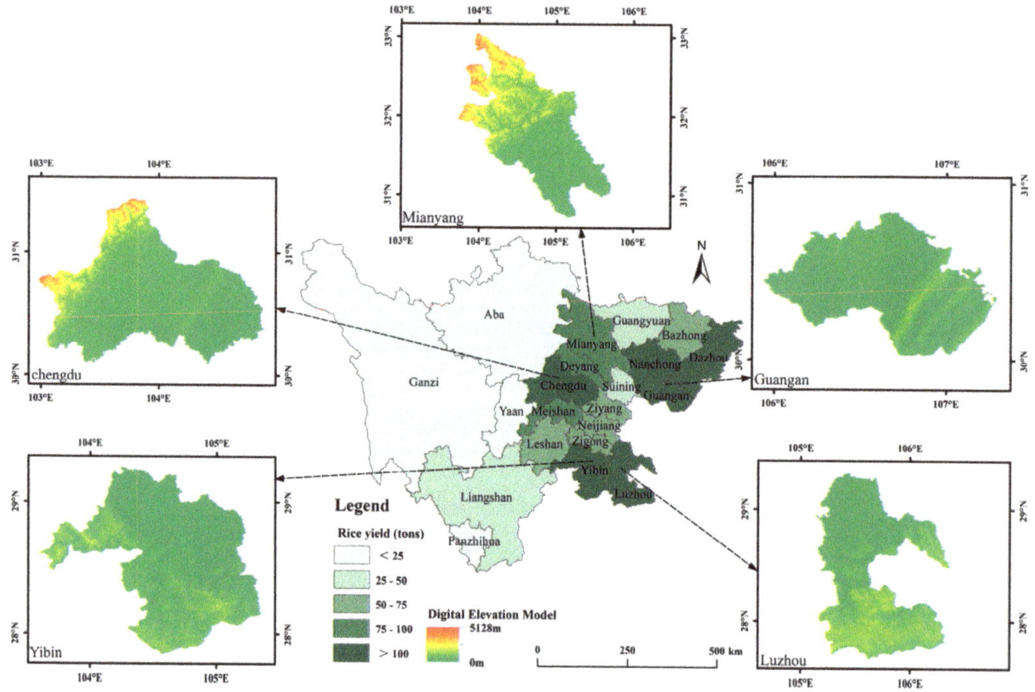

Figure 4. Distribution of Survey Regions.

4.3. Model Specification and Variable Selection

4.3.1. Model Specification

The degree of adoption of GATs by farmers is a ranking variable and can be estimated using the Ordered Probit model. Therefore, this paper constructs the following model:

$$GTA_i^* = \beta_0 + \beta_1 DTU_i + \beta_2 Control + \varepsilon_i \quad (1)$$

In Equation (1), GTA_i^* represents the extent to which farmer i adopts GATs; DTU_i represents the digital technology usage of farmer i; $Control$ refers to the control variables; β_0 denotes the constant term; ε_i denotes the random error term.

GTA_i^* is a latent variable. Assume that r_0, r_1, r_2, r_3, r_4 are unknown segmentation points of farmers' GATs adoption behavior, and $r_0 < r_1 < r_2 < r_3 < r_4$. Equation (2) can be defined as follows:

$$GTA_i = \begin{cases} 0, \ GTA_i^* \le r_0 \\ 1, r_0 < GTA_i^* \le r_1 \\ 2, r_1 < GTA_i^* \le r_2 \\ 3, r_2 < GTA_i^* \le r_3 \\ 4, r_3 < GTA_i^* \le r_4 \\ 5, \ GTA_i^* > r_4 \end{cases} \quad (2)$$

Accordingly, the conditional probabilities of farmers adopting different numbers of GATs can be expressed as:

$$P(GTA_i = 0) = \Phi(r_0 - \beta_1 DTU_i - \beta_2 Control) \quad (3)$$

$$P(GTA_i = 1) = \Phi(r_1 - \beta_1 DTU_i - \beta_2 Control) - \Phi(r_0 - \beta_1 DTU_i - \beta_2 Control)$$

$$P(GTA_i = 2) = \Phi(r_2 - \beta_1 DTU_i - \beta_2 Control) - \Phi(r_1 - \beta_1 DTU_i - \beta_2 Control)$$

$$P(GTA_i = 3) = \Phi(r_3 - \beta_1 DTU_i - \beta_2 Control) - \Phi(r_2 - \beta_1 DTU_i - \beta_2 Control)$$

$$P(GTA_i = 4) = \Phi(r_4 - \beta_1 DTU_i - \beta_2 Control) - \Phi(r_3 - \beta_1 DTU_i - \beta_2 Control)$$

$$P(GTA_i = 5) = 1 - \Phi(r_4 - \beta_1 DTU_i - \beta_2 Control)$$

In Equation (3), $\Phi(\cdot)$ denotes the cumulative distribution function of the standard normal distribution. The parameters of the model are estimated using the method of maximum likelihood.

Meanwhile, the Control Function Method (CFM) is applied in this study to mitigate potential endogeneity issues within the model [78]. This approach consists of two steps. First, the DTU variable is regressed on its instrumental variables and control variables to estimate the generalized residuals. Second, the predicted residuals are included as an additional regressor in the second-stage estimation. If the residual term is statistically significant, it indicates that DTU is endogenous, and incorporating the residual helps correct estimation bias. Conversely, if the residual is not significant, it suggests that DTU is exogenous, and excluding the residual leads to more efficient and unbiased estimation results. Accordingly, the model is constructed as follows:

$$DGT_i = \delta_0 + \delta_1 HCE_i + \delta_2 Control + \omega_i \quad (4)$$

$$GTA_i = b_0 + b_1 DTU_i + b_2 \sigma_i + b_3 Control + \epsilon_i \quad (5)$$

In Equations (4) and (5), HCE_i denotes the household Internet communication expenditure of farmer i, and σ_i represents the generalized residual obtained from the regression in Equation (4). δ_0 and b_0 are constant terms, δ_1, δ_2, b_1, b_2, and b_3 are parameters to be estimated. ω_i and ϵ_i denote random error terms.

4.3.2. Variable Selection

The dependent variable in this study is farmers' adoption of GATs. Following the approaches of Kuang et al. (2023) and Liu et al. (2024) [3,79], we consider five categories of GATs to comprehensively capture adoption behavior: green cultivation, green pest and disease control, green fertilization, green irrigation, and green waste management. Each category is measured as a binary variable, where a value of 1 indicates adoption and 0 otherwise. The overall adoption level is represented by the cumulative number of technologies adopted, ranging from 0 to 5.

The core explanatory variable is the use of digital technologies. Following the methodology of Zheng et al. (2022) and Kai et al. (2023) [80,81], this study assesses farmers' digital technology usage across various aspects of production and daily life, including social interaction, information access, agricultural input procurement, and agricultural product sales. During the survey, farmers were asked to rate their level of DTU in these areas using five-point Likert scales. An entropy-weighted average was then applied to construct a composite index measuring the overall extent of DTU.

Technology perception and the regional soft environment are identified as mediating variables in this study. Farmers' perceptions of GATs are primarily reflected in three dimensions: perceived usefulness, perceived risk, and perceived ease of use. Following the approach of Huang et al. (2022) [82], perceived usefulness is measured by the degree to which farmers recognize the economic benefits (e.g., increasing income, reducing input costs, and saving labor) and social benefits (e.g., improving product quality and promoting social development) of green technologies. Perceived ease of use is assessed by asking farmers about the level of difficulty they experience when adopting GATs. Perceived risk is captured by farmers' concerns regarding potential technical risks (e.g., yield reduction) and market risks (e.g., failure to achieve premium pricing despite quality improvements). The regional soft environment primarily consists of three dimensions: the social environment, policy environment, and market environment. The social environment is measured by the frequency of communication among farmers, the prevalence of GATs adoption among peers, and the farmer's sense of community identity. The policy environment is assessed across three aspects: government publicity, incentives, and constraints. The market environment is evaluated based on market recognition and expectations. All responses to the related questions are rated using five-point Likert scales, ranging from 1 (strongly disagree) to 5 (strongly agree). A comprehensive index is then constructed using the entropy weighting method.

Following the approach of Luo et al. (2022) [83], this study uses household expenditure on Internet and communication services as an instrumental variable for DTU. Such expenditure is closely related to the extent of farmers' digital technology adoption but does not directly influence their decisions regarding GATs' adoption. Therefore, it serves as a theoretically sound and appropriate instrumental variable.

To ensure reliable estimation, the analysis accounts for individual, household, and environmental characteristics as control variables. Individual characteristics include the respondent's gender, age, education level, political affiliation, off-farm employment status, and risk preference. Household operational characteristics encompass total household size, number of migrant workers, highest education level among household members, political affiliation of the main household member, and farm size. Environmental characteristics include the distance to the county seat, topographical features, and regional dummy variables. Detailed variable definitions are presented in Table 3.

Table 3. Variable selection and assignment description.

Variables	Definition and Assignment	Mean	S.D.
Dependent variable			
Green cultivation technologies	Whether to adopt green cultivation techniques, No = 0, Yes = 1	0.610	0.488
Green pest control technologies	Whether to adopt green pest control techniques, No = 0, Yes = 1	0.497	0.500
Green fertilization technologies	Whether to adopt green fertilization techniques, No = 0, Yes = 1	0.594	0.492
Green waste utilization technologies	Whether to adopt green waste utilization techniques, No = 0, Yes = 1	0.770	0.421
Green irrigation technologies	Whether to adopt green irrigation techniques, No = 0, Yes = 1	0.015	0.121
Green agricultural technologies (GATs)	Number of green production technologies adopted (items)	2.485	1.413
Core explanatory variable			
Digital technology use (DTU)	Entropy-based composite score	0.285	0.300
Mediating variables			
Technology perception	Entropy-based composite score	0.455	0.218
Regional soft environment	Entropy-based composite score	0.453	0.205
Control variables			
Gender	Gender of farmer, Female = 0, Male = 1	0.877	0.329
Age	Age of farmer (years)	54.984	10.291
Education	Years of Education (years)	8.498	3.191
Village cadre	Whether serving as a village cadre, No = 0, Yes = 1	0.148	0.355
Part-time farming	Whether engaged in part-time farming, No = 0, Yes = 1	0.286	0.452
Risk preference	Risk preference of farmers, Low = 1, Medium = 2, High = 3	2.125	0.832
Total household size	Number of all household members	4.684	1.802
Number of household members working away from home	Calculated in real terms, in number of persons	1.053	1.192
Highest education level in household	Highest education level among household members (years)	11.928	3.461
Number of cadre members	Number of cadre members in the household	0.194	0.420
Planting scale	Logarithmic value of rice planting area	3.264	1.884
Distance to county seat	Measured by actual distance (km)	23.681	12.769
Topography	Plain = 1, Hilly = 2, Mountainous = 3	1.691	0.507
Region	Chengdu Plain Economic Zone = 1, Northeastern Sichuan Economic Zone = 2, Southern Sichuan Economic Zone = 3	2.031	0.903
Instrumental variable			
Household communication and Internet expenditure	Logarithm of household communication and Internet expenditure	7.025	0.882

4.4. Results

4.4.1. Benchmark Regression Results

As shown in Table 4, Models (1) and (2) report estimates based on the Ordered Probit model, while Models (3) and (4) address potential endogeneity using the CFM. Regional dummy variables are included in Models (2) and (4). In Models (3) and (4), the residual terms are statistically significant at the 1% level, indicating the presence of endogeneity in DTU. Incorporating the residuals into the regression helps correct estimation bias. The results from Model (2) show that DTU has a significantly positive effect on farmers' adoption of GATs at the 1% level, suggesting that DTU significantly promotes adoption. This finding is consistent with the conclusions of Zhang et al. (2024) and Musajan et al. (2024) [58,84]. The estimates in Model (4), which correct for endogeneity, remain consistent with those in Model (2), confirming the robustness of the baseline findings and supporting Hypothesis 1.

Table 4. Benchmark regression results.

Variable Name	(1) Ordered Probit	(2) Ordered Probit	(3) CFM	(4) CFM
DTU	2.179 ***	2.223 ***	1.864 ***	1.902 ***
	(0.285)	(0.287)	(0.286)	(0.288)
Residual			9.576 ***	9.796 ***
			(1.428)	(1.492)
Gender	−0.063	−0.123	0.152	0.093
	(0.121)	(0.115)	(0.124)	(0.117)
Age	−0.017 **	−0.018 **	0.103 ***	0.105 ***
	(0.007)	(0.008)	(0.019)	(0.019)
Education	0.041 *	0.056 **	−0.186 ***	−0.176 ***
	(0.022)	(0.022)	(0.039)	(0.040)
Village Cadre	0.242	0.217	1.311 ***	1.309 ***
	(0.191)	(0.191)	(0.230)	(0.236)
Part-time farming	−0.171 *	−0.120	−0.429 ***	−0.382 ***
	(0.103)	(0.103)	(0.108)	(0.107)
Risk preference	0.136 *	0.164 **	−0.295 ***	−0.275 ***
	(0.075)	(0.075)	(0.101)	(0.102)
Total household size	0.122 ***	0.100 ***	0.104 ***	0.080 **
	(0.030)	(0.032)	(0.031)	(0.032)
Number of household members working away from home	0.056	0.074	0.023	0.041
	(0.050)	(0.052)	(0.053)	(0.054)
Highest education level in household	0.023	0.028 *	0.067 ***	0.073 ***
	(0.017)	(0.017)	(0.017)	(0.017)
Number of cadre members	0.073	0.111	−0.602 ***	−0.576 ***
	(0.160)	(0.163)	(0.187)	(0.190)
Planting scale	0.086 ***	0.183 ***	0.070 **	0.173 ***
	(0.032)	(0.039)	(0.033)	(0.039)
Distance to county seat	−0.001	−0.000	0.030 ***	0.032 ***
	(0.004)	(0.004)	(0.006)	(0.007)
Topography	0.063	−0.040	0.351 ***	0.251 **
	(0.108)	(0.109)	(0.116)	(0.114)
Region	No	Yes	No	Yes
Pseudo R^2	0.2087	0.2196	0.2375	0.2495
Wald test	354.48 ***	336.02 ***	406.43 ***	360.91 ***
Observations	608	608	608	608

Note: *, **, and *** indicate statistical significance at the 10%, 5%, and 1% levels, respectively.

4.4.2. Analysis of the Mechanism Pathways

The grounded theory analysis indicates that technology perception and the regional soft environment are key pathways through which DTU promotes the adoption of GATs among farmers. To further validate this finding, we quantify both technology perception and the regional soft environment using micro-level data and conduct an empirical analysis using a mediation effect model. The results are presented in Table 5.

Table 5. Results of the mediation effect test.

Variable Name	(1) GATs	(2) Technology Perception	(3) GATs	(4) GATs	(5) Regional Soft Environment	(6) GATs
DTU	2.223 *** (0.287)	0.303 *** (0.030)	1.245 *** (0.317)	2.223 *** (0.287)	0.263 *** (0.031)	1.409 *** (0.309)
Technology perception			3.959 *** (0.394)			
Regional soft environment						4.014 *** (0.388)
Control variable	Yes	Yes	Yes	Yes	Yes	Yes
Region	Yes	Yes	Yes	Yes	Yes	Yes
Constant term		0.297 *** (0.088)			0.277 *** (0.090)	
R^2		0.5540			0.5217	
Pseudo R^2	0.2196		0.2955	0.2196		0.2942
F value		60.40 ***			52.88 ***	
Wald test	336.02 ***		353.97 ***	336.02 ***		396.30 ***
Observations	608	608	608	608	608	608

Note: *** indicates statistical significance at the 1% level.

Models 1 to 3 examine the mediating role of technology perception, while Models 4 to 6 test the mediating effect of the regional soft environment. Model 2 shows that DTU has a significantly positive effect on farmers' technology perception at the 1% significance level, indicating that DTU can significantly enhance farmers' awareness and understanding of agricultural technologies. Model 3 reveals that both DTU and technology perception positively influence the adoption of GATs at the 1% level. This suggests that DTU promotes adoption by improving farmers' perception of technology, providing empirical support for Hypothesis 2. This finding is consistent with the conclusions of Yu et al. (2024) [85].

Similarly, Models 5 and 6 demonstrate that DTU significantly improves the regional soft environment at the 1% level, and that both DTU and the regional soft environment have significantly positive effects on farmers' adoption of GATs. These findings confirm that DTU facilitates adoption by enhancing the regional soft environment, thereby supporting Hypothesis 3. This finding aligns with the perspective of Musajan et al. (2024) [84], despite notable differences in study subjects, methods, and regions between the two studies.

4.4.3. Further Analysis

To examine how DTU influences farmers' adoption of different types of GATs, following the approach of Xiong et al. (2025) [86], this study categorizes the GATs involved in rice production into five types: green cultivation, green pest control, green fertilization, green waste utilization, and green irrigation. A series of regression analyses are conducted accordingly. As shown in Table 6, DTU is significantly positively associated with farmers' adoption of green cultivation techniques, green pest and disease management, green fertilization techniques, and green waste utilization, with coefficients of 1.224, 2.524, 2.519, and 2.319, respectively, all significant at the 1% level. This indicates that DTU has a stronger effect on the adoption of green pest and disease management, green fertilization, and green waste utilization techniques. A possible explanation is that China's agricultural development has long relied heavily on conventional chemical pesticides and fertilizers, resulting in severe agricultural non-point source pollution [5,7,8], while agricultural waste is often burned by farmers, causing serious air pollution [87]. Consequently, the Chinese government has placed particular emphasis on promoting green pest and disease management, green fertilization, and green waste utilization technologies. Digital technology facilitates the rapid and widespread dissemination of relevant policies among farmers, thereby incentivizing the adoption of these green production practices. In contrast, adopting green cultivation techniques requires investment in new farmland machinery, which entails high costs that some smallholder farmers cannot afford, resulting in a relatively lower effect of DTU on this adoption. Interestingly, DTU does not have a significant effect on farmers' adoption of green irrigation techniques. One possible explanation is that the surveyed area lies in a subtropical monsoon climate zone with abundant rainfall throughout the year. Rice production in this region typically relies on natural precipitation, resulting in relatively low irrigation demand. Additionally, the high investment and maintenance costs of water-saving irrigation facilities may further discourage farmers from adopting green irrigation technologies.

Table 6. Effects of DTU on Farmers' Adoption of Different GATs.

Variable Name	(1) Green Cultivation	(2) Green Pest Control	(3) Green Fertilization	(4) Green Waste Utilization	(5) Green Irrigation
DTU	1.224 ***	2.524 ***	2.519 ***	2.319 ***	−1.564
	(0.372)	(0.339)	(0.379)	(0.499)	(1.189)
Control variable	Yes	Yes	Yes	Yes	Yes
Region	Yes	Yes	Yes	Yes	Yes
Constant term	−2.101 **	−3.241 ***	−1.157	0.439	−5.574 ***
	(0.963)	(0.857)	(0.857)	(1.091)	(1.875)
Pseudo R^2	0.3026	0.3205	0.2715	0.3252	0.3721
Wald test	167.42 ***	221.02 ***	159.24 ***	122.35 ***	48.01 ***
Observations	608	608	608	608	608

Note: ** and *** indicate statistical significance at the 5% and 1% levels, respectively.

Considering the significant heterogeneity among sampled farmers, this study follows the approach of Guo et al. (2022) [88] and Chen et al. (2025) [89] to conduct subgroup regressions based on farmers' average age and average education level. Farmers with age less than or equal to the average are classified as the younger group, while those above the average are classified as the older group. Similarly, farmers with education level less than or equal to the average are grouped as the lower-education group, and those above the average as the higher-education group. The regression results are presented in Table 7. DTU has a significantly positive effect on the adoption of GATs among younger and more educated farmers, with larger coefficients. This may be because younger and more educated farmers tend to use digital technologies more intensively and are more capable of learning and adopting new agricultural technologies.

Table 7. Effects of DTU on GATs Adoption Among Different Farmer Groups.

Variable Name	Age		Education	
	(3) Younger Farmers	(4) Older Farmers	(7) Lower Education Group	(8) Higher Education Group
DTU	2.947 ***	1.306 **	2.484 ***	2.756 ***
	(0.342)	(0.539)	(0.514)	(0.352)
Control variable	Yes	Yes	Yes	Yes
Region	Yes	Yes	Yes	Yes
Pseudo R^2	0.2504	0.1302	0.1627	0.2463
Wald test	153.62 ***	125.96 ***	140.27 ***	175.94 ***
Observations	303	305	248	360

Note: ** and *** indicate statistical significance at the 5% and 1% levels, respectively.

5. Conclusions and Policy Implications

The emergence of digital technologies has exerted a subtle yet profound influence on farmers' production and daily lives. Some scholars even argue that digital technologies may usher in a new wave of industrial revolution in agriculture [90]. Against this backdrop, this study combines qualitative and quantitative methods to investigate the pathways through which DTU facilitates farmers' adoption of GATs. Based on grounded theory analysis, the study identifies that DTU significantly affects the adoption of GATs among farmers. Specifically, digital infrastructure, farmers' capital endowment, and practical needs serve as prerequisite conditions; technology perception and the regional soft environment represent two key transmission pathways; and achieving economic, social, and ecological benefits constitutes the ultimate goal and outcome. To further verify the findings from the grounded theory analysis, this study conducts an empirical test using survey data from 608 rice farmers in Sichuan Province, China. Employing an Ordered Probit model and the CFM, the results show that DTU has a significantly positive impact on farmers' adoption of GATs. This effect is primarily mediated by improvements in farmers' technology perception and enhancements in the regional agricultural soft environment—findings that are consistent with the grounded theory analysis. Moreover, the study finds considerable heterogeneity in the effects of DTU across different types of GATs and farmer groups.

Based on the above analysis, DTU can effectively promote farmers' adoption of GATs. Driven by strong government support, the digitalization of agriculture and rural areas in China is advancing rapidly. However, the study area in western China has relatively lower socio-economic development, digital infrastructure, and farmers' digital literacy. To further facilitate farmers' adoption of GATs in this region, this study proposes four policy recommendations: (1) Provide digital skills training for farmers to enhance their awareness and application of digital technologies, guiding them to access and learn more about green agricultural production knowledge and techniques, thereby improving their understanding of GATs. (2) Strengthen government leadership and support by increasing financial investment in the research and application of agricultural digital technologies. Support agricultural research institutions in developing technologies in areas such as the Internet of Things, big data, and artificial intelligence, and promote the broad application of advanced technologies in agricultural production, further improving the regional soft environment for farmers' adoption of green technologies GATs. (3) Establish dedicated platforms for promoting GATs to provide more targeted guidance and publicity, leveraging digital technologies to disseminate practical, easily adoptable green production techniques that meet farmers' immediate needs. (4) Give priority to digitally marginalized rural groups by offering customized digital support and training. Emerging technologies, such as smartphone applications and agricultural IoT systems, can be targeted at younger farmers, while simplified devices or applications combined with individualized guidance should be provided for older or less-educated farmers.

This study adopts a mixed-methods approach to provide a more comprehensive and in-depth analysis, addressing some of the gaps in existing research. It offers useful insights for future studies on how DTU influences farmer behavior. Nevertheless, the study has certain limitations. First, both the case materials and survey data were collected exclusively from rice farmers in Sichuan Province, without nationwide data or consideration of other typical crops. Given the considerable variation in agricultural production environments across regions and the differing GATs required for various crops, the effect of DTU on farmers' adoption of green technologies may vary across regions and crop types. Second, the level of agricultural digitalization within the study area is heterogeneous, which may affect the precision of the estimates. For instance, Chengdu and Mianyang exhibit higher overall economic and technological development and more advanced digital infrastructure, resulting in relatively higher agricultural digitalization and a stronger effect of DTU on farmers' adoption of green technologies. Third, while grounded theory is employed to explore the antecedents and consequences of DTU in promoting farmers' adoption of GATs, empirical testing is limited to the effect of DTU itself due to constraints in data availability and article length. The causal mechanisms related to the antecedents and consequences identified through qualitative analysis have not yet been empirically verified. This represents a promising direction for future research.

Author Contributions: Conceptualization, X.Y. and W.L.; methodology, P.T. and W.L.; software, X.Y. and Y.L.; validation, X.Y., P.T. and W.L.; formal analysis, X.Y. and Y.L.; investigation, S.T. and J.Z.; resources, P.T.; data curation, X.Y.; writing—original draft preparation, X.Y., P.T. and W.L.; writing—review and editing, S.T. and J.Z.; funding acquisition, P.T. All authors have read and agreed to the published version of the manuscript.

Funding: This work is supported by the Youth Project of National Social Science Fund of China "Research on the mechanism and path of digital Empowering rural revitalization to help low-income population stabilize employment in key counties", grant number 24CJY095; the Youth Project of Sichuan Rural Development Research Center "Research on the mechanism and path of digital economy promoting rural labor transfer: A case study of Liangshan Yi Area", grant number CR2417.

Institutional Review Board Statement: This study is waived for ethical review by Sichuan Agricultural University, as this research does not involve any biological or human experiments and does not require ethical review.

Informed Consent Statement: Informed consent for participation was obtained from all subjects involved in the study.

Data Availability Statement: The associated dataset of the study is available from the corresponding author upon request.

Acknowledgments: The authors would like to thank the anonymous reviewers for their comments and suggestions, which contributed to the further improvement of this paper.

Conflicts of Interest: The authors declare no conflicts of interest.

Abbreviations

The following abbreviations are used in this manuscript:

DTU	Digital Technology Use
GATs	Green Agricultural Technologies
CFM	Control Function Method

References

1. Gawel, A.; Benesova, I.; Kotyza, P. The Green Transformation and Gender Equality in Agricultural Entrepreneurship: Insights from the European Union. *J. Rural Stud.* **2024**, *105*, 103202. [CrossRef]
2. Han, C.; Lyu, J.; Zhong, D. The Impact of Financial Support and Innovation Awareness on Farmers' Adoption of Green Production Technology. *Financ. Res. Lett.* **2025**, *79*, 107259. [CrossRef]
3. Kuang, F.; Li, J.; Jin, J.; Qiu, X. Do green production technologies improve household income? Evidence from rice farmers in China. *Land* **2023**, *12*, 1848. [CrossRef]
4. Hussain, F.; Ahmed, S.; Naqvi, S.M.Z.A.; Awais, M.; Zhang, Y.; Zhang, H.; Hu, J. Agricultural Non-Point Source Pollution: Comprehensive Analysis of Sources and Assessment Methods. *Agriculture* **2025**, *15*, 531. [CrossRef]
5. Sui, Y.; Gao, Q. Farmers' endowments, technology perception and green production technology adoption behavior. *Sustainability* **2023**, *15*, 7385. [CrossRef]
6. Wen, S.; Li, J.; Xiao, Q.; Wu, H.; Li, J. Aging, Intergenerational Ties, and Adoption of Green Agricultural Technologies: Survey Evidence from China's Yellow River Basin. *Arid Land Resour. Environ.* **2024**, *38*, 80–91.
7. Yu, F. Driving Forces, Core Elements, and Policy Recommendations for Green Agricultural Development in the New Era. *China Rural Econ.* **2018**, *5*, 19–34.

8. Chai, Y.; Pannell, D.J.; Pardey, P.G. Nudging Farmers to Reduce Water Pollution from Nitrogen Fertilizer. *Food Policy* **2023**, *120*, 102525. [CrossRef]
9. Huang, Y.; Zhang, X.; Li, Z. Analysis of Nationwide Soil Pesticide Pollution: Insights from China. *Environ. Res.* **2024**, *252*, 118988. [CrossRef]
10. Feder, G.; Just, R.E.; Zilberman, D. Adoption of Agricultural Innovations in Developing Countries: A Survey. *Econ. Dev. Cult. Change* **1985**, *33*, 255–298. [CrossRef]
11. Murage, A.; Midega, C.; Pittchar, J.; Pickett, J.; Khan, Z. Determinants of Adoption of Climate-Smart Push-Pull Technology for Enhanced Food Security through Integrated Pest Management in Eastern Africa. *Food Secur.* **2015**, *7*, 709–724. [CrossRef]
12. He, X.; Zhu, K. Farmers' Information Level, Elite Capture, and the Targeting of Rural Minimum Living Security. *Econ. Res. J.* **2019**, *54*, 150–164.
13. Li, C.; Zhang, Y.; Wang, X.; Liu, J. Impact Effects of Cooperative Participation on the Adoption Behavior of Green Production Technologies by Cotton Farmers and the Driving Mechanisms. *Agriculture* **2024**, *14*, 213. [CrossRef]
14. Qiao, H.; Wu, S. Research on the Empowerment of China's Rural Residents' Income Growth by the Digital Economy. *J. Xi'an Univ. Financ. Econ.* **2025**, 1–21. [CrossRef]
15. Ping, W.; Wang, Y.; Zhang, Y. Digital Village Construction, Rural Residents' Income Growth, and the Urban–Rural Income Gap: A Quasi-Natural Experiment Based on the National Digital Village Pilot Program. *J. Agrotech. Econ.* **2025**, *8*, 80–104.
16. Chang, Y. The Urban–Rural Digital Divide: Manifestations, Causes, and Countermeasures. *Agric. Econ. Manag.* **2025**, *1*, 45–54.
17. State Council of China. China Digital Village Development Report (2022) Released. 2023. Available online: https://www.gov.cn/xinwen/2023-03/01/content_5743969.htm (accessed on 20 December 2024).
18. Liu, G.; Li, D. Research on the Interactive Development of Digital Technology and Agricultural and Rural Informatization. *Inf. Sci.* **2023**, *41*, 127–132+190.
19. Du, Y.; Feng, H.; Zhang, Q.; Zheng, S. Influence of Information Literacy on Farmers' Green Production Technology Adoption Behavior: The Moderating Role of Risk Attitude. *Agriculture* **2025**, *15*, 701. [CrossRef]
20. Wei, J.; Zheng, S.; Zhang, Q. Use of Digital Technology, Green Cognition, and Farmers' Adoption of Green Pest Control Technologies: Evidence from Pear Growers in Shanxi and Hebei Provinces. *World Agric.* **2024**, *3*, 99–112.
21. Zhu, D. Methodology, Research Strategies, and Applications of Mixed Methods: A Case Study of Consumption Pattern Research. *Sociol. Stud.* **2012**, *27*, 146–166+244–245.
22. Brannen, J. The Practice of a Mixed Methods Research Strategy: Personal, Professional and Project Considerations. In *Advances in Mixed Methods Research: Theories and Applications*; Bergman, M.M., Ed.; SAGE Publications: London, UK, 2008; pp. 53–65.
23. Gao, R.; Zhang, H.; Gong, C.; Wu, Z. The Role of Farmers' Green Values in Creation of Green Innovative Intention and Green Technology Adoption Behavior: Evidence from Farmers Grain Green Production. *Front. Psychol.* **2022**, *13*, 980570. [CrossRef]
24. Hong, X.; Chen, Y.J.; Gong, Y.; Wang, H. Farmers' Green Technology Adoption: Implications from Government Subsidies and Information Sharing. *Naval Res. Logist.* **2024**, *71*, 286–317. [CrossRef]
25. Wu, Q.; Zhou, J. Need for Cognitive Closure, Information Acquisition and Adoption of Green Prevention and Control Technology. *Ecol. Chem. Eng.* **2021**, *28*, 129–143. [CrossRef]
26. Gao, F.; Xu, J.; Li, D.; Chen, F. The Impact of Adopting Green Control Technologies on Farmers' Income: Based on Research Data from Tea Farmers in 16 Provinces of China. *Pol. J. Environ. Stud.* **2024**, *33*, 3. [CrossRef] [PubMed]
27. Simon, H.A. Rational Choice and the Structure of the Environment. *Psychol. Rev.* **1956**, *63*, 129–138. [CrossRef]
28. Eheazu, C.L.; Ezeala, J.I. Environmental Adult Education for Mitigating the Impacts of Climate Change on Crop Production and Fish Farming in Rivers State of Nigeria. *J. Educ. Pract.* **2017**, *8*, 98–107.
29. Adnan, N.; Nordin, S.M.; Ali, M. A Solution for the Sunset Industry: Adoption of Green Fertiliser Technology amongst Malaysian Paddy Farmers. *Land Use Policy* **2018**, *79*, 575–584. [CrossRef]
30. Gailhard, I.; Unay, A.; Bojnec, Š. Farm Size and Participation in Agri-Environmental Measures: Farm-Level Evidence from Slovenia. *Land Use Policy* **2015**, *46*, 273–282. [CrossRef]
31. Gao, Y.; Zhang, L.; Wang, Y.; Li, S. Influence of a New Agricultural Technology Extension Mode on Farmers' Technology Adoption Behavior in China. *J. Rural Stud.* **2020**, *76*, 173–183. [CrossRef]
32. Nordlund, A.M.; Garvill, J. Value Structures Behind Pro-Environmental Behavior. *Environ. Behav.* **2002**, *34*, 740–756. [CrossRef]
33. Bougherara, D.; Gassmann, X.; Piet, L.; Reynaud, A. Structural Estimation of Farmers' Risk and Ambiguity Preferences: A Field Experiment. *Eur. Rev. Agric. Econ.* **2017**, *44*, 782–808. [CrossRef]
34. Davis, F.D.; Venkatesh, V. A Critical Assessment of Potential Measurement Biases in the Technology Acceptance Model: Three Experiments. *Int. J. Hum.-Comput. Stud.* **1996**, *45*, 19–45. [CrossRef]
35. Lee, D.Y.; Lehto, M.R. User Acceptance of YouTube for Procedural Learning: An Extension of the Technology Acceptance Model. *Comput. Educ.* **2013**, *61*, 193–208. [CrossRef]
36. Matsumura, Y.; Minowa, T.; Yamamoto, H. Amount, Availability, and Potential Use of Rice Straw (Agricultural Residue) Biomass as an Energy Resource in Japan. *Biomass Bioenergy* **2005**, *29*, 347–354. [CrossRef]

37. Goyal, M.; Netessine, S. Strategic Technology Choice and Capacity Investment under Demand Uncertainty. *Manag. Sci.* **2007**, *53*, 192–207. [CrossRef]
38. Du, S.; Luo, X.; Tang, L. The Nonlinear Impact of Premium Expectation on Farmers' Adoption of Green Pest Control Technology: From the Perspective of Quality Certification. *J. Huazhong Agric. Univ. (Soc. Sci. Ed.)* **2024**, *4*, 95–108.
39. Putnam, R.D. *Making Democracy Work: Civic Traditions in Modern Italy*; Princeton University Press: Princeton, NJ, USA, 1993.
40. Fisher, R. A Gentleman's Handshake: The Role of Social Capital and Trust in Transforming into Usable Knowledge. *J. Rural Stud.* **2013**, *31*, 13–22. [CrossRef]
41. Schultz, T.W. *Transforming Traditional Agriculture*; Yale University Press: New Haven, CT, USA, 1964.
42. Midingoyi, S.-K.G.; Kassie, M.; Muriithi, B.; Diiro, G.; Ekesi, S. Do Farmers and the Environment Benefit from Adopting Integrated Pest Management Practices? Evidence from Kenya. *J. Agric. Econ.* **2019**, *70*, 452–470. [CrossRef]
43. Martey, E.; Etwire, P.M.; Abdoulaye, T. Welfare Impacts of Climate-Smart Agriculture in Ghana: Does Row Planting and Drought-Tolerant Maize Varieties Matter? *Land Use Policy* **2020**, *95*, 104622. [CrossRef]
44. Meena, R.S.; Kumar, S.; Datta, R.; Lal, R.; Vijayakumar, V.; Brtnicky, M.; Holatko, J.; Sharma, M.P.; Yadav, G.S. Impact of Agrochemicals on Soil Microbiota and Management: A Review. *Land* **2020**, *9*, 34. [CrossRef]
45. Bharadwaj, A.; El Sawy, O.A.; Pavlou, P.A.; Venkatraman, N. Digital business strategy: Toward a next generation of insights. *MIS Q.* **2013**, *37*, 471–482. [CrossRef]
46. von Briel, F.; Davidsson, P.; Recker, J. Digital technologies as external enablers of new venture creation in the IT hardware sector. *Entrepren. Theory Pract.* **2018**, *42*, 47–69. [CrossRef]
47. Boccia, F.; Leonardi, R. *The Challenge of the Digital Economy: Markets, Taxation and Appropriate Economic Models*; Springer International Publishing: Cham, Switzerland, 2016.
48. Cheng, Q.; Lu, S. Digital technology application, economic uncertainty, and green innovation. *Soft Sci.* **2023**, *37*, 1–7+30.
49. Wang, W.; Huang, Z.; Fu, Z.; Jia, L.; Li, Q.; Song, J. Impact of digital technology adoption on technological innovation in grain production. *J. Innov. Knowl.* **2024**, *9*, 100520. [CrossRef]
50. Liu, H.; Chen, Z.; Wen, S.; Zhang, J.; Xia, X. Impact of Digital Literacy on Farmers' Adoption Behaviors of Green Production Technologies. *Agriculture* **2025**, *15*, 303. [CrossRef]
51. Mao, H.; Chai, Y.; Shao, X.; Chang, X. Digital Extension and Farmers' Adoption of Climate Adaptation Technology: An Empirical Analysis of China. *Land Use Policy* **2024**, *143*, 107220. [CrossRef]
52. Dong, Y.; Jia, C.; Su, L. The Impact of Digital Agricultural Insurance on Farmers' Fertilizer Reduction Technology Adoption: Evidence from China. *China Agric. Econ. Rev.* **2025**, *17*, 379–397. [CrossRef]
53. Aker, J.C.; Ghosh, I.; Burrell, J. The Promise (and Pitfalls) of ICT for Agriculture Initiatives. *Agric. Econ.* **2016**, *47*, 35–48. [CrossRef]
54. Karanasios, S.; Slavova, M. Understanding the Impacts of Mobile Technology on Smallholder Agriculture. In *Digital Technologies for Agricultural and Rural Development in the Global South*; Baumüller, H., Ed.; CAB International: Wallingford, UK, 2018; pp. 111–122.
55. Misaki, E.; Apiola, M.; Gaiani, S.; Tedre, M. Challenges Facing Sub-Saharan Small-Scale Farmers in Accessing Farming Information through Mobile Phones: A Systematic Literature Review. *Electron. J. Inf. Syst. Dev. Ctries.* **2018**, *84*, e12034. [CrossRef]
56. Yuan, Y.; Sun, L.; She, Z.; Chen, S. Influence of Digital Literacy on Farmers' Adoption Behavior of Low-Carbon Agricultural Technology: Chain Intermediary Role Based on Capital Endowment and Adoption Willingness. *Sustainability* **2025**, *17*, 2187. [CrossRef]
57. Rosen, S. Hedonic Prices and Implicit Markets: Product Differentiation in Pure Competition. *J. Polit. Econ.* **1974**, *82*, 34–55. [CrossRef]
58. Zhang, J.; Xie, S.; Li, X.; Xia, X. Adoption of Green Production Technologies by Farmers through Traditional and Digital Agro-Technology Promotion—An Example of Physical versus Biological Control Technologies. *J. Environ. Manag.* **2024**, *370*, 122813. [CrossRef]
59. Shen, Z.; Zhang, Y.; Huang, J.; Li, X.; Yu, X. Digital Transition and Green Growth in Chinese Agriculture. *Technol. Forecast. Soc. Change* **2022**, *181*, 121742. [CrossRef]
60. Harou, A.P.; van den Broek, K.; Bulte, E.; Hoffmann, V. The Joint Effects of Information and Financing Constraints on Technology Adoption: Evidence from a Field Experiment in Rural Tanzania. *J. Dev. Econ.* **2022**, *155*, 102707. [CrossRef]
61. Glaser, B.G.; Strauss, A.L. *The Discovery of Grounded Theory: Strategies for Qualitative Research*; Aldine Transaction: New Brunswick, NJ, USA, 1967.
62. Strauss, A.; Corbin, J.M. *Grounded Theory in Practice*; Sage Publications: Thousand Oaks, CA, USA, 1997.
63. Charmaz, K. *Constructing Grounded Theory*; Sage Publications: Thousand Oaks, CA, USA, 2006.
64. Strauss, A.; Corbin, J.M. *Basics of Qualitative Research: Grounded Theory Procedures and Techniques*; Sage Publications: Thousand Oaks, CA, USA, 1990.
65. Peng, P.; Liu, D. The Operation Mechanism of Land Management Rights Mortgage Financing under the Three Rights Separation: A Multi-Case Study Based on Grounded Theory. *China Rural Econ.* **2019**, *11*, 32–50.

66. Yang, F.; Wang, X.H. The Mechanism of Digital Empowerment in Matching the Supply and Demand of Rural Public Cultural Services: A Qualitative Study Based on Grounded Theory. *J. Libr. Inf. Sci.* **2021**, *1*, 62–69.
67. Zhang, J.J.; Liu, T.S. Does Internet Use Make Rural Residents Happier? Evidence from CFPS 2018. *Dongyue Luncong* **2020**, *41*, 172–179.
68. Du, S.X.; Luo, X.F.; Huang, Y.Z.; Tang, L.; Yu, W.Z. Risk Perception, Agricultural Socialized Services, and Paddy Farmers' Adoption of Biological Pesticide Technology. *Yangtze River Basin Res. Environ.* **2021**, *30*, 1768–1779.
69. Huang, Y.Z.; Luo, X.F. Eat and Sell: An Analysis of Differences in Paddy Farmers' Use of Biological Pesticides. *China Rural Econ.* **2018**, *7*, 63–78.
70. Zhou, Y.X.; Gao, R.M. The Impact of Digital Literacy on the Green Production Transformation of Facility Vegetable Growers. *J. Arid Land Res. Environ.* **2025**, *39*, 79–93.
71. Qiu, H.; Tang, W.; Huang, Y.; Deng, H.; Liao, W.; Ye, F. E-Commerce Operations and Technology Perceptions in Promoting Farmers' Adoption of Green Production Technologies: Evidence from Rural China. *J. Environ. Manag.* **2024**, *370*, 122628. [CrossRef] [PubMed]
72. Tan, Y.; Lu, Q. The Influence of Risk Aversion and Social Learning on Farmers' Adoption of Modern Irrigation Technologies: An Empirical Analysis Based on the Heckman Sample Selection Model. *Yangtze River Basin Res. Environ.* **2021**, *30*, 234–245.
73. Davis, F.D. Perceived Usefulness, Perceived Ease of Use, and User Acceptance of Information Technology. *MIS Q.* **1989**, *13*, 319–340. [CrossRef]
74. Peng, X.; Yan, X.; Wang, H. Study on the Effect of Digital Technology Adoption and Farmers' Cognition on Fertilizer Reduction and Efficiency Improvement Behavior. *Agriculture* **2024**, *14*, 973. [CrossRef]
75. Tversky, A.; Kahneman, D. The Framing of Decisions and the Psychology of Choice. *Science* **1981**, *211*, 453. [CrossRef]
76. Chen, Q.; Yang, Q.; Ye, D. Regional Environment, Household Endowments, and Straw Disposal Behavior: Evidence from Dry Farming Areas in Gansu Province. *Chin. J. Appl. Ecol.* **2020**, *31*, 563–572.
77. Hu, Z.D. Study on the Impact of Regional Soft Environment on Farmers' Technology Adoption Behavior. Ph.D. Thesis, Hunan Agricultural University, Changsha, China, 2011.
78. Wooldridge, J.M. Control Function Methods in Applied Econometrics. *J. Hum. Resour.* **2015**, *50*, 420–445. [CrossRef]
79. Liu, B.; Li, N.; Liao, C. Effects of Social Capital on the Adoption of Green Production Technologies by Rice Farmers: Moderation Effects Based on Risk Preferences. *Sustainability* **2024**, *16*, 8879. [CrossRef]
80. Zheng, Y.-Y.; Zhu, T.-H.; Wei, J. Does Internet Use Promote the Adoption of Agricultural Technology? Evidence from 1449 Farm Households in 14 Chinese Provinces. *J. Integr. Agric.* **2022**, *21*, 282–292. [CrossRef]
81. Kai, L.; Yu, J.; Zhou, J.-H. Are Vulnerable Farmers More Easily Influenced? Heterogeneous Effects of Internet Use on the Adoption of Integrated Pest Management. *J. Integr. Agric.* **2023**, *22*, 3220–3233. [CrossRef]
82. Huang, X.; Yang, F.; Fahad, S. The Impact of Digital Technology Use on Farmers' Low-Carbon Production Behavior under the Background of Carbon Emission Peak and Carbon Neutrality Goals. *Front. Environ. Sci.* **2022**, *10*, 1002181. [CrossRef]
83. Luo, M.Z.; Liu, Z.Y. Digital Technology Adoption, Social Network Expansion, and Farmers' Common Prosperity. *South. Econ.* **2022**, *3*, 1–16.
84. Musajan, A.; Lin, Q.; Wei, D.; Mao, S. Unveiling the mechanisms of digital technology in driving farmers' green production transformation: Evidence from China's watermelon and muskmelon sector. *Foods* **2024**, *13*, 3926. [CrossRef]
85. Yu, X.; Sheng, G.; Sun, D.; He, R. Effect of digital multimedia on the adoption of agricultural green production technology among farmers in Liaoning Province, China. *Sci. Rep.* **2024**, *14*, 13092. [CrossRef] [PubMed]
86. Xiong, F.; Peng, Y.; Liu, Y.; Zhu, S. Impact of digital information capability on farmers' adoption of green production technologies: The mediating effect of technological ecological cognition. *Changjiang Liuyu Ziyuan Yu Huanjing* **2025**, *34*, 216–225.
87. He, J.; Zhou, W.; Guo, S.; Deng, X.; Song, J.; Xu, D. Effect of land transfer on farmers' willingness to pay for straw return in Southwest China. *J. Clean. Prod.* **2022**, *369*, 133397. [CrossRef]
88. Guo, Z.; Chen, X.; Zhang, Y. Impact of Environmental Regulation Perception on Farmers' Agricultural Green Production Technology Adoption: A New Perspective of Social Capital. *Technol. Soc.* **2022**, *71*, 102085. [CrossRef]
89. Chen, C.; Peng, R.; Wu, Z.; Ye, F. The Impact of High-Standard Farmland Construction on Farmers' Agricultural Green Production Technologies Adoption Behavior. *Front. Sustain. Food Syst.* **2025**, *9*, 1581894. [CrossRef]
90. Ingram, J.; Maye, D.; Bailey, C.; Barnes, A.; Bear, C.; Bell, M.; Cutress, D.; Davies, L.; de Boon, A.; Dinnie, L. What Are the Priority Research Questions for Digital Agriculture? *Land Use Policy* **2022**, *114*, 105962. [CrossRef]

sustainability

MDPI

Article

APD-YOLOv7: Enhancing Sustainable Farming through Precise Identification of Agricultural Pests and Diseases Using a Novel Diagonal Difference Ratio IOU Loss

Jianwen Li [1,2,†], Shutian Liu [3,4,†], Dong Chen [1,2], Shengbang Zhou [1,2,*] and Chuanqi Li [1,2,*]

1 Guangxi Key Laboratory of Functional Information Materials and Intelligent Information Processing, Nanning Normal University, Nanning 530001, China; ljwaassslw@163.com (J.L.)
2 Smart Logistics Industry School of Nanning Normal University, Demonstrative Modern Industrial School of Guangxi University, Nanning 530001, China
3 Guangxi Geographical Indication Crops Research Center of Big Data Mining and Experimental Engineering Technology, Nanning Normal University, Nanning 530001, China
4 Guangxi Key Laboratory of Earth Surface Processes and Intelligent Simulation, Nanning Normal University, Nanning 530001, China
* Correspondence: zhoushengbang@foxmail.com (S.Z.); lchq@nnnu.edu.cn (C.L.)
† These authors contributed equally to this work.

Abstract: The diversity and complexity of the agricultural environment pose significant challenges for the collection of pest and disease data. Additionally, pest and disease datasets often suffer from uneven distribution in quantity and inconsistent annotation standards. Enhancing the accuracy of pest and disease recognition remains a challenge for existing models. We constructed a representative agricultural pest and disease dataset, FIP6Set, through a combination of field photography and web scraping. This dataset encapsulates key issues encountered in existing agricultural pest and disease datasets. Referencing existing bounding box regression (BBR) loss functions, we reconsidered their geometric features and proposed a novel bounding box similarity comparison metric, DDRIoU, suited to the characteristics of agricultural pest and disease datasets. By integrating the focal loss concept with the DDRIoU loss, we derived a new loss function, namely Focal-DDRIoU loss. Furthermore, we modified the network structure of YOLOV7 by embedding the MobileViTv3 module. Consequently, we introduced a model specifically designed for agricultural pest and disease detection in precision agriculture. We conducted performance evaluations on the FIP6Set dataset using mAP75 as the evaluation metric. Experimental results demonstrate that the Focal-DDRIoU loss achieves improvements of 1.12%, 1.24%, 1.04%, and 1.50% compared to the GIoU, DIoU, CIoU, and EIoU losses, respectively. When employing the GIoU, DIoU, CIoU, EIoU, and Focal-DDRIoU loss functions, the adjusted network structure showed enhancements of 0.68%, 0.68%, 0.78%, 0.60%, and 0.56%, respectively, compared to the original YOLOv7. Furthermore, the proposed model outperformed the mainstream YOLOv7 and YOLOv5 models by 1.86% and 1.60%, respectively. The superior performance of the proposed model in detecting agricultural pests and diseases directly contributes to reducing pesticide misuse, preventing large-scale pest and disease outbreaks, and ultimately enhancing crop yields. These outcomes strongly support the promotion of sustainable agricultural development.

Keywords: loss function design; MobileViTv3; YOLOv7; agricultural diseases and pests; sustainable agricultural development

1. Introduction

Globally, agriculture plays a crucial role in contributing to the health of the population, maintaining social stability, and even contributing to national security. Therefore, there has been a consistent demand for continuous innovation in agricultural technology to enhance the efficiency of the agricultural industry and maximize food production to meet the needs of a growing population [1]. However, due to the abundance of harmful organisms and microorganisms in the cultivation environment, crops are increasingly susceptible to pests and diseases. Infestations of pests and diseases seriously threaten agricultural production safety and the sustainable supply of food. Therefore, accurate identification of crop pests and diseases and effective early warning of their outbreaks can help prevent agricultural disasters and ensure the quality and yield of farmland [2,3].

However, traditional methods for pest and disease identification typically depend on the extensive experience and professional knowledge of agricultural experts [4], which can pose efficiency challenges when handling large-scale datasets. While expert diagnosis is crucial for ensuring accuracy, the increasing diversity and complexity of crop diseases may strain the availability of experts' time and resources. Therefore, it is particularly important to explore more efficient and accurate methods for pest and disease identification, in order to assist agricultural experts in enhancing overall identification efficiency, support agricultural practitioners in boosting crop yields, and promote sustainable agricultural development.

With the advancement of image processing technology, crop pest and disease identification methods based on machine learning have emerged [5,6]. However, these methods still struggle with extracting abstract features and improving accuracy to a level suitable for practical applications. Moreover, improving accuracy to a practical application level remains challenging. Subsequently, with the development of graphics processing units (GPUs), deep learning technology, which relies on the powerful processing capabilities of GPUs, has rapidly evolved [7] and is widely used in agricultural pest and disease identification [8]. For large-scale agricultural pest and disease datasets, deep learning technologies have demonstrated superior accuracy in identifying crop pests and diseases compared to traditional methods and other machine learning-based approaches [9]. Therefore, exploring deep learning models that are tailored to agricultural pest and disease identification can significantly enhance the accuracy and efficiency of diagnosis. This, in turn, has the potential to prevent large-scale outbreaks of crop pests and diseases, reduce excessive pesticide usage, foster healthy crop growth, increase crop yields, and ultimately steer agriculture towards sustainable development.

Although many studies provide references and demonstrate the feasibility of using deep learning neural networks to identify agricultural pests and diseases, existing deep learning models still face challenges in addressing issues such as high computational costs and ensuring high accuracy in agricultural pest and disease identification.

The bounding box regression (BBR) module is crucial for determining object locations in advanced object detectors. During training, BBR adjusts predicted bounding box coordinates to better align with the ground truth bounding box, thereby improving object localization accuracy—a pivotal aspect of object detection tasks. The design of the loss function is crucial for BBR's success, as it directly affects training efficiency and the model's overall performance. So far, BBR-based loss functions are mostly classified into two categories: ℓn-norm-based loss functions and intersection-over-union (IoU)-based loss functions.

In the field of object detection, the YOLO series [10–16] models, particularly YOLOv7 [15], have occupied a mainstream position in real-time object detection tasks due to their excellent inference speed and high detection accuracy. We selected YOLOv7 as the foundation for our research due to its demonstrated balance between high speed and accuracy across multiple public datasets, particularly its proficiency in detecting small targets and managing complex scenes. Additionally, YOLOv7 offers a rich array of configuration options and good hardware compatibility, facilitating model portability and subsequent improvements.

From YOLOv1 [10] to YOLOv7, each generation of the YOLO series has meticulously optimized its network structure and loss function in pursuit of higher detection efficiency and accuracy. Meanwhile, successors to the RCNN [17] model, such as Fast R-CNN [18], Mask R-CNN [19], and Dynamic R-CNN [20], have each improved their loss functions according to their respective functional emphases. Furthermore, the DETR [21] model is a direct set prediction object detection model based on the transformer architecture and bipartite matching loss, with one of its core designs being the computation of loss through a bipartite matching process. Despite their application in various domains, the loss function remains a key component common to all these models. Enhancing the loss function is thus of paramount importance for improving the overall performance.

However, existing BBR loss functions sometimes have the same value under different prediction results, which reduces the convergence speed and accuracy of BBR. Additionally, the current BBR loss functions are not fully suitable for precise identification of agricultural pests and diseases. A well-designed loss function can assist the model in performing agricultural pest and disease recognition tasks with greater accuracy and efficiency.

Therefore, building upon the YOLOv7 model, we introduce Agricultural Pest Detection-YOLOv7 (APD-YOLOv7), a model specifically tailored for the detection of agricultural pests and diseases. The core aim of this study is to devise a novel model through the development of a new BBR loss function and the enhancement of YOLOv7's network architecture. This approach aims to attain precise identification of crop pests and diseases, thereby contributing to the sustainable development of agriculture.

(1) Existing BBR loss functions have limitations because they sometimes assign identical values to distinct prediction outcomes. To mitigate this issue, we propose a novel BBR loss function, termed Diagonal Difference Ratio IoU (DDRIoU), which exploits the geometric properties inherent in BBR. Our aim is to utilize DDRIoU as a metric to assess the similarity between predicted and ground truth bounding boxes during the BBR procedure.
(2) Furthermore, during the BBR process, an imbalance exists between high-quality and low-quality anchor boxes. To address this, we intend to integrate the DDRIoU loss with the focal loss concept, resulting in a regression-focused loss, termed Focal-DDRIoU loss. We expect that this combined loss function will enhance the contribution of high-quality anchor boxes to model optimization while simultaneously minimizing the impact of irrelevant anchor boxes.
(3) In order to refine the YOLOv7 network architecture for more precise identification of agricultural pests and diseases, we plan to modify the network structure of the YOLOv7 model and incorporate the MobileViTv3 module. We anticipate that this innovative network design will not only reduce the number of model parameters and computational complexity but also enhance its feature extraction and fusion capabilities, ultimately leading to improved model accuracy.

2. Materials and Methods

2.1. Bounding Box Regression Loss Function

2.1.1. Commonly Used Boundary Box Regression Loss Functions

The YOLO series of real-time detectors has been widely accepted by researchers and applied in various scenarios since its inception. For instance, YOLOv1 constructed a loss function weighted by BBR loss, classification loss, and objectness loss. Until now, this approach has remained one of the most effective loss function paradigms in object detection tasks, as the BBR loss directly determines the localization performance of the model. To further enhance the model's localization accuracy, a carefully designed BBR loss is crucial.

Initially, the ℓn-norm [22] loss function was extensively used for BBR. This function is straightforward but sensitive to various scales. Later, to more accurately calculate the differences between predicted and ground truth bounding boxes, the IoU loss function was introduced in Unitbox [22] and has been widely employed since. This is a loss function designed to optimize the position and size of bounding boxes in object detection tasks. To ensure training stability, the Bounded-IoU [23] loss function introduces an upper bound for

IoU. In deep learning-based object detection and instance segmentation, IoU-based metrics are considered more consistent than the ℓn-norm.

$$IoU = \frac{A_{gt} \cap A_{pr}}{A_{gt} \cup A_{pr}} \tag{1}$$

Here, A_{gt} denotes the ground truth bounding box and A_{pr} the predicted bounding box.

The original intersection-over-union (IoU) metric computes the union area of two bounding boxes, but it fails to differentiate between nonoverlapping scenarios of the two boxes. To address this limitation, Generalized-IoU (GIoU) [24] was introduced, incorporating a penalty term within its loss function. This ensures that, in cases of nonoverlap, the predicted bounding box converges towards the target box.

$$GIoU = IoU - \frac{|C - (A_{gt} \cap A_{pr})|}{|C|}, \tag{2}$$

where C represents the minimum bounding rectangle enclosing both A_{gt} and A_{pr}, and $|C|$ denotes the area of C.

Nevertheless, GIoU becomes ineffective when the predicted bounding box is completely covered by the ground truth bounding box. To overcome this challenge, Distance-IoU (DIoU) [25] was proposed, considering the distance between the center points of the predicted and ground truth bounding boxes.

$$DIoU = IoU - \frac{\rho^2(b_{gt}, b_{pr})}{h_c^2 + w_c^2} \tag{3}$$

Here, b_{gt} and b_{pr} are the center points of the ground truth and predicted bounding boxes, respectively, and $\rho(b_{gt}, b_{pr})$ is the Euclidean distance between b_{gt} and b_{pr}.

Yet, in situations where the center points of the predicted and ground truth bounding boxes overlap, the metric reduces to the original IoU. To rectify this, Complete-IoU (CIoU) [26] was introduced, which integrates both the center point distance and the aspect ratio.

$$CIoU = IoU - \frac{\rho^2(b_{gt}, b_{pr})}{h_c^2 + w_c^2} - \alpha v, \tag{4}$$

$$v = \frac{4}{\pi^2}(arctan\frac{h_{gt}}{w_{gt}} - arctan\frac{h_{pr}}{w_{pr}})^2, \tag{5}$$

$$\alpha = \frac{v}{1 - IoU - v} \tag{6}$$

h_{gt} and w_{gt} can be defined as the height and width of the ground truth bounding box, respectively, and h_{pr} and w_{pr} as the corresponding dimensions of the predicted bounding box.

In CIoU, however, the aspect ratio is defined based on a relative value, not an absolute one. To address this, Efficient-IoU (EIoU) [27] was proposed, building upon DIoU.

$$EIoU = IoU - \frac{\rho^2(b_{gt}, b_{pr})}{h_c^2 + w_c^2} - \frac{\rho^2(w_{gt}, w_{pr})}{w_c^2} - \frac{\rho^2(h_{gt}, h_{pr})}{h_c^2}, \tag{7}$$

Here, h_c and w_c denote the height and width of C, respectively. Similarly, h_{pr} and w_{pr} are the height and width of the predicted bounding box, and h_{gt} and w_{gt} are the corresponding dimensions of the ground truth bounding box.

Notwithstanding, EIoU could exhibit limitations in scenarios where the predicted bounding box and the ground truth bounding box possess an identical aspect ratio yet differ in terms of their widths and heights. Such circumstances have the potential to hinder convergence speed and accuracy. To mitigate this issue, further improvements or alternative metrics might be necessary.

Figure 1 is a schematic diagram of commonly used BBR loss functions.

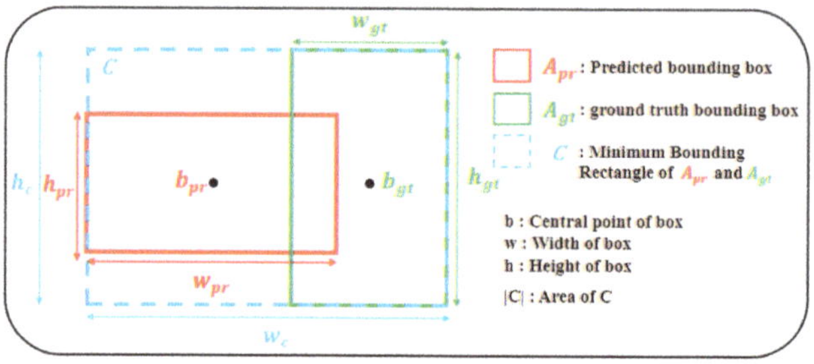

Figure 1. Commonly used BBR loss functions.

2.1.2. Analysis of BBR Loss Function

The existing loss functions fail to fully leverage the geometric attributes of BBR. In scenarios where the predicted and ground truth bounding boxes share an aspect ratio but diverge in terms of their widths and heights, these loss functions prove ineffective, thereby hindering convergence speed and accuracy.

We have observed a significant imbalance in datasets pertaining to agricultural pest and disease identification, as compared to deep learning datasets in other domains. This imbalance contributes to inconsistencies in the quality of agricultural pest and disease datasets currently available.

As depicted in Figure 2, numerous deep learning datasets for agricultural pests and diseases are compiled using varying image acquisition techniques and annotation standards. Consequently, scenarios arise where the predicted and ground truth bounding boxes exhibit matching aspect ratios but differ in their dimensions.

Figure 2. Current status of annotations in agricultural pest and disease datasets: (**a**) Citrus canker disease. (**b**) Black rot of grapes. (**c**) Cabbage caterpillar on snow peas. (**d**) Acridoidea insect. (**e**) Legume blister beetle. (**f**) Leaf blight of grapes.

The aforementioned scenarios can lead to existing BBR loss functions producing uniform values for disparate prediction outcomes, thereby diminishing the convergence rate and precision of BBR. The recurrence of this problem undermines the accurate identification of agricultural pests and diseases. Presently, BBR loss functions possess constraints that hinder precise identification, particularly in datasets bearing similar traits.

Hence, considering the merits of loss functions like GIoU, DIoU, CIoU, and EIoU, we strive to devise a novel loss function for BBR, termed DDRIoU loss, aimed at enhancing the efficacy and precision of BBR.

2.2. Construction of the FIP6Set Dataset

The identification of agricultural pests and diseases faces certain challenges, particularly the imbalance in the dataset caused by the varying availability of data. Some types of data are easy to obtain, resulting in a large quantity, while others are difficult to acquire, leading to a smaller amount. This imbalance affects the training and generalization capabilities of the model. To address this issue, we have constructed a crop pest and disease dataset called FIP6Set.

The FIP6Set dataset originates from three primary sources: Firstly, photos taken by us in Nanning, Guangxi Zhuang Autonomous Region, including images of citrus canker disease and cabbage caterpillars on snow peas; secondly, images from the public dataset IP102 [28], specifically, those of the Acridoidea insect and legume blister beetle; and thirdly, images sourced from the internet, featuring black rot of grapes and leaf blight of grapes.

The construction of the FIP6Set dataset involved three main components. Firstly, on several sunny days and under the guidance of agricultural experts, we captured 821 images of citrus canker disease and 800 images of cabbage caterpillars on snow peas. These images were subsequently cleaned based on clarity and realism, resulting in a selection of 440 images of citrus canker disease and 440 images of cabbage caterpillars on snow peas. These selected images were then annotated under the direction of agricultural experts and reviewed by a second expert. Secondly, we selected a portion of images from the public dataset IP102. Out of 533 images of Acridoidea insects, 470 high-quality images were chosen. Due to the limited number of legume blister beetle images available, all 435 images were included in our dataset. Lastly, we collected 1043 images of black rot of grapes and 1007 images of leaf blight of grapes from the internet. These images were also cleaned based on quality, with 460 images of black rot of grapes and 475 images of leaf blight of grapes being incorporated into our dataset. Similar to the previous components, these images were annotated under the guidance of agricultural experts and reviewed by a second expert. Thus, the new dataset, FIP6Set, was constructed.

During the image selection process, we deliberately selected a combination of complex lesions suitable for broad annotation, as seen in Figure 2d,f as well as simpler lesions amenable to precise annotation, exemplified by Figure 2a. Furthermore, we incorporated pests that bear similar colors to the leaves, as shown in Figure 2c,d and those that contrast with the leaf color, as in Figure 2e.

Based on Saleh Shahinfar's [29] research, which suggests that a minimum of 150–500 images are required for training, and considering the current state of agricultural pest and disease datasets, we maintained the image count for each category in our dataset between 435 and 475. This approach not only ensures the consistency of the dataset but also prevents the quantity of data from adversely affecting training effectiveness.

2.3. Diagonal Difference Ratio Intersection over Union

As shown in Figure 3, the red box represents the predicted bounding box, the green box represents the ground truth bounding box, and the blue box represents the minimum enclosing rectangle encompassing both the red and green boxes. When the predicted bounding box and the ground truth bounding box possess the same aspect ratio but differ in widths and heights, the traditional IoU loss function becomes less effective. Figure 3a,b depict two distinct scenarios; however, the traditional IoU loss function assigns an equivalent loss value in both instances.

To tackle the aforementioned issues and inspired by the geometric characteristics of the bounding box, we introduce DDRIoU, which is based on CIoU.

Figure 4 shows the key components of our proposed DDRIoU.

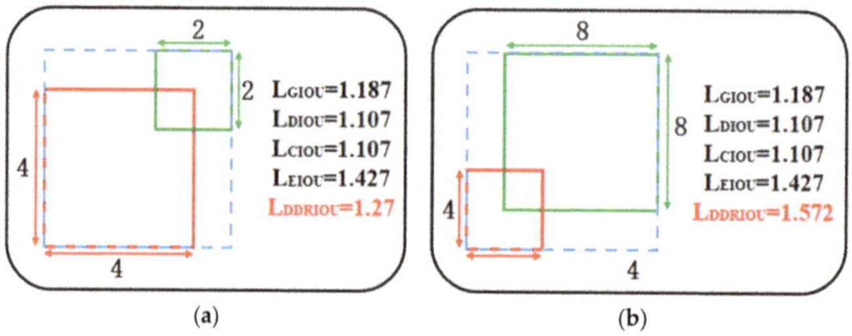

Figure 3. Calculation results across different scenarios, displaying diverse BBR losses such as GIoU, DIoU, CIoU, EIoU, and DDRIoU. Specifically, Figure (**a**) illustrates the values of various BBR losses for a scenario where the predicted bounding box is a 4 × 4 rectangle, while the ground truth bounding box is a 2 × 2 rectangle. Similarly, Figure (**b**) depicts the BBR loss values for a case where the predicted bounding box remains a 4 × 4 rectangle, but the ground truth bounding box is an 8 × 8 rectangle.

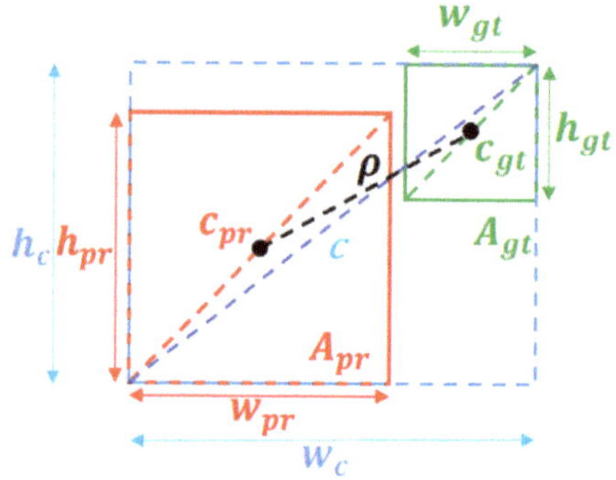

Figure 4. Our DDRIoU component. Specifically, the red box represents the predicted bounding box, the green box represents the ground truth bounding box, and the blue box represents the minimum enclosing rectangle encompassing both the red and green boxes.

Algorithm 1 outlines the computational process for DDRIoU:

Algorithm 1: Intersection over Union with Diagonal Difference Ratio.

Input: Two arbitrary convex shapes: A_{gt}, A_{pr}, width and height of input image: w, h
Output: DDRIoU
1 : Let h_{gt} and w_{gt} represent the height and width of A_{gt}, respectively.
Let h_{pr} and w_{pr} represent the height and width of A_{pr}, respectively.
2 : Let ρ represent the Euclidean distance between the center points of A_{gt} and A_{pr}.
3 : C represents the minimum bounding rectangle enclosing both A_{gt} and A_{pr}, where h_c and w_c denote the height and width of C, respectively.
4 : $IoU = \frac{A_{gt} \cap A_{pr}}{A_{gt} \cup A_{pr}}$
5 : $c_{gt}^2 = h_{gt}^2 + w_{gt}^2$
6 : $c_{pr}^2 = h_{pr}^2 + w_{pr}^2$
7 : $c^2 = h_c^2 + w_c^2$
8 : $DDRIoU = IoU - \frac{\rho^2}{c^2} - \frac{1}{\pi} * arctan\left(\left(\frac{c_{gt}^2 - c_{pr}^2}{c_{pr}^2}\right)^2\right)$
9 : return DDRIoU.

---

where c_{gt} represents the length of the hypotenuse of the ground truth bounding box, and c_{pr} represents the length of the hypotenuse of the predicted bounding box.

Therefore, the DDRIoU loss is defined as follows:

$$L_{DDRIoU} = 1 - IoU - \frac{\rho^2}{c^2} - \frac{1}{\pi} * \arctan((\frac{c_{gt}^2 - c_{pr}^2}{c_{pr}^2})^2), \tag{8}$$

The core component is the penalty term $\frac{1}{\pi} * \arctan((\frac{c_{gt}^2 - c_{pr}^2}{c_{pr}^2})^2)$. We utilize a function, $\frac{1}{\pi} * \arctan$, to confine the penalty term within the range of $\left[0, \frac{1}{2}\right)$. This confinement aids in stabilizing the training process and prevents gradient explosion due to excessively large loss values. The expression $\frac{c_{gt}^2 - c_{pr}^2}{c_{pr}^2}$ reflects the relative difference in proportion between the predicted bounding box and the ground truth bounding box. This is advantageous for object detection tasks involving significant variations in scale, such as detecting objects of different sizes.

Compared to traditional loss functions, utilizing the square of the hypotenuse of the bounding box as a metric enhances the stability of the loss function. This is because it provides a more consistent and robust measure of the bounding box size, regardless of its aspect ratio. Traditional metrics based on aspect ratio (or width/height difference) can cause instability during training, especially in tasks that require frequent detection of large and small targets.

In our approach, the expression $\frac{c_{gt}^2 - c_{pr}^2}{c_{pr}^2}$ plays a crucial role, where $c_{gt}^2 - c_{pr}^2$ represents the difference or deviation and c_{pr}^2 serves as a normalizing factor. Meanwhile, we specifically use c_{pr}^2 as the denominator in this expression. This approach not only avoids potential instability issues during training, especially when the ground truth bounding box size is too small, but also makes the loss function more sensitive to changes in the predicted bounding box size. This sensitivity encourages the model to better adjust the size of the predicted bounding box to closely approximate the ground truth bounding box.

In this way, we retain the characteristics of CIoU loss. DDRIoU loss, however, goes beyond CIoU by considering both the predicted and actual bounding boxes. This approach effectively addresses the issue of identical loss that arises when the predicted and actual bounding boxes share the same aspect ratio but differ in widths and heights.

2.4. Focal-DDRIoU

In BBR, there exists an issue of imbalance in training examples, where due to the sparsity of target objects in the image, the number of high-quality examples with small regression errors is far fewer than low-quality examples. Outliers can produce excessively large gradients, which are detrimental to the training process. Hence, it is crucial to allow high-quality examples to provide a greater contribution to training. Therefore, we propose the Focal-DDRIoU loss to enhance the performance of the DDRIoU loss.

Based on the research by Yi-Fan Zhang et al. [27], we similarly adopt the strategy of weighting the DDRIoU loss using the IoU value to obtain the Focal-DDRIoU loss, as shown in Formula (9):

$$L_{Focal-DDRIoU} = IoU^{\gamma} L_{DDRIoU}, \tag{9}$$

where $IoU = \frac{A_{gt} \cap A_{pr}}{A_{gt} \cup A_{pr}}$, and γ is a parameter that controls the degree of outlier suppression.

2.5. Overview of the YOLOv7 Network Structure

There are seven different variants of the YOLOv7 network structure, including the standard YOLOv7 and YOLOv7-tiny, among others. In this paper, we focus on the standard YOLOv7 structure for our research.

Figure 5 is a schematic diagram of the YOLOv7 network structure.

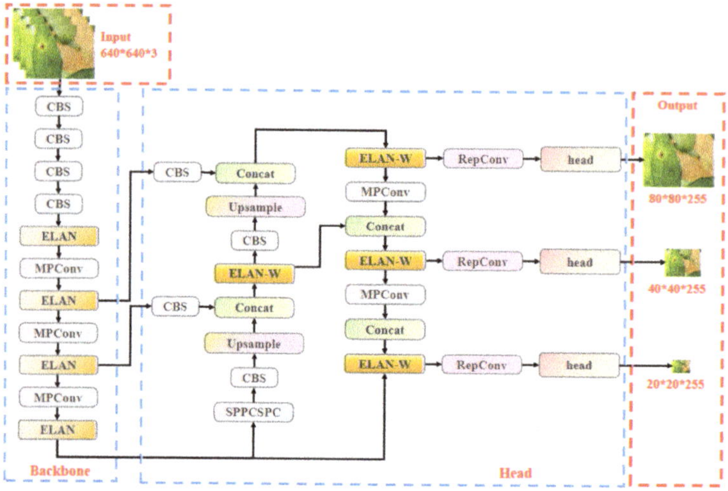

Figure 5. YOLOv7 network structure.

The overall framework of YOLOv7 is largely inherited from YOLOv5 [15], encompassing the overall network architecture and configuration file settings, as well as the training, inference, and validation processes. Furthermore, it incorporates certain aspects from YOLOR [30], such as the design of distinct networks, hyperparameter configurations, and the integration of implicit knowledge learning. Additionally, the SimOTA approach from YOLOX [31] is adopted for positive sample matching.

Apart from the elements adopted from other YOLO iterations, YOLOv7 also integrates several notable advancements, including efficient aggregated networks (e.g., ELAN and E-ELAN), re-parameterized convolution, auxiliary head detection, and model scaling.

2.6. APD-YOLOv7 Network Structure

We have made the following improvements to the YOLOv7 network structure:

(1) Introduction of C3 and MobileViTv3 Modules to Enhance Feature Extraction and Scalability.

The C3 module, a pivotal component in YOLOv5, employs a hierarchical structure for feature extraction. Due to its design, the C3 module is capable of more effectively capturing contextual information, thereby achieving superior performance in handling complex backgrounds.

MobileViT, originally known as MobileViTv1, represents a lightweight model tailored for mobile vision tasks, seamlessly integrating convolutional neural networks (CNNs) and vision transformers (ViTs).

The MobileViTv3 module, proposed by WADEKAR ShaktiN et al. [32], addresses scalability issues and simplifies learning tasks by enhancing the fusion module. As shown in Figure 6, this module seamlessly integrates features from both local and global representation blocks within the fusion block, utilizing 1×1 convolution. This approach not only simplifies the learning task but also prevents significant increases in parameters and FLOPs during scaling. Within the local representation block, 3×3 depthwise separable convolution is employed to further reduce parameters. Before the module's final output, input features are combined with the output of the fusion block, leveraging residual connections to enhance accuracy.

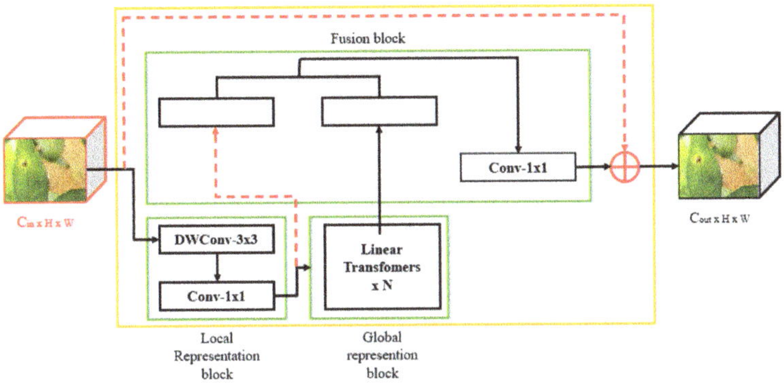

Figure 6. Structure of the MobileViTv3 module.

The incorporation of the MobileViTv3 module significantly contributes to the reduction in model parameters and the enhancement of model accuracy.

(2) Adjustments to the YOLOv7 Network Structure.

We have refined the YOLOv7 network structure with the aim of further minimizing computational complexity, parameter count, and floating-point operations, all while boosting the model's recognition accuracy and generalization capability.

As depicted in Figure 7, we initially removed two ELAN modules from the backbone section and introduced C3 modules to compensate for the network's feature extraction capabilities. Subsequently, to further elevate model performance, we embedded MobileViTv3 modules between the remaining two ELAN modules. In the head section, we substituted all ELAN-W modules with C3 modules, culminating in the new APD-YOLOv7 network structure.

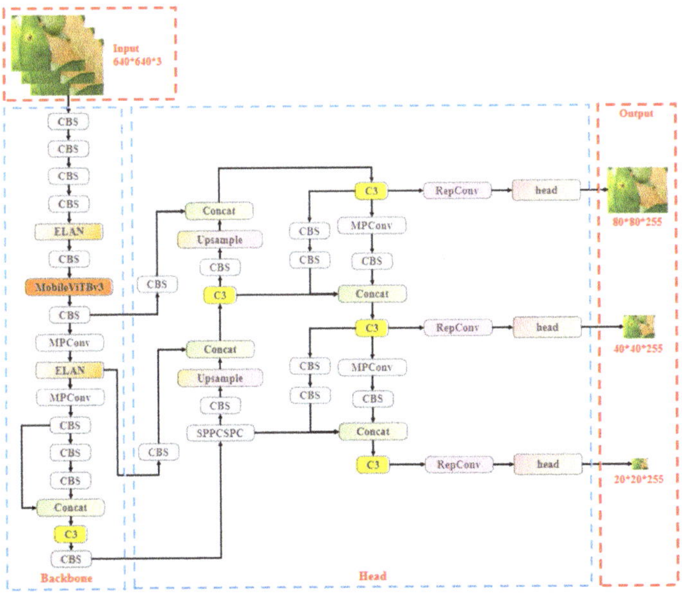

Figure 7. APD-YOLOv7 network structure.

3. Experiment and Result Analysis

3.1. Experimental Setup

In this study, we employed the open-source PyTorch deep learning framework for model development and enhancement. Additionally, the training and evaluation of these models were conducted on a system running Ubuntu 20. The hardware configuration comprised an Intel (R) Core (TM) i9-11900K processor and an NVIDIA GeForce RTX 3090 graphics card. The GPU was equipped with 24 GB of video memory and used CUDA version 11.3. During the experiments, consistent training parameters were applied as follows:

- The input image size was consistently adjusted to 640 $\times$ 640 pixels.
- The batch size parameter was set to 16.
- The epochs parameter was set to 300.
- The initial training weights were set to empty.
- All other parameters were left at their default settings.

3.2. Experimental Evaluation Metrics

When identifying targets of agricultural pests and diseases in complex environments, it is crucial to consider both the accuracy and real-time performance of the model. To evaluate the model's performance in identifying agricultural pests and diseases, this paper uses mean average precision as the primary metric, with single-image inference time and model parameter size serving as secondary evaluation criteria. It is worth noting that mean average precision is related to both precision and recall, and its detailed calculation formulas are provided below:

1. P represents precision, indicating the proportion of actually positive samples among those predicted as positive by the model.

$$P = \frac{TP}{TP + FP} \tag{10}$$

2. R represents recall, indicating the proportion of positive samples correctly predicted as positive by the model among all actual positive samples.

$$R = \frac{TP}{TP + FN} \tag{11}$$

3. AP represents average precision for a single category.

$$AP = \int_0^1 P(R)dR \tag{12}$$

4. mAP represents the mean of average precisions for all categories.

$$mAP = \frac{1}{N}\sum_{i=1}^{N} AP_i \tag{13}$$

In this context, TP denotes the count of samples accurately classified as positive, FP signifies the count of samples erroneously classified as positive, and FN indicates the count of samples mistakenly identified as negative.

Hence, AP75 refers to the AP computed with an IoU threshold of 0.75, while mAP75 signifies the mAP calculated at the same IoU threshold. For the purpose of this study, mAP75 serves as the primary evaluation metric.

In the subsequent experiments, we utilized the FIP6Set dataset, developed in this study, and implemented a 5-fold cross-validation approach across all experimental trials.

3.3. Experimental Analysis of Focal-DDRIoU Loss Function

We conducted experiments utilizing the YOLOv7 model, establishing baseline results by training the model with the default CIoU loss function. To assess the influence of various

IoU loss functions, we trained YOLOv7 using GIoU, DIoU, CIoU, EIoU, DDRIoU, and Focal-DDRIoU. To achieve this, we substituted the BBR IoU loss with L_{GIoU}, L_{DIoU}, L_{CIoU}, L_{EIoU}, L_{DDRIoU}, and $L_{Focal-DDRIoU}$, respectively.

As shown in Table 1, under the FIP6Set dataset, using a 5-fold cross-validation method and averaging the results of five experiments, DDRIoU loss and Focal-DDRIoU loss demonstrated superior performance compared to GIoU loss, DIoU loss, CIoU loss, and EIoU loss. Specifically, DDRIoU loss enhanced the mAP75 value by 0.76% over the default CIoU loss used in YOLOv7, while Focal-DDRIoU loss achieved a notable improvement of 1.04%.

Table 1. Performance comparison of YOLOv7 model trained with default loss (L_{CIoU}) and L_{GIoU}, L_{DIoU}, L_{EIoU}, L_{DDRIoU}, and $L_{Focal-DDRIoU}$ using 5-fold cross-validation method.

mAP75 / Loss	Rounds					5-Fold Average	5-Fold Variance
	1	2	3	4	5		
L_{GIoU}	0.561	0.548	0.536	0.555	0.588	0.5576 (−0.08%)	0.000300
L_{DIoU}	0.571	0.536	0.549	0.560	0.566	0.5564 (−0.20%)	0.000158
L_{CIoU}	0.544	0.542	0.560	0.560	0.586	0.5584 (0)	0.000249
L_{EIoU}	0.546	0.531	0.573	0.546	0.573	0.5538 (−0.46%)	0.000276
L_{DDRIoU}	0.566	0.556	0.555	0.571	0.582	0.5660 (+0.76%)	0.000100
$L_{Focal-DDRIoU}$	0.560	0.561	0.561	0.569	0.593	0.5688 (+1.04%)	0.000157

The improved performance can be primarily attributed to the introduction of the diagonal difference ratio in DDRIoU loss. This innovation addresses a limitation of other IoU loss functions, which produce the same loss value when the predicted and ground truth bounding boxes have the same aspect ratio but differ in size. DDRIoU loss more accurately captures these differences, guiding the model towards more precise optimization and enhancing detection accuracy.

Beyond improving accuracy, the diagonal difference ratio in DDRIoU loss also contributes to a more nuanced reflection of disparities between predicted and ground truth bounding boxes. This refinement aids in enhancing the stability and consistency of prediction results, thereby reducing outcome variance.

Compared to DDRIoU loss, Focal-DDRIoU loss exhibits a slight increase in variance, yet it maintains a relatively low level. This suggests that incorporating the focal loss concept allows the model to focus more effectively on challenging samples during training. As a result, it boosts performance without significantly compromising result stability. This design fosters a more robust and reliable model, especially in complex scenarios.

3.4. Experimental Analysis of Network Structure

In this section, a comparative experiment was conducted between the APD-YOLOv7 network structure devised in this study (hereinafter referred to as the APD network structure) and the original YOLOv7 network structure. The experimental steps are outlined below:

1. Initially, we modified only the IoU loss function. Specifically, we trained the YOLOv7 model using GIoU loss, DIoU loss, CIoU loss, EIoU loss, and Focal-DDRIoU loss, respectively, to evaluate the performance of the original YOLOv7 network structure.
2. Next, we substituted the original YOLOv7 network structure with the APD network structure.
3. Finally, we replicated the first step, adjusting the loss function to assess the performance of the APD network structure.

As demonstrated in Table 2, the performance of the APD network structure exceeds that of the YOLOv7 network structure when evaluated under equivalent IoU loss conditions. Following the substitution of the YOLOv7 network structure with the APD network structure, we conducted experiments employing GIoU loss, DIoU loss, CIoU loss, EIoU loss, and Focal-DDRIoU loss, respectively. The results showed an improvement in mAP75 by 0.68%, 0.68%, 0.78%, 0.6%, and 0.56%, respectively. In summary, the YOLOv7 model, when trained with Focal-DDRIoU loss and incorporating the APD-YOLOv7 network structure, achieved a 2.06% improvement in mAP75 compared to the model trained with EIoU loss and the standard YOLOv7 architecture under identical conditions.

Table 2. Performance comparison between the YOLOv7 network structure and the APD network structure trained using L_{GIoU}, L_{DIoU}, L_{CIoU}, L_{EIoU}, and $L_{Focal-DDRIoU}$, respectively, under the YOLOv7 model.

Loss \ Evaluation	YOLOv7 5-Fold Average mAP75	APD-YOLOv7 5-Fold Average mAP75
L_{GIoU}	0.5576	0.5644 (+0.68%)
L_{DIoU}	0.5564	0.5632 (+0.68%)
L_{CIoU}	0.5584	0.5662 (+0.78%)
L_{EIoU}	0.5538	0.5598 (+0.60%)
$L_{Focal-DDRIoU}$	0.5688	0.5744 (+0.56%)

Table 3 illustrates that, in comparison to the YOLOv7 network structure, the APD network structure achieves a 3.9MB parameter reduction and a 9.74% decrease in giga floating-point operations per second (GFLOPs), effectively lowering the model's computational demands and parameter count.

Table 3. Comparison of parameters between YOLOv7 network structure and APD network structure (batch = 1).

Model	Parameters (MB)	GFLOPs	Inference Time (ms)
YOLOv7	36.9	104.7	7.2
APD-YOLOv7	33.0	94.5	10.8

This significant improvement is primarily attributed to modifications made to the YOLOv7 network structure. We adjusted the YOLOv7 architecture by removing redundant modules, appropriately supplementing with C3 modules, and embedding MobileViTv3 modules. These modifications successfully reduced the model's parameter count and computational complexity while further enhancing the model's accuracy.

Although the inference time for a single image has increased relatively, in agricultural pest and disease identification, strict real-time performance is not typically required. As long as the identification results can be provided within a few seconds, it is sufficient for most application scenarios, thus still meeting the need for rapid identification of agricultural pests and diseases.

3.5. *Experimental Analysis of the APD-YOLOv7 Model*

We conducted a performance evaluation of other widely used object detection models on the FIP6Set dataset.

Referring to Table 4, the APD-YOLOv7 model presented in this study demonstrates significant superiority over the YOLOv7 and YOLOv5-L models in terms of detection accuracy. Specifically, the APD-YOLOv7 model achieves an mAP75 metric that is 1.86% and 1.6% higher than the YOLOv5-L and YOLOv7 models, respectively. Additionally, it reduces the model parameter count by 13.5 MB and 3.9 MB compared to YOLOv5-L and YOLOv7, respectively. In terms of model detection speed, the inference time for a single image using the APD-YOLOv7 model is 10.8 ms. Although the inference time of the APD-YOLOv7 model is slightly longer compared to the YOLOv7 and YOLOv5-L models, it still satisfies the requirements for swift identification of agricultural pests and diseases.

Table 4. Performance comparison between APD-YOLOv7 and other object detection models.

Model	5-Fold Average mAP75	Parameters (MB)	Inference Time (ms)
YOLOv5-L	0.5558	46.5	9.2
YOLOv7	0.5584	36.9	7.2
APD-YOLOv7	0.5744	33.0	10.8

4. Conclusions and Future Directions

In this study, we present a model, termed APD-YOLOv7, for the detection of agricultural pests and diseases. Firstly, we design a novel loss function, termed Focal-DDRIoU loss, to enhance the efficiency and accuracy of BBR. Secondly, we introduce C3 and MobileViTv3 modules to modify the YOLOv7 network structure. Specifically, the MobileViTv3 module is integrated into the 17th layer of the backbone, thereby reducing the computational complexity, parameter count, and floating-point operations of the model, while simultaneously enhancing its recognition accuracy.

The performance evaluation is conducted on the FIP6Set dataset, employing mAP75 as the evaluation metric. Experimental results indicate that the APD-YOLOv7 model surpasses mainstream models, such as YOLOv7 and YOLOv5-L, by achieving improvements of 1.6% and 1.86%, respectively. Additionally, our proposed Focal-DDRIoU loss outperforms popular loss functions, including GIoU, DIoU, CIoU, and EIoU, yielding improvements of 1.12%, 1.24%, 1.04%, and 1.5%, respectively.

Although the inference time for a single image is increased in the APD-YOLOv7 model, it still fulfills the criteria for swift identification of agricultural pests and diseases.

While the APD-YOLOv7 model strikes a good balance between recognition accuracy and speed, its recognition diversity remains to be further explored. Future research efforts should focus on training the APD-YOLOv7 model with additional pest and disease samples to ascertain its robustness. Moreover, as the investigation of Focal-DDRIoU loss is currently confined to the FIP6Set dataset, future studies should seek to apply it across diverse datasets to assess its generalizability.

Author Contributions: Conceptualization, J.L., S.L., and C.L.; methodology, J.L., S.L., C.L., and S.Z.; software, J.L. and S.Z.; validation, J.L. and D.C.; formal analysis, J.L. and D.C.; investigation, J.L., S.L., and S.Z.; resources, J.L. and C.L.; data curation, J.L. and S.L.; writing—original draft preparation, J.L. and D.C.; writing—review and editing, J.L. and S.L.; visualization, J.L., S.L., and S.Z.; supervision, J.L., S.L., D.C., S.Z., and C.L.; project administration, J.L., S.L., S.Z., and C.L.; funding acquisition, S.L., S.Z., and C.L. All authors have read and agreed to the published version of the manuscript.

Funding: This work was funded in part by The Basic Ability Enhancement Program for Young and Middle-aged Teachers of Guangxi (2022KY0381), Guangxi Key Research and Development Project (Application Number: 2023AB19026, Research and demonstration of rapid monitoring technology and early warning system for major citrus diseases and pests based on deep learning).

Institutional Review Board Statement: Not applicable.

Informed Consent Statement: Not applicable.

Data Availability Statement: The public IP102 dataset can be downloaded from https://github.com/xpwu95/IP102 (accessed on 15 May 2023). Publicly available web data were used in this study and can be accessed without restrictions. The FIP6Set dataset, which is a combination of the IP102 dataset, publicly available web data, and private data, used and/or analyzed during the current study is available from the corresponding author upon reasonable request.

Conflicts of Interest: The authors declare no conflicts of interest.

References

1. Manavalan, R. Automatic identification of diseases in grains crops through computational approaches: A review. *Comput. Electron. Agric.* **2020**, *178*, 105802. [CrossRef]
2. Kong, J.; Wang, H.; Wang, X.; Jin, X.; Fang, X.; Lin, S. Multi-stream hybrid architecture based on cross-level fusion strategy for fine-grained crop species recognition in precision agriculture. *Comput. Electron. Agric.* **2021**, *185*, 106134. [CrossRef]

3. Zheng, Y.Y.; Kong, J.L.; Jin, X.B.; Wang, X.Y.; Su, T.L.; Zuo, M. CropDeep: The crop vision dataset for deep-learning-based classification and detection in precision agriculture. *Sensors* **2019**, *19*, 1058. [CrossRef] [PubMed]
4. Li, Y.; Nie, J.; Chao, X. Do we really need deep CNN for plant diseases identification? *Comput. Electron. Agric.* **2020**, *178*, 105803. [CrossRef]
5. Shruthi, U.; Nagaveni, V.; Raghavendra, B.K. A review on machine learning classification techniques for plant disease detection. In Proceedings of the 2019 5th International Conference on Advanced Computing & Communication Systems (ICACCS), Coimbatore, India, 15–16 March 2019; IEEE: Piscataway, NJ, USA, 2019; pp. 281–284.
6. Shrivastava, V.K.; Pradhan, M.K. Rice plant disease classification using color features: A machine learning paradigm. *J. Plant Pathol.* **2021**, *103*, 17–26. [CrossRef]
7. Kim, K.G. Book review: Deep learning. *Healthc. Inform. Res.* **2016**, *22*, 351. [CrossRef]
8. Li, L.; Zhang, S.; Wang, B. Plant disease detection and classification by deep learning—A review. *IEEE Access* **2021**, *9*, 56683–56698. [CrossRef]
9. Oppenheim, D.; Shani, G.; Erlich, O.; Tsror, L. Using deep learning for image-based potato tuber disease detection. *Phytopathology* **2019**, *109*, 1083–1087. [CrossRef] [PubMed]
10. Chen, Q.; Wang, Y.; Yang, T.; Zhang, X.; Cheng, J.; Sun, J. You only look one-level feature. In Proceedings of the IEEE/CVF Conference on Computer Vision and Pattern Recognition, Nashville, TN, USA, 19–25 June 2021; IEEE: Piscataway, NJ, USA, 2021; pp. 13039–13048.
11. Redmon, J.; Farhadi, A. YOLO9000: Better, faster, stronger. In Proceedings of the IEEE Conference on Computer Vision and Pattern Recognition, Honolulu, HI, USA, 21–26 July 2017; IEEE: Piscataway, NJ, USA, 2017; pp. 7263–7271.
12. Redmon, J.; Farhadi, A. Yolov3: An incremental improvement. *arXiv* **2018**, arXiv:1804.02767.
13. Bochkovskiy, A.; Wang, C.Y.; Liao, H.Y.M. Yolov4: Optimal speed and accuracy of object detection. *arXiv* **2020**, arXiv:2004.10934.
14. YOLOv5: An Open-Source Object Detection Algorithm. Available online: https://github.com/ultralytics/yolov5 (accessed on 15 May 2023).
15. Wang, C.Y.; Bochkovskiy, A.; Liao, H.Y.M. YOLOv7: Trainable bag-of-freebies sets new state-of-the-art for real-time object detectors. In Proceedings of the IEEE/CVF Conference on Computer Vision and Pattern Recognition, Vancouver, BC, Canada, 18–22 June 2023; IEEE: Piscataway, NJ, USA, 2023; pp. 7464–7475.
16. Wang, C.Y.; Yeh, I.H.; Liao, H.Y.M. YOLOv9: Learning What You Want to Learn Using Programmable Gradient Information. *arXiv* **2024**, arXiv:2402.13616.
17. Girshick, R.; Donahue, J.; Darrell, T.; Malik, J. Rich feature hierarchies for accurate object detection and semantic segmentation. In Proceedings of the IEEE Conference on Computer Vision and Pattern Recognition, Columbus, OH, USA, 23–28 June 2014; IEEE: Piscataway, NJ, USA; pp. 580–587.
18. Ren, S.; He, K.; Girshick, R.; Sun, J. Faster R-CNN: Towards real-time object detection with region proposal networks. *IEEE Trans. Pattern Anal. Mach. Intell.* **2016**, *39*, 1137–1149. [CrossRef] [PubMed]
19. He, K.; Gkioxari, G.; Dollár, P.; Girshick, R. Mask r-cnn. In Proceedings of the IEEE International Conference on Computer Vision, Venice, Italy, 22–29 October 2017; IEEE: Piscataway, NJ, USA, 2017; pp. 2961–2969.
20. Zhang, H.; Chang, H.; Ma, B.; Wang, N.; Chen, X. Dynamic R-CNN: Towards high quality object detection via dynamic training. In Proceedings of the European Conference on Computer Vision, Glasgow, UK, 23–28 August 2020; Springer: Berlin, Germany, 2020; pp. 260–275.
21. Carion, N.; Massa, F.; Synnaeve, G.; Usunier, N.; Kirillov, A.; Zagoruyko, S. End-to-end object detection with transformers. In Proceedings of the European Conference on Computer Vision, Glasgow, UK, 23–28 August 2020; Springer: Berlin, Germany, 2020; pp. 213–229.
22. Yu, J.; Jiang, Y.; Wang, Z.; Cao, Z.; Huang, T. Unitbox: An advanced object detection network. In Proceedings of the 24th ACM international conference on Multimedia, Amsterdam, The Netherlands, 15–19 October 2016; ACM: New York, NY, USA, 2016; pp. 516–520.
23. Tychsen-Smith, L.; Petersson, L. Improving object localization with fitness nms and bounded iou loss. In Proceedings of the IEEE conference on computer vision and pattern recognition., Salt Lake City, UT, USA, 18–22 June 2018; IEEE: Piscataway, NJ, USA, 2018; pp. 6877–6885.
24. Rezatofighi, H.; Tsoi, N.; Gwak, J.; Sadeghian, A.; Reid, I.; Savarese, S. Generalized intersection over union: A metric and a loss for bounding box regres-sion. In Proceedings of the IEEE/CVF Conference on Computer Vision and Pattern Recognition, Long Beach, CA, USA, 16–20 June 2019; IEEE: Piscataway, NJ, USA, 2019; pp. 658–666.
25. Zheng, Z.; Wang, P.; Liu, W.; Li, J.; Ye, R.; Ren, D. Distance-IoU loss: Faster and better learning for bounding box regression. In Proceedings of the AAAI Conference on Artificial Intelligence, New York, NY, USA, 7–12 February 2020; AAAI: Menlo Park, CA, USA, 2020; Volume 34, pp. 12993–13000.
26. Zheng, Z.; Wang, P.; Ren, D.; Liu, W.; Ye, R.; Hu, Q.; Zuo, W. Enhancing geometric factors in model learning and inference for object detection and instance segmentation. *IEEE Trans. Cybern.* **2021**, *52*, 8574–8586. [CrossRef] [PubMed]
27. Zhang, Y.-F.; Ren, W.; Zhang, Z.; Jia, Z.; Wang, L.; Tan, T. Focal and efficient IOU loss for accurate bounding box regression. *arXiv* **2021**, arXiv:2101.08158. [CrossRef]

28. Wu, X.; Zhan, C.; Lai, Y.K.; Cheng, M.M.; Yang, J. Ip102: A large-scale benchmark dataset for insect pest recognition. In Proceedings of the IEEE/CVF Conference on Computer Vision and Pattern Recognition, Long Beach, CA, USA, 16–20 June 2019; IEEE: Piscataway, NJ, USA, 2019; pp. 8787–8796.
29. Shahinfar, S.; Meek, P.; Falzon, G. "How many images do I need?" Understanding how sample size per class affects deep learning model performance metrics for balanced designs in autonomous wildlife monitoring. *Ecol. Inform.* **2020**, *57*, 101085. [CrossRef]
30. Wang, C.Y.; Yeh, I.H.; Liao, H.Y.M. You only learn one representation: Unified network for multiple tasks. *arXiv* **2021**, arXiv:2105.04206.
31. Ge, Z.; Liu, S.; Wang, F.; Li, Z.; Sun, J. Yolox: Exceeding yolo series in 2021. *arXiv* **2021**, arXiv:2107.08430.
32. Wadekar, S.N.; Chaurasia, A. Mobilevitv3: Mobile-friendly vision transformer with simple and effective fusion of local, global and input features. *arXiv* **2022**, arXiv:2209.15159.

sustainability

MDPI

Article

Sustainable and Traditional Irrigation and Fertigation Practices for Potato and Zucchini in Dry Mediterranean Regions

Talal Darwish [1,*], Amin Shaban [1], Ghaleb Faour [1], Ihab Jomaa [2], Peter Moubarak [3] and Roula Khadra [3]

[1] National Council for Scientific Research, Beirut 11072260, Lebanon; geoamin@cnrs.edu.lb (A.S.); gfaour@cnrs.edu.lb (G.F.)
[2] Lebanese Agronomic Research Institute, Rayak 901965, Lebanon; ijomaa@lari.gov.lb
[3] Mediterranean Agronomic Institute of Bari, 70010 Bari, Italy; moubarak.p@gmail.com (P.M.); khadra@iamb.it (R.K.)
* Correspondence: tlldarwish@gmail.com

Abstract: Transforming irrigation practices is essential to address aquifer depletion and food security in Mediterranean regions facing climate change and water scarcity. Developing local and national resilience to climate change requires capacity building to boost soil health and adaptation to drought. Recent attempts undertaken by the SEALACOM Project reduced irrigation rates in protected agriculture. The purpose of this work is to enhance traditional farmer's practices and promote the potential of advanced fertigation of field crops (i.e., potato and zucchini) cultivated under two different pedo-climatic conditions to improve water and nutrient use efficiency. Results showed the yield of zucchini and potato on SEALACOM plots with continuous fertigation was 22% and 17.8%, respectively, which was higher than the yield with traditional irrigation and fertilization practices. Elite potato tuber size was 40% higher in SEALACOM plots ($p < 0.05$). The farmer applied 359 L of water to produce 1 kg of fresh zucchini compared to 225 L by the SEALACOM Project, indicating a significant, 60% water saving in the SEALACOM practice. Compared to farmer's practices of potato production, the SEALACOM Project achieved more than 50% higher water productivity. In zucchini production, farmers applied 19.5% more nitrogen and 19.6% more phosphorus fertilizers. Compared to 58 kg of N applied by the farmers, the SEALACOM Project applied 38 kg of N to produce 1 ton of Zucchini, showing a 34% saving in major nutrient application. To cultivate 1 kg of fresh potato tubers, SEALACOM utilized 4.06 g of nitrogen and 1.34 g of phosphorus, compared to the traditional practice, which required 13.2 g of nitrogen and 2.25 g of phosphorus. Water and nutrient saving and higher productivity and commerciality of the final product have a high positive impact on the farmer's income and positive attitude towards the adoption of modern, sustainable practices.

Keywords: fertilization; continuous feeding; field crops; water productivity; nutrient use efficiency

1. Introduction

Climate change is expected to raise Lebanon's mean annual temperature by 1.45 °C by mid-century and reduce precipitation by 2.7% by 2025 [1,2]. If current trends in water scarcity continue, the direct costs to the Lebanese economy caused by higher temperatures, fluctuating precipitation, and drought events are estimated to reach USD 138,900 million in 2080 [3], equivalent to 1.71% of current national gross domestic product. The resulting reduction in agricultural productivity will certainly threaten food security. Water scarcity

https://doi.org/10.3390/su17051860

causes yield reductions that can be offset by controlled irrigation to maintain sustainable crop production [4].

However, climate change has diminished the positive effect of soil organic carbon sequestration, reduced wheat yield up to 39%, and increased CO_2 and N_2O emissions from the soil [5]. Efforts to enhance on-farm irrigation management are crucial to addressing the growing challenges of a finite resource and maintaining food production [6]. To avoid excessive irrigation, which causes depleted water resources, controlled or deficit irrigation is recommended. This approach restricts excessive leaf growth and enhances fruit development, positively affecting water consumption and improving water use efficiency [7].

In Lebanon, emissions from energy, agriculture, and forestry-land use accounted for 58% of total emissions in 2019 [8]. The majority of these emissions were attributed to energy consumption, particularly from diesel engines powering pumps to extract water from individual wells, costing approximately USD 0.86 per liter. This issue is especially prevalent in the Bekaa Plain, where collective irrigation schemes are lacking [9]. Water scarcity has led to the chaotic drilling of unlicensed (i.e., illegal) wells and uncontrolled groundwater abstraction for irrigation to meet increasing demands for food and feed. This has resulted in aquifer depletion and the salinization of soil and water resources [10,11]. A study done 10 years ago in Lebanon reported more than 18,000 private, unlicensed wells in the Bekaa Plain, primarily used for irrigation with fossil energy [12].

Integrating climate-proof agriculture into a comprehensive management approach involves monitoring soil and climatic conditions and considering crop water and nutrient demands in relation to crop growth and development. This can support pro-active decision-making to anticipate economic and environmental losses. Improving water management in the country can reduce yield gaps and meet the challenges of food security and sustainable development goals (SDGs), specifically SDG 2. But this outcome is impossible to achieve without increased water productivity at the farm level.

A comprehensive multiscale analysis that couples plot scale tests with district-wide assessments can be useful to identify the technical and socio-economic driving forces behind low water productivity and yield gaps and to support the adoption of modern irrigation technology. The three-year national development priorities and strategy to revive the Lebanese national economy through targeted investments in various sectors [13], along with the Ministry of Agriculture's five-year strategic plan [14], proposed investing in capacity building and upgrading the technical basis and productivity of irrigated agriculture. They also call for innovation campaigns in fertilization and irrigation practices, which can contribute significantly to national economic growth and recovery. Developing local and national resilience to climate change requires a thorough analysis of prevailing irrigation and fertilization practices at the farmer level to boost adaptation capacity to drought and ensure food security. Since agro-ecosystems largely depend on both natural and climatic processes, resilience must be combined with innovation, adaptation, and the development of sustainable agricultural practices [15].

A sustainable adaptation strategy can be achieved through information technology (IT) intervention and the adoption of permanent and simple field monitoring tools to control water application based on soil moisture and crop demands [16]. This practice maintains optimal soil health conditions and achieves higher yields with less water and nutrient input, thereby reducing the footprint of irrigated agriculture. In addition to other adaptation practices, the use of regulated irrigation can decrease the risks of nitrogen losses, nitrate leaching, and gas emissions from agricultural lands [17].

With net exploitable surface and groundwater estimated at around 2 billion m^3 (Bm^3), the 2012 National Water Sector Strategy assessed the available water resources at 2.7 Bm^3, of which 2.2 Bm^3 are surface water and 0.50 Bm^3 are derived from groundwater recharge [18].

Since total groundwater abstraction by wells is estimated at 0.70 Bm^3, under normal climatic conditions, the yearly deficit in groundwater is equivalent to 0.2 Bm^3, causing a drop in the water table and decreased pumping. Recurring drought and depleted groundwater have caused local and regional drops in water table level [19]. However, in semi-arid regions, crop evapotranspiration (ET) is negatively correlated with groundwater depth, indicating a higher contribution of irrigation to crop water requirements in deeper aquifers. This is associated with better yields and more efficient water use in the root zone [20].

Over irrigation in field crop production has led to the intensive leaching of soluble pollutants, resulting in economic and environmental losses due to the deterioration of soil and groundwater quality [21]. Current irrigation practices in the Bekaa Plain threaten sustainable food production and challenge the implementation of SDGs. An optimized irrigation schedule and water accounting, combined with adequate nitrogen application, increased soil organic carbon, boosted soil fertility, and enhanced crop production and water use efficiency [22].

Despite governmental efforts to disseminate good practices among water users and farmers, adopt climatic information, and consider water balance [23], the increase in water use efficiency remains insufficient [24]. One of the main challenges hindering the expansion of such improvement is the lack of cost-effective and reliable data monitoring systems [6]. Field experiments based on climate-soil-crop-water smart agriculture practices have shown good potential to enhance the sustainability of local farming. These experiments indicate substantial water saving potential and improved agronomic productivity through a shift from traditional fertilization and irrigation practices to well managed modern fertigation systems [21].

Recent regional progress in determining crop water demands using data from a typical climatic year, correlated to crop growth cycles [25], did not significantly influence farmers' attitudes towards effective irrigation management in the Mediterranean region. Conservation measures in the USA, such as better correlation between pumping and precipitation and adjusting the area of irrigated lands, did not close the gaps between groundwater recharge and pumping rates [26]. However, the application of deficit irrigation in major potato-growing regions in Lebanon [21] and Northern China [27] significantly increased potato water use efficiency (WUE) by 10% and irrigation water use efficiency (IWUE) by 31.6%, with a considerable reduction in ET by 26.3%.

Alternatively, greater success in reducing irrigation water use per irrigated area was achieved by using locally based climatic stations to estimate potential evapotranspiration (ET0) and link it to the crop phenological development [28]. However, previous research in the region did not consider the total amount of water applied, starting from land preparation, amelioration of pre-sowing soil moisture, and post-sowing water application to maintain optimal conditions for good sprouting. Published studies mainly count water application related to evapotranspiration and crop coefficient (Kc) after seedling establishment, i.e., they account for effectively applied water covering the crop production cycle from plant emergence to physiological maturity.

A weak extension service is an additional constraint to the dissemination of good agricultural practices and effective water management in the semi-arid areas. Therefore, it is crucial to create and sustain living labs and demonstration (and experimental) sites in the main agroclimatic zones of the East and South Mediterranean. These sites ensure continuous interaction and participatory learning with local farmers. To our knowledge, this is the first time effectively applied water is considered in conjunction with the total applied water, with both parameters integrally analyzed within the concept of water accounting and water productivity under field conditions.

Meeting the objectives of the SDGs and addressing challenges to boost national capacities in the food-water-energy nexus, a project named "SEALACOM" was launched in 2023. It aimed to sustainably manage sea and land resources by the community, support development, sustain the environment, and strengthen farmer resilience. Accordingly, the purpose of this paper is to compare locally prevailing unsustainable fertilization and irrigation practices with science-based techniques and climate-smart practices proposed by the project. It assesses the impact in terms of water and nutrient saving and efficient use for two open field crops (potato and zucchini) cultivated under two different pedoclimatic conditions, following the traditional and advanced fertigation methods.

2. Materials and Methods

To reach the objectives of the SEALACOM Project and address the sustainable use of water in agriculture with a low energy and environmental footprint, two experimental sites in two different soil zones of Lebanon were selected. The first Demo Sites (DS1) is located in Serein-Cental Bekaa (Casa of Zahle) at 33°52′38″ N and 36°02′59″ E, and the second Demo Sites (DS2) is located in Sultan Yacoub (Casa of West Bekaa) at 33°40′21″ N and 35°51′40″ E. The study was run in fall 2023 in the two DS representing two agroecological zones of the Bekaa Plain of Lebanon with a total area of 86251.8 ha, of which 45.8% represent cultivated agricultural lands (Figure 1).

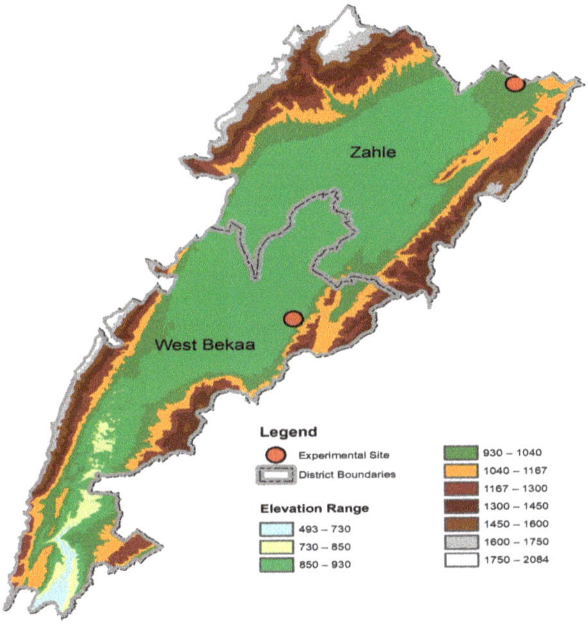

Figure 1. Topographic map of the SEALACOM Project Demo Sites showing the elevation range derived from DEM.

The topography of the demo sites varies between level plains in Serein and undulating foot slopes in Sultan Yacoub. Both areas are surrounded from east and west by the sloping lands of Mount Lebanon and the Anti-Lebanon mountain chain. The average altitude of the first demo site (DS1), located at Serein, is 950 m, while that of the second demo site (DS2), located at Sultan Yacoub, is 850 m.

The annual climatic data were received from the climatic stations of Tal-Amara, Lebanese Agricultural Research Institute. DS1 is located within the semi-arid climatic zone with 566 mm of average annual precipitation falling in one season between November and March (Figure 2). The rest of the year is dry with mean high and low annual temperatures reaching 25.6 °C and 8.5 °C, respectively.

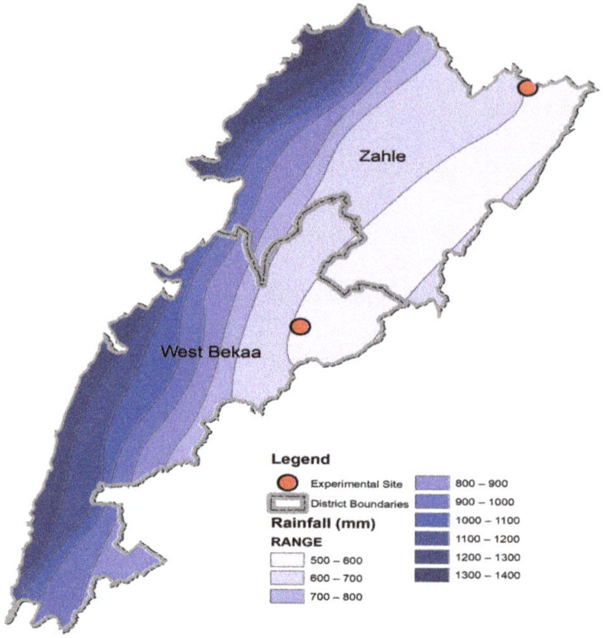

Figure 2. Average precipitation map of the SEALACOM Demo Sites showing the 100-precipitation interval.

DS2 belongs to the dry-subhumid climatic zone with an average annual rainfall of 700 mm and mean high and low annual temperatures reaching 24.6 °C and 10.8 °C, respectively. According to the World Reference Base for Soil Classification [29], the area of the two DSs is represented by the dominance of Cambisols, Regosols, and Luvisols with the inclusion of Vertisols, Calcisols, Leptosols, Arenosols, and Andosols (Figure 3).

The soil cover of DS1 was classified as deep Eutric Regosols. The soil is of neutral pH and non-saline (Table 1), clay texture with negligible content of $CaCO_3$. The soil has an average low organic matter content and is poor in total nitrogen, enriched with phosphorous, and moderately high in available potassium.

Table 1. Physio-chemical characteristics of soils of the two DSs before the experiment.

Soil Type	Depth, cm	Total Sand	Silt	Clay	$CaCO_3$ Total	$CaCO_3$ Active	Organic Matter	EC	pH H_2O	Available Nutrients, ppm		
		%						ds m^{-1}		N	P_2O_5	K_2O
Eutric Regosols DS1-Serein	0–20	36	18	46	5.6	1.9	2.74	0.49	7.6	23.8	354	280
	20–40	30	20	50	3.8	1.3	1.55	0.33	7.4	23.1	305	268
Eutric Cambisols DS2-Sultan Yacoub	0–20	26	12	62	5.8	1.9	1.25	0.34	7.7	32.2	206	578
	20–40	24	16	60	5.8	1.9	1.31	0.30	7.8	27.0	185	627

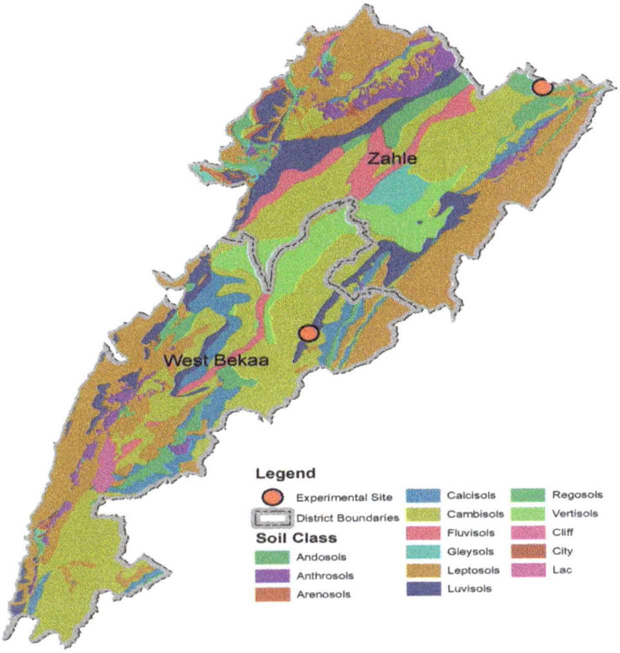

Figure 3. Soil map of the SEALACOM Project area of study [27].

The soil cover of DS2 was classified as Eutric Cambisols. Soil texture is clay with low organic matter content and weakly basic pH (Table 1). The soil is non-saline, poor in nitrogen, enriched with available phosphorous, and highly enriched with potassium.

Farmers traditionally apply nutrients and water by fertigation using closed tanks, managing the types, amounts, and timing of applications themselves. This approach relied on traditional irrigation management tools with intermittent nutrient application. The SEALACOM Project adopted an advanced smart climate system, estimating irrigation water needs based on ET0 calculations from a nearby climatic station. The daily crop evapotranspiration (ETc) was defined by multiplying ETc by the crop coefficient (Kc). Crop coefficient (Kc) was proposed by Jensen [30] and used together with the FAO guidelines to estimate crop water requirements [31] to relate the evapotranspiration of a specific crop (ETc) to the calculated potential evapotranspiration (ET0) according to the equation:

$$ETc = ET0 \times Kc \tag{1}$$

where ETc is the Daily Crop Evapotranspiration [mm/day]
ET0 is the Daily Reference Evapotranspiration [mm/day];
Kc is crop coefficient.

Since Kc is crop dependent and unaffected by climate, it follows the crop's growth and development dynamics, regardless of location. Kc values increased throughout the cropping season from 0.2 to 0.4, 0.6, 0.8, and 1.0 at full growth, then decreased to 0.7 by the end of the irrigation season, corresponding to the physiological maturity in potatoes and the onset of low night temperature (<10 °C) for zucchini.

Fertigation schedules usually divide the portion of the total recommended rate of nutrient to be injected into varying amounts or percentages for each application over the growing season. Small amounts (low concentration) are usually injected in the early plant growth stage, larger amounts at mid-season, and then it is diminished toward the end of the season. The overall picture is similar to the dynamic of the crop fraction (Kc). Therefore, nutrient concentrations (i.e., N, P, K, and Mg) were adjusted according to the crop's age, starting from 50 mg/L and increasing to a maximum of 100 mg/L, with proportional increases in major and minor nutrients to meet the crop's growing demands.

The basic factors in fertigation procedures to be taken into account are:

a. Injection rate (injection pump capacity)
b. Amount of material to be injected
c. Time available for injection

The injection rate (a = b/c) must be calibrated using a graduated cylinder of 1000 mL as a container and a watch to control how much of the nutritive solution is injected in one minute. To calculate the mass of dry fertilizer (m) to be dissolved for the preparation of stock solution, we used the following equation.

$$M = \frac{b \times df \times v \times 100}{c} \quad (2)$$

where M is the mass of soluble fertilizer (kg).
b is the concentration of the final nutritive solution (mg L^{-1} or kg m^{-3})
df is the dilution factor (unitless)
v is the volume of the barrel to prepare the stock solution (m^3)
c is the concentration of the given fertilizer

Therefore, to prepare 100 L or 0.1 m^3 of stock solution, with a desired concentration (b) of the final solution equivalent to 100 mg L^{-1} (100 g m^{-3} or 0.1 kg m^{-3}) of nitrogen (N), at a dilution factor (df) = 100, i.e., one liter of stock solution is injected against 99 L of irrigation water, we proceeded as follows: The used fertilizer is Ammonium Sulfate (c = 21.5% NH_4-N). To calculate the needed amount of this fertilizer to be dissolved in a 100 L barrel, the calculation was done as follows:

$$M = \frac{0.1\ \text{kg m}^{-3} \times 100 \times 0.1\ \text{m}^3 \times 100\%}{21.5\%} = 4.651\ \text{kg}$$

Instead of using the relatively expensive, high-precision dosatron injectors (DOSATRON, USA), the project used the feasible venturi system (Venturi Pump, 1 inch Elysee – Cyprus) to modify the concentration of the final solution and control the ratio between nutrients. Venturi is recommended for the fertigation of large areas. This system allows for the homogeneous and regulated distribution of water and nutrients in a continuous feeding mode compared to the farmer's intermittent fertigation using the traditional closed tank (Figure 4a,b).

Water-meters (2 inch - Flange Type- Solid- China) were installed in both the SEALACOM plot and the farmer's plot to control the amount of applied water (Figure 5).

Zucchini (Cucurbita pepo) was cultivated in Serein (DS1) in an open field on a total area of 6000 m^2 starting also from August 2023. The field was subdivided into two equal parts, 3000 m^2 each. The SEALACOM project managed 3000 m^2 and the farmer managed the other 3000 m^2. The soil and irrigation water were similar in both treatments. Land preparation and time and density of sowing of the same Zucchini variety were also identical. SEALACOM used full fertigation with drip lines following the approach of continuous application of fertilizers and water. The farmer also used the drip system, but he used the traditionally followed intermittent application of nutrients.

(a) (b)

Figure 4. (**a**). Fertigation of field crops using differential and proportional injectors. (**b**). Closed tanks (local production) are used by local farmers to mix, solubilize, and apply fertilizers with irrigation.

Figure 5. Water-meter used to control water application both at the farmer's plot and project plot.

Potato (Solanum Tuberosum), variety Spunta, was grown in Sultan Yacoub (DS2) on a total area of 2000 m^2. The farmer managed the control plot of 1000 m^2, followed intermittent application of nutrients using the closed tank for fertilizer injection through macro-sprinklers. The SEALACOM Project managed the neighboring plot of 1000 m^2. The project applied continuous feeding with the same type of fertilizers using a venturi system, complemented by improved water management and application using mini-sprinklers.

A comparison of crop production, fertilizer, and water use efficiency between the project plot and traditional farmer practices was conducted to identify gaps and improve efficiency in drylands. Farmers' water application was monitored through water-meter readings and coordination meetings held face to face two to three times a week. These meetings included field visits to monitor and compare crop health conditions.

Additionally, a comparison of crop production, total yield performance, and fertilizer recovery (i.e., productivity in kg product per 1 g of applied nutrient) and water productivity (kg product per 1 m^3 of applied water) was conducted. This aimed to explore the potential for sustainable agriculture and disseminate good practices to meet the requirements of SDGs, particularly access to water and food security.

3. Statistical Analysis

Statistical analysis for potato crops was done using the descriptive statistics on Excel 2016, Window 10 (USA). The means of yields, tuber calibration, and dry matter content, randomly harvested at physiological maturity from 1 m^2 in four replicates, were calculated, followed by the confidence intervals, considering a 95% significance level. The comparison of the means could be done through the confidence ranges obtained as [mean ± confidence interval]. If two confidence ranges are overlapped, then the means were not considered different ($p < 0.05$).

4. Results

4.1. Zucchini Yield

Seedlings were planted on the same day (16 August 2023) on nested plots with similar soil conditions and the same history of land use and management. It is noteworthy that the periodically harvested and cumulative yield of zucchini on SEALACOM plots with continuous fertigation was consistently higher compared to that achieved by farmers using traditional irrigation and fertilization practices (Figure 6).

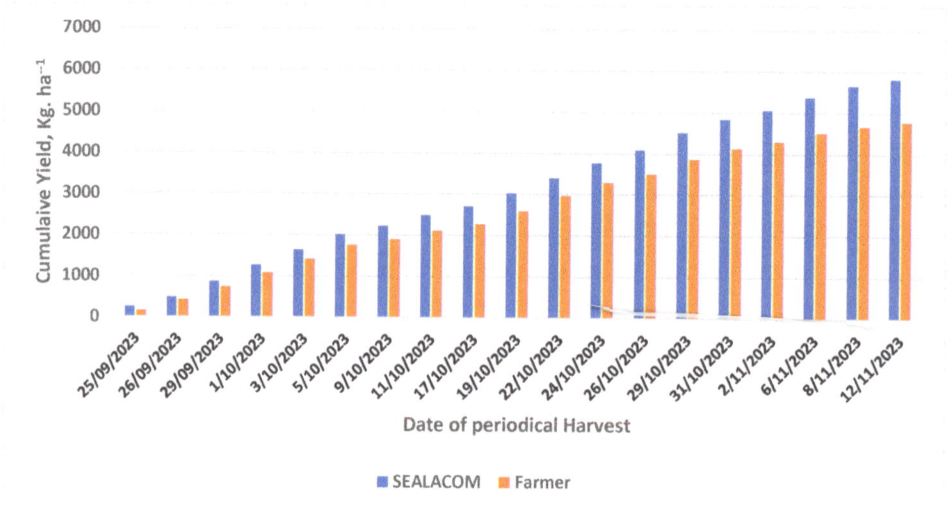

Figure 6. Comparison of cumulative yield of harvested commercial Zucchini (kg ha^{-1}) by continuous fertigation (SEALACOM) against traditional (Farmer) practices.

At the beginning of the experiment, the gap between yields in the two systems was minimal, but it increased linearly throughout the vegetation season. The total cumulative yield from 3000 m^2 in each treatment was 1729 kg for the SEALACOM approach and 1417 kg for the farmer's approach (Table 2), corresponding to a 22% difference in harvested and realized product. Since the demonstration trial was conducted directly on the farmer's land, with separate harvest on SEALACOM and the Farmer's plots undertaken by the farmer's workers the following day, the produce was weighed and sent to the local wholesale market. The farmer counted the cumulative yield in each treatment and periodically reported to us in writing. Following the principle that seeing is better than hearing, the project staff did not interfere with the harvest and realization processes. Therefore, as there were no yield replicates in the zucchini trial, statistical analysis could not be performed, and we only compared the total cumulative yield and related parameters.

Table 2. Commercial yield of zucchini harvested in traditional and advanced irrigation and fertilization practices.

Total Zucchini Yield (kg from 3000 m^2)	Farmer	SEALACOM
Sum for the 2023 season	1417	1729
Difference (kg)		+312
%		22.02

4.2. Water Application to Zucchini

The basic amount of water applied under zucchini during the preparation phase was 100 mm. Subsequently, the effectively applied water by the farmer and the SEALACOM Project amounted to 1697.8 m^3 ha^{-1} and 1294.4 m^3 ha^{-1}; respectively (Table 3). The difference in total and effective water application resulted in higher water use efficiency, which was equivalent to 1.75–2.51 kg m^{-3} for traditional farmer's practice and 2.78–4.45 kg m^{-3} for the SEALACOM practice.

Table 3. Water application and water use efficiency in zucchini by traditional and advanced practices.

Indicator	Farmer	SEALACOM	Difference
Basic application m^3 ha^{-1}	1000.3	1000.3	No
Effectively applied m^3 ha^{-1}	1697.8	1294.4	−403.40
Total applied water m^3 ha^{-1}	2698.1	2294.7	−403.40
Tuber fresh yield, kg ha^{-1}	4723.3	5763.3	+1040.00
Applied water productivity, kg m^{-3}	1.75	2.51	+0.76
Effective water productivity, kg m^{-3}	2.78	4.45	+1.67

4.3. Fertilization Policy in Traditional and Advanced Practices

The farmer applied 19.5% more nitrogen and 19.6% more phosphorus than the SEALACOM project did on its plot. However, the farmer applied 286.8% less potassium and 142% less magnesium compared to the project approach. The project's emphasis on sufficient application of K and Mg is justified by the significant contribution of K to plant resilience against climate-induced stresses [32] and crop yield [33]. Additionally, due to the local high soil pH and soil saturation with exchangeable Ca, researchers from the region highlighted the depletion of Mg from the soils and stressed the importance of Mg and the necessity of adding magnesium fertilizers [34] to maintain nutrient balance and boost soil fertility (Table 4).

Table 4. Amount of total applied nutrients under zucchini (kg Deca^{-1}) in Serein (DS1).

Input	Farmer				SEALACOM			
	N	P_2O_5	K_2O	MgO	N	P_2O_5	K_2O	MgO
Nutrients	27.52	11.39	6.05	2.01	22.15	9.16	23.40	4.87
Fertilizers	131.02	18.66	12.11	13.39	103.02	15.02	46.80	32.43

Difference in K and Mg application might have reflected on crop photosynthetic activity. Based on low nocturne temperature prevailing in highlands during the fall season, the color of plant leaves in the farmer's plot turned into yellow-brown, compared to the green color in SEALACOM treatment reflecting better plant health and fruit shape. This treatment was less affected by decreasing temperature in late October and early

November. These physiological changes were reflected in the measurements of normalized differential vegetation index (NDVI) values for zucchini (Figure 7), detected from Sentinel-2 Satellite Images throughout the season from 1 September 2023 to 11 December 2023. At the SEALACOM plot, the canopy of zucchini showed higher vegetation index values throughout the season, reflecting better plant health conditions and higher harvest potential. Monitoring crop NDVI can provide valuable, real time, information on crop water and nutrient status and allow for timely intervention to apply corrective measures.

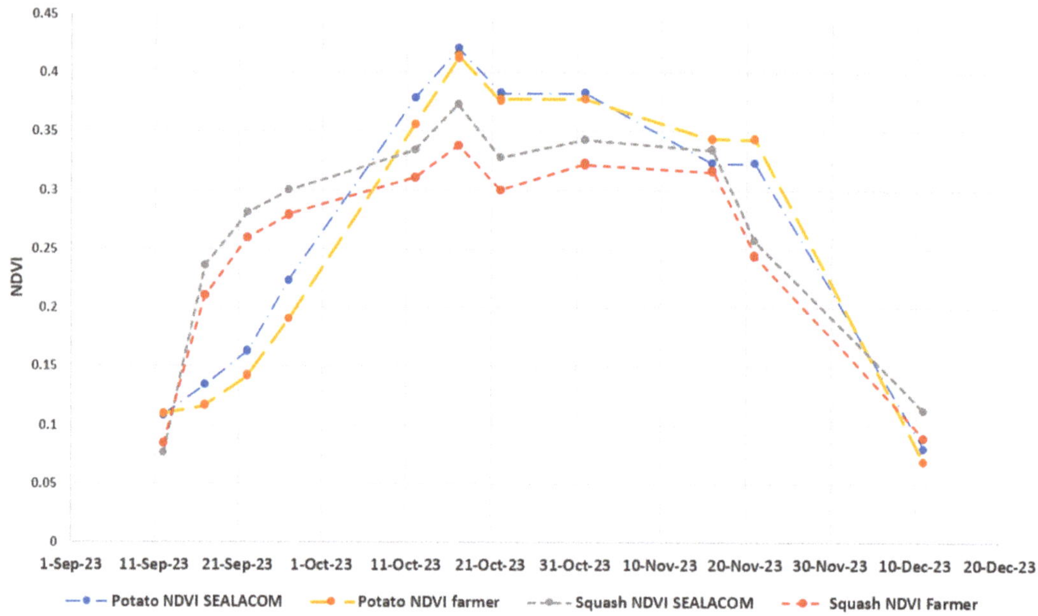

Figure 7. Comparative monitoring of crop canopy using NDVI data from Sentinel-2 during vegetation and production cycles demonstrates that continuous nutrient feeding enhances canopy tolerance to cold temperatures, while intermittent nutrient application induces stress in plants.

Improved crop resilience to low temperature was also observed in the color of the longitudinal scratches along zucchini fruits, with the SEALACOM plot showing completely green fruits compared to the whitish green color in the farmer's practice. This outcome has an important impact on the commercial value of the product.

4.4. Nutrient Use Efficiency in Zucchini Production

The SEALACOM Project achieved higher nutrient use efficiency, applying 38.43 kg of N to produce 1 ton of zucchini, compared to 58.25 kg N applied by the farmer (Table 5). This represents a 34% saving in nitrogen and phosphorus application, which can prevent salt buildup and reduce the hazards of nitrate accumulation in the soil and groundwater.

The SEALACOM approach required 27.8 kg more potassium (K_2O) and 4.19 kg more magnesium (MgO) to produce 1 ton of zucchini, corresponding to 216.8% and 98.5% higher nutrient use efficiency compared to the farmer's practice (Figure 8). However, the improved commercial quality of the product justifies the higher input of potassium and magnesium. Adequate potassium provision has been shown to boost crop resistance to water stress, particularly in drought- and temperature-sensitive crops [32].

Table 5. Ratio of applied nutrients in the traditional and advanced fertilization practice of zucchini in the two demo sites.

Ratio of Nutrients							
SEALACOM				Farmer			
N	P_2O_5	K_2O	MgO	N	P_2O_5	K_2O	MgO
1.00	0.41	1.06	0.22	1.00	0.41	0.22	0.07

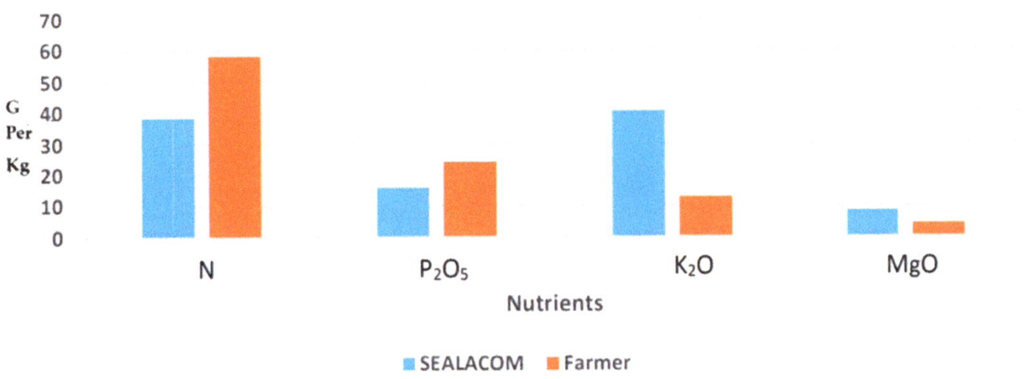

Figure 8. Amounts of macronutrient application (g) to produce 1 kg of fresh zucchini product in the West Bekaa demo site.

The alternative fertilization policy for the late summer–fall season, which favors the application of potassium and magnesium due to their contributions to the metabolism and turgor pressure in the leave's guard cells, has resulted in a different nutrient ratio compared to the standard practice. This discrepancy requires further investigation to verify the corresponding fertilization recommendations of late-season zucchini (Table 5).

Figure 8 clearly illustrates the different outputs of both practices, specifically the amount of nutrients (in grams) to produce 1 kg of fresh, healthy, zucchini. While the farmer applied 58.25 g of nitrogen to produce 1 kg of fresh zucchini, SEALACOM achieved a nitrogen saving of 9.82 g by applying only 38.48 g.

4.5. Water Application to Zucchini in the Farmer's and SEALACOM Practices

Water savings in zucchini cultivation were achieved by using data from a climatic station to estimate water demand. This approach proved to be more efficient compared to traditional irrigation practices, which relied on farmers' skills and subjective interaction with weather and soil conditions. The SEALACOM plot used 224.59 L of water to produce 1 kg of fresh zucchini, against the application of 359.46 L by the farmer. This indicates a significant water saving of 60% in the SEALACOM practice.

4.6. Potato Performance Under Traditional and Improved Practices

4.6.1. Tuber Yield

Monitoring the NDVI showed higher values for the SEALACOM potato plot from seedling until the period of maximum canopy development (Figure 7). After this stage, the potato plant starts senescing and transferring dry matter and nutrients towards the tubers. At the farmer plot, NDVI values continued to show a higher range during the tuber development stage, witnessing continued potato canopy growth triggered by excess irrigation and N application in the traditional practice. The difference in NDVI was proved by the higher potato dry matter accumulation and yield in the SEALACOM plot.

In the late potato fall season, the reduced length of daylight might have affected the low tuber production of 19,000 kg ha^{-1} using the farmer's practice compared to 22,000 kg ha^{-1} with the SEALCOM fertilization and irrigation practice (Table 6), indicating a 17.8% increase in favor of the SEALCOM practice. Despite the late sowing period (10 August), the obtained results are comparable with previous studies on summer potato crops sown in Bekaa during the first half of July, providing an average yield of 2.5 tons deca^{-1}.

Table 6. Potato tuber yield * following different fertilization and irrigation policies.

Indicator	Farmer	SEALACOM
Tuber fresh yield production, kg ha^{-1}	1900.00 b	2200.00 a

* Differences (a, b) between average values, within the same rows, are statistically significant at <0.05.

Tubers were sampled at physiological maturity from a representative area of 2.8 m^2, and the fresh tuber yield and dry matter content were measured, and results were transferred to 1 Deca (1000 m^2). Results showed no overlapping between confidence ranges, which means a significant difference exists between harvested potato yields from the SEALACOM plot and Farmer's plot (Table 7).

Table 7. Statistical analysis of tuber fresh yield collected a physiological maturity from plots fertigated continuously (SEALACOM) and plots fertigated with the traditional farmer's approach in DS1 and DS2.

Treatment	Yield kg Deca^{-1}				Mean *	SD	Confidence Interval	Confidence Range	
	R1	R2	R3	R4				Min	Max
SEALACOM	2415.10	2376.30	2293.40	2379.80	2366.10 a	±51.58	35.74	2330.39	2401.88
Farmers	2157.50	2065.40	1982.90	2017.10	2055.70 b	±75.82	56.17	1999.50	2111.90

* Differences (a, b) between average values, within the same rows, are statistically significant at <0.05.

4.6.2. Tuber Calibration

The calibration of tuber size, harvested in four replicates, at full maturity, from 2 m length (equivalent to 2.8 m^2), corresponding to an average of 5748.5 g and 6625.2 g from the farmer's and SEALACOM plots, respectively, is presented in Table 8. The statistical analysis of calibrated tubers into small diameter size (<1.5 cm), medium size (1.5–3.5 cm), large size (3.5–5.0 cm) and elite size (5.5–7.5 cm), showed a significant difference with the advantage of the SEALACOM approach to produce a larger quantity of elite size potato tubers (Table 8). It can be concluded that the yield of elite-size tubers was 40% higher in the SEALACOM plot.

Table 8. Statistical analysis of tuber calibration collected at full maturity stage from advanced fertigation (i.e., SEALACOM) and framer's practice.

Tuber Size *, cm	SEALACOM					Farmer				
	Average	SD ±	Confidence Interval	Confidence Range		Average	SD ±	Confidence Interval	Confidence Range	
				Min	Max				Min	Max
<1.5	63.70 b	1.28	1.25	62.45	64.95	284.00 a	15.10	14.80	269.20	298.80
1.5–3.5	1403.30 a	15.29	14.98	1388.32	1418.28	1148.60 b	57.28	56.15	1092.40	1204.70
3.5–5.5	2169.00 a	66.35	65.02	2103.98	2234.02	2182.90 a	100.98	98.98	2083.90	2281.90
5.5–7.5	2989.20 a	139.50	136.71	2852.49	3125.91	2133.00 b	118.49	116.13	2016.90	2249.10

* Differences (a, b) between average values of tuber size, within the same rows, are statistically significant at <0.05.

4.6.3. Tuber Dry Matter Content

Since dry matter (DM) content in potato tubers is one of the main criteria for the cooked product and culinary preference, we tested the tuber dry matter content and compared the two practices based on four replicates of subsamples dried in the oven at 72 °C over three days. The statistical analyses of DM content showed an advantage in the SEALACOM approach, providing significantly higher DM content (Table 9).

Table 9. Comparative dry matter content of potato tubers produced by continuous fertigation and traditional practices.

SEALACOM					Farmer				
Average *	SD ±	Confidence Interval	Confidence Range		Average	SD ±	Confidence Interval	Confidence Range	
			Min	Max				Min	Max
20.350 a	0.551	0.540	19.810	20.890	18.400 b	0.540	0.524	17.880	18.920

* Differences (a, b) between average values within the same rows are statistically significant at <0.05.

4.6.4. Water Application Under Potato in Farmer's and SEALACOM Practices

The initial water application for land preparation in both treatments was substantial (357 mm). This considerable amount was necessary due to the extreme soil dryness resulting from a long, dry summer following the early June wheat harvest, which typically depletes soil moisture to a depth of over 2 m. Therefore, the amount of effectively applied water in the SEALACOM approach irrigated by mini-sprinklers was 111.2 m^3 versus 521.33 m^3 in the farmer's practice with irrigation using macro sprinklers (Table 10). Therefore, the total applied water was 468.2 m^3 and 878.3 m^3 in both practices, respectively.

Table 10. Water application and water use efficiency in potato by traditional and advanced practices.

Indicator	Farmer	SEALACOM
Basic water application, m^3 ha^{-1}	3570.00	3570.00
Effectively applied water *, m^3 ha^{-1}	5210.33	1110.20
Total applied water, m^3 ha^{-1}	8780.33	4680.20
Applied water productivity fresh tubers, kg m^{-3}	2.16	4.70
Effectively applied water productivity DM ** tubers, kg m^{-3}	0.51	1.15
Increase in applied water productivity (kg DM per unit applied water)	1.15 − 0.51 = 0.64	
% increase in applied water productivity	(0.642 × 100)/0.513 = 125.10	
Effective water productivity Fresh Tubers, kg m^{-3}	3.64	19.78
Effective water productivity DM Tubers, kg m^{-3}	0.86 ***	4.86
Increase in effective water productivity (kg DM per unit applied water)	4.86−0.86 = 4.00	
% increase in effective water productivity	(4.00 × 100)/0.86 = 465.12	

* Since the start of fertigation and differential water application; ** DM content in potato tubers in SEALACOM = 24.55%, in Farmer's plot = 23.76%; *** The base line for summer potato is 0.95 kg DM m^{-3}. In this trial, the sowing of potato was late (15 August 2023). Autumn potato season is very short with low night T and provides 30–50% from the potential harvest of the normal summer season.

Compared to the farmer's practice, this outcome, in the SEALACOM approach, allowed for more than 50% better applied water productivity and more than doubled the productivity of effectively applied water. Our results show that to produce 1 kg of fresh potato, the farmer effectively applied 274.38 L of water against 50.61 L of water applied by the SEALACOM practice.

4.6.5. Nutrient Application Under Potato in the Farmer's and SEALACOM Practices

Nutrient application under potatoes showed a similar trend like in zucchini, highlighting a pattern in traditional Lebanese farming practices. There is an overestimation of nitrogen and phosphorous inputs, while the importance of potassium and magnesium in the management of soil fertility and crop production is underestimated (Table 11).

Table 11. Amount of applied nutrients under potato by advanced fertigation (i.e., SEALACOM) and framer's practice (kg ha^{-1}).

Type of Nutrients	Farmer				SEALACOM			
	N	P_2O_5	K_2O	MgO	N	P_2O_5	K_2O	MgO
Pure elements	250.8	42.7	93.0	3.0	89.4	29.5	118.8	6.4
Total Fertilizer equivalent	1166.5	70.0	186.0	18.7	415.6	48.3	237.6	40.1

Results of the experiment show the application of 250 kg ha^{-1} and 42.7 kg ha^{-1} versus 89.36 kg ha^{-1} and 29.5 kg ha^{-1} of nitrogen and phosphorous in the farmer's practice and SEALACOM approach, respectively. Due to high potato demands in K and Mg, SEALACOM applied 30% and 50% more potassium and magnesium.

4.6.6. Nutrient Use Efficiency in Farmer's and SEALACOM Practices of Late Season Potato Production

To produce 1 kg of fresh potato tubers, SEALACOM applied 4.06 g of nitrogen and 1.34 g of phosphorous, whereas the farmer applied 13.2 g and 2.25 g; respectively (Figure 9). Conversely, the SEALACOM practice applied 10.4% and 81.2% more potassium and magnesium, which resulted in the dominance of large tubers in the harvest using the project's fertilization method.

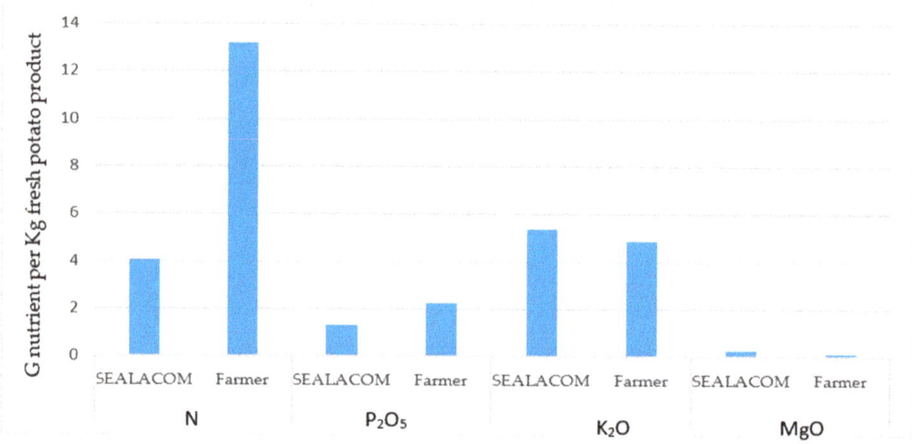

Figure 9. Amounts of macronutrients applied to produce 1 kg fresh potato tubers in traditional and advanced practices.

The fertilization policy of late summer potato followed by the SEALACOM Project and the farmer shows different understandings of the role and importance of each element. This discrepancy indicates poor nutrient formulation in the traditional practice (Table 12). While potassium should typically exceed nitrogen, as commonly recommended, the nitrogen level in the farmer's potato production practice was three times higher than the applied potassium level.

Table 12. Ratio of applied nutrients in the traditional and advanced fertilization practice under potato.

Ratio of Applied Nutrients							
Farmer's Practice				SEALACOM Practice			
N	P_2O_5	K_2O	MgO	N	P_2O_5	K_2O	MgO
1.00	0.17	0.37	0.01	1.00	0.33	1.33	0.07

5. Discussion

SEALACOM practice was more efficient in N and P utilization by the potato crop (Figure 9). Similarly, squash grown in Jordan demonstrated higher yields with lower N rates and better water and fertilizer utilization through full fertigation compared to the traditional practices. This approach can minimize production costs and reduce nitrate pollution in the soil-groundwater system [35]. Case studies from the region have reported nitrate accumulation and leaching, leading to significant groundwater contamination with nitrates [34–36].

Given the deficiency of organic and mineral nitrogen in the soils of the area [21,32], it is recommended to develop a farming practice with an adequate dose of nitrogen for successful crop production. This policy must take into account the soil and climate conditions, plant response to nutrients, and crop tolerance to drought to improve carbon and nitrogen metabolism and achieve production goals and water use efficiency [37–39].

The approach used to define crop water requirements has limitations, as the tested crop coefficients (Kc) for some field crops might not align with the FAO-56 Kc recommended values when determined for the same crops in different locations [40]. Therefore, to properly regulate nutrient ratios and amounts, it is advisable to determine the crop coefficient in situ using both climatic stations and lysimeters to coordinate the lower and higher ranges of Kc values. This method allows for more precise water and nutrient application by matching the Kc range with crop genotypes and specific water demands, as well as local soil and climatic conditions. Consequently, it enables the modification and control of nutrient concentrations through continuous fertigation.

Using the same genotypes of potato and zucchini, cultivated simultaneously in the same locations under similar soil and climate conditions, reduced variability due to the smoothing effect of systematic error. Therefore, it is crucial to upgrade environmental performance, sustainable agricultural production, and precision farming in Lebanon at both farm and national levels. Capacity building will enhance the requirement for mass, energy, and human labor inputs that support food security through sustainable production and environmental practices [41]. This aligns with the adoption of a multi-criteria approach to sustainable irrigated agriculture and crop endurance, addressing the challenges of food security and water scarcity in arid environments [42].

In China, moderate and precise water and nitrogen application within the targeted root zone of sweet potatoes, considering soil and water pools, improved tuber yield and quality and enhanced water and nitrogen use efficiency [43]. Three categories of farmers were identified in the Mediterranean: proactive, skeptical, and reluctant farmers, with a significant portion showing negative attitudes toward using digital technologies in agri-farming and environmental policies [44]. However, adopting feasible tools and simple techniques that demonstrate both economic and environmental resilience can lead to a breakthrough.

Given the significant average global increase in ET over the last three decades, at an annual rate of 1.33 (±0.84) mm per year, largely due to an average summer ET increase of 2.06%, the adoption of sustainable irrigation strategies is a prime concern [45]. Advanced soil and crop management, along with improved fertilization and irrigation practices in semi-arid zones, are among the key alternative farming options at all decision-making levels. These practices aim to improve carbon and nitrogen balance, enhance soil fertility, and boost crop yields [46].

Since 1970, a significant decrease in groundwater levels, with a drop between 20 and 25 m in the major aquifers of Lebanon, has been attributed to chaotic drilling of water wells and over exploitation [47]. Observations for Lebanon based on land subsidence and space information showed a total water storage decline in the Bekaa plain at a rate of 1.10 cm per year, caused by groundwater depletion [19].

Evidence of uncontrolled expansion of illegal wells and excess pumping calls for immediate interventions and the activation of water governance to improve water accounting and water productivity in Lebanon and the Eastern Mediterranean. A good example of groundwater governance comes from Jordan, where acts of violence and mismanaged irrigation practices from private wells, along with incorrect metering of pumped water and inefficient water conveyance, were detected and corrected to promote the wise use of limited water resources [10].

Adaptive responsive measures, such as the development and efficient use of water resources, were recommended in Jordan to reduce the impacts of drought and increased public demands on water resources and food security [48]. In the face of water scarcity and recurrent drought events in semi-arid Mediterranean conditions, implementing sustainable agricultural practices for more efficient use of limited water and soil resources becomes a national and regional priority. These practices aim to enhance nitrogen and water use efficiency and address the challenges of food security [49].

Quantifying evapotranspiration (ET) is crucial for a valid understanding of the global water cycle and precise resource management. However, accurately estimating ET, especially at local scales, has always been challenging [50]. Substantial progress has been made in estimating regional crop water demands using low-resolution platforms such as the FAO Water Productivity Portal-WAPOR [51], and AgSAT [52]. These platforms provide real-time and lagged information to support decision-making on regional water use. However, unless these remote sensing platforms are applicable at local scales and integrated into a land-based smart climate and agricultural management system, achieving sustainable irrigated agriculture remains challenging for medium and small farmers.

The results obtained in the SEALACOM approach using mini sprinklers showed trends similar to those of earlier studies on potato performance and water use efficiency conducted in the country using a drip system [21,53,54]. Our results indicated better performance of the Spunta variety tested in Lebanon compared to its performance in the Jordanian desert [55]. However, full fertigation of potatoes ensured better nutrient distribution within the root zone on the fine textured Mediterranean soils of Jordan and prevented rapid immobilization of applied nitrogen by soil microorganisms [55].

Similarly, with the drip-irrigated squash in Syria [56], our results consistently showed higher water productivity, i.e., better biomass production per unit of applied water. This was explained by the higher irrigation frequency and the reduced amount of water applied per each irrigation event.

Adopting a practical, field-validated, simple, low-cost, and efficient continuous fertigation system that considers crop water and nutrient requirements and soil conditions and integrates weather-based irrigation with crop performance under dry Mediterranean climate is crucial. This approach can significantly maintain high yields and crop water productivity, which are essential for food security in water-scarce regions [57]. Structural models that combine precision irrigation with slow-release nitrogen fertilizers and modern irrigation systems have significantly improved crop yield, N use efficiency, and water productivity [58].

Compared to traditional farmer irrigation practices, the use of good agricultural practices with strong deficit irrigation mitigated water stress through expanded root density and better water interception [59]. However, with the complex factors affecting water saving, crop performance, and water productivity in dry regions [60,61], local soil and climatic conditions as well as farmers' attitudes and skills must always be considered.

6. Conclusions

The SEALACOM Project's advanced methodology for assessing crop water demands and applying metered irrigation achieved increased yields and higher water use efficiency. This is crucial in an era of increasing drought incidences, which exacerbate water scarcity and put pressure on limited water resources. Compared to traditional practices, the improved SEALACOM fertigation method resulted in over a 22% increase in higher-quality zucchini yield, from 1417 kg ha^{-1} to 1729 kg ha^{-1}, more than 23% water savings, and over 43% higher water productivity. Farmers expressed their intention to follow the SEALACOM approaches and adopt continuous feeding instead of intermittent fertilization, using a modern injector rather than a closed tank. Local farmers were trained in advanced fertilization and irrigation techniques.

In continuous feeding, higher yields and better quality zucchini and potatoes were obtained with doubled nitrogen and phosphorous use efficiency. For potato production, the improved practice using mini-sprinklers and continuous feeding resulted in a 17.8% yield increase, from 1900 kg ha^{-1} to 2200 kg ha^{-1}, and a 40% increase in large tuber size ($p < 0.05$). The project methodology provided a better nutrient ratio formulation, aligning with soil conditions and crop demands to major nutrients. The efficient use of water and nutrients will reduce energy consumption for water pumping and application. Continuous feeding's water and fertilizer savings have significant economic and environmental impacts, particularly in drylands.

Demonstrating water accounting and productivity based on totally and effectively applied water can help disseminate good, proactive practices, enhancing food production with a lower environmental footprint, higher sustainability, reduced production costs, and better commercial value of the final product. The findings from the SEALACOM Project are becoming part of local agricultural and water policies. Turning these results into an action plan requires strong commitment, decision-making, and capacity building to conserve and properly manage water resources, with special emphasis on groundwater, to enhance crop production. This is particularly important for the Bekaa Plain, where agriculture is the major source of income.

Author Contributions: Conceptualization, T.D.; Methodology, I.J. and R.K.; Validation, P.M.; Investigation, T.D. and A.S.; Resources, G.F. and P.M.; Writing—original draft, T.D.; Writing—review and editing, T.D. and R.K.; Visualization, A.S. and I.J.; Supervision, G.F. and R.K.; Project administration, A.S. and G.F.; Funding acquisition, R.K. All authors have read and agreed to the published version of the manuscript.

Funding: This research was conducted as part of the SEALACOM project, funded by the Italian Ministry of Foreign Affairs and International Cooperation and the Italian Agency for Development Cooperation. We extend our deepest gratitude to the farmers of the Serein and Sultan Yacoub cooperatives in Lebanon for generously providing their fields for experimental trials and for their cooperation in following the project team's instructions. We also wish to acknowledge Professor Therese Atallah for her invaluable support with the statistical analysis.

Institutional Review Board Statement: Not applicable.

Informed Consent Statement: Not applicable.

Data Availability Statement: Data are contained within the article.

Conflicts of Interest: Authors Talal Darwish, Amin Shaban and Ghaleb Faour were employed by the company National Council for Scientific Research. The remaining authors declare that the research was conducted in the absence of any commercial or financial relationships that could be construed as a potential conflict of interest.

References

1. MoE; UNDP. Third National Communication to the UNFCCC. 2016. Available online: https://unfccc.int/sites/default/files/resource/lbnnc3.pdf (accessed on 13 January 2025).
2. WB. Lebanon Country Climate and Development Report 2024. Available online: https://documents1.worldbank.org/curated/en/099030624151542784/pdf/P179617111e46d05c187b91daa689f7b351.pdf (accessed on 13 January 2025).
3. MoE; UNDP; GEF. Economic Costs to Lebanon from Climate Change: A First Look; 2015, Beirut, Lebanon. Available online: https://www.unescwa.org/sites/default/files/event/materials/09_aboujaoude_economic_cost_of_cc_on_lebanon.pdf (accessed on 13 January 2025).
4. Garg, N.K.; Dadhich, S.M. A proposed method to determine yield response factors of different crops under deficit irrigation using inverse formulation approach. *Agric. Water Manag.* **2014**, *137*, 68–74. [CrossRef]
5. Wang, S.; Sun, N.; Zhang, X.; Hu, C.; Wang, Y.; Xiong, W.; Zhang, S.; Colinet, G.; Xu, M.; Wu, L. Assessing the impacts of climate change on crop yields, soil organic carbon sequestration and N2O emissions in wheat–maize rotation systems. *Soil Tillage Res.* **2024**, *240*, 106088. [CrossRef]
6. Abdelmoneim, A.A.; Khadra, R.; Derardja, B.; Dragonetti, G. Internet of Things (IoT) for Soil Moisture Tensiometer Automation. *Micromachines* **2023**, *14*, 263. [CrossRef]
7. Wen, S.; Cui, N.; Wang, Y.; Gong, D.; Xing, L.; Wu, Z.; Zhang, Y.; Wang, Z. Deficit irrigation enhances yield and water productivity of apples by inhibiting excessive vegetative growth and improving photosynthetic performance. *Agric. Water Manag.* **2025**, *307*, 109220. [CrossRef]
8. MoE/UNDP/GEF. *Lebanon's First Biennial Transparency Report (BTR) on Climate Change, Beirut, Lebanon*. 2024. Available online: https://www.undp.org/lebanon/publications/lebanons-biennial-transparency-report-climate-change (accessed on 13 January 2025).
9. CEDRO. Solar-Driven Water Pumping: An Untapped Resource for Lebanon. Project Funded by the EU. *Exchange Issue 15*. 8 pages. 2015. Available online: https://www.cedro-undp.org/Library/Assets/Gallery/PDFfiles/Solar%20Water%20Pumping.pdf (accessed on 13 January 2025).
10. Al-Bakri, J.T.; Shawash, S.; Ghanim, A.; Abdelkhaleq, R. Geospatial Techniques for Improved Water Management in Jordan. *Water* **2016**, *8*, 132. [CrossRef]
11. Molle, F.; Nassif, M.H.; Jaber, B.; Closas, A.; Baydoun, S. Groundwater Governance in Lebanon. The case of Central Beqaa, *A Policy White Paper*; IWMI, USAID. 2017. Available online: https://issuu.com/bauwebsiteoffice/docs/groundwater-governance-in-lebanon-t/1 (accessed on 13 January 2025).
12. UNDP. *Groundwater Assessment and Database Project*; Final Output; UNDP: New York, NY, USA, 2014.
13. UNDP. *Climate-Proofing Lebanon's Development Plans*; UNDP: Beirut, Lebanon, 2021.
14. FAO. *Lebanon National Agriculture Strategy 2020–2025 (NAS)*; FAO: Rome, Italy, 2023. Available online: https://www.fao.org/faolex/results/details/en/c/LEXFAOC202167/ (accessed on 13 January 2025).
15. El Chami, D.; El Moujabber, M. Sustainable Agriculture and Climate Resilience. *Sustainability* **2024**, *16*, 113. [CrossRef]
16. Dhanaraju, M.; Chenniappan, P.; Ramalingam, K.; Pazhanivelan, S.; Kaliaperumal, R. Smart Farming: Internet of Things (IoT)-Based Sustainable Agriculture. *Agriculture* **2022**, *12*, 1745. [CrossRef]
17. Carbonell-Bojollo, R.M.; Veroz-González, Ó.; González Sánchez, E.J.; Ordóñez-Fernández, R.; Moreno-García, M.; Repullo Ruibérriz de Torres, M.A. Soil Management, Irrigation and Fertilisation Strategies for N_2O Emissions Mitigation in Mediterranean Agricultural Systems. *Agronomy* **2022**, *12*, 1349. [CrossRef]
18. MoEW. *National Water Sector Strategy Report*; Ministry of Energy and Water: Beirut, Lebanon, 2010. Available online: https://faolex.fao.org/docs/pdf/leb166572E.pdf (accessed on 13 January 2025).
19. Massoud, E.C.; Liu, Z.; Shaban, A.; El Hage, M. Groundwater Depletion Signals in the Beqaa Plain, Lebanon: Evidence from GRACE and Sentinel-1 Data. *Remote Sens.* **2021**, *13*, 915. [CrossRef]
20. Gul, N.; Mangrio, M.A.; Shaikh, I.A.; Siyal, A.G.; Semiromi, M.T. Quantifying the impacts of varying groundwater table depths on cotton evapotranspiration, yield, water use efficiency, and root zone salinity using lysimeters. *Agric. Water Manag.* **2024**, *301*, 108933. [CrossRef]

21. Darwish, T.; Atallah, T.; Hajhasan, S.; Haidar, A. Nitrogen and water use efficiency of fertigated processing potato. *Agric. Water Manag.* **2006**, *85*, 95–104. [CrossRef]
22. Shi, J.; Jiang, L.; Wang, L.; Wang, C.; Li, R.; Pan, L.; Jia, T.; Hou, S.; Jia, Z. Effects of Seven-Year-Optimized Irrigation and Nitrogen Management on Dynamics of Soil Organic Nitrogen Fractions, Soil Properties, and Crop Growth in Greenhouse Production. *Agriculture* **2024**, *14*, 2319. [CrossRef]
23. Karam, F.; Breidy, J.; Stephan, C.; Rouphael, J. Evapotranspiration, yield and water use efficiency of drip irrigated corn in the Bekaa Valley of Lebanon. *Agric. Water Manag.* **2003**, *63*, 125–137. [CrossRef]
24. Karaa, K.; Karam, F.; Tarabey, N. Participatory water saving management and water cultural heritage: Lebanon country report. *Options Médit.* **2004**, *48*, 185–198.
25. Cano, A.; Pardo, J.J.; Montero, J.; Domínguez, A. Determining Irrigation Requirements of Extensive Crops Using the Typical Meteorological Year Adjusted to the Growing Cycle Period. *Agronomy* **2022**, *12*, 2208. [CrossRef]
26. Whittemore, D.O.; Butler, J.J.; Bohling, G.C.; Wilson, B.B. Are we saving water? Simple methods for assessing the effectiveness of groundwater conservation measures. *Agric. Water Manag.* **2023**, *287*, 108408. [CrossRef]
27. Niu, Y.; Zhang, K.; Khan, K.S.; Fudjoe, S.K.; Li, L.; Wang, L.; Luo, Z. Deficit Irrigation as an Effective Way to Increase Potato Water Use Efficiency in Northern China: A Meta-Analysis. *Agronomy* **2024**, *14*, 1533. [CrossRef]
28. Darwish, T.; Shaban, A.; Faour, G.; Jomaa, I.; Moubarak, P.; Khadra, R. Transforming Irrigated Agriculture in Semi-Arid and Dry Subhumid Mediterranean Conditions. A case of Protected Cucumber cultivation. *Sustainability* **2024**, *16*, 10050. [CrossRef]
29. IUSS Working Group WRB. *World Reference Base for Soil Resources 2006*; World Soil Resources Reports No. 103; FAO: Rome, Italy, 2007.
30. Jensen, M.E. Water consumption by agricultural plants. In *Water Deficits and Plant Growth*; Kozlowski, T.T., Ed.; Academic Press, Inc.: New York, NY, USA, 1968; Volume II, pp. 1–22.
31. FAO. *Crop Evapotranspiration—Guidelines for Computing Crop Water Requirements—FAO Irrigation and Drainage Paper 56*; FAO: Rome, Italy, 1998. Available online: https://www.avwatermaster.org/filingdocs/195/70653/172618e_5xAGWAx8.pdf (accessed on 13 January 2025).
32. Darwish, T.; Fadel, A.; Baydoun, S.; Jomaa, I.; Awad, M.; Hammoud, Z.; Halablab, O.; Atallah, T. *Potato Performance Under Different Potassium Levels and Deficit Irrigation in Dry Sub-Humid Mediterranean Conditions*; e-ifc N 43; International Potash Institute: Zug, Switzerland, 2015; pp. 14–20.
33. Wang, Z.; Hassan, M.U.; Nadeem, F.; Wu, L.; Zhang, F.; Li, X. Magnesium Fertilization Improves Crop Yield in Most Production Systems: A Meta-Analysis. *Front. Plant Sci.* **2020**, *10*, 1727. [CrossRef]
34. El-Fouly, M.M.; Rezk, A.I.; Nofal, O.A.; Abou El-Nour, E.A.A. Depletion of magnesium in Egyptian soils, its content in crops and estimated needs. *Afr. J. Agric. Res.* **2010**, *5*, 1060–1067. [CrossRef]
35. Mohammad, M.J. Squash yield, nutrient content and soil fertility parameters in response to methods of fertilizer application and rates of nitrogen fertigation. *Nutr. Cycl. Agroecosyst.* **2004**, *68*, 99–108. [CrossRef]
36. Moeller, A.; Altfelder, S.; Moeller, H.W.; Darwish, T.; Abdelgawad, G. *A Guide to Sustainable Nitrogen Management in Agricultural Practices*; ACSAD: Yafour, Syria; BGR: Hannover, Germany; CNRS: Beirut, Lebanon, 2023; Volume 8, 90p. [CrossRef]
37. Olšovská, K.; Sytar, O.; Kovácik, P. Optimizing Nitrogen Application for Enhanced Barley Resilience: A Comprehensive Study on Drought Stress and Nitrogen Supply for Sustainable Agriculture. *Sustainability* **2024**, *16*, 2016. [CrossRef]
38. Sun, J.; Li, W.; Li, C.; Chang, W.; Zhang, S.; Zeng, Y.; Zeng, C.; Peng, M. Effect of Different Rates of Nitrogen Fertilization on Crop Yield, Soil Properties and Leaf Physiological Attributes in Banana Under Subtropical Regions of China. *Front. Plant Sci.* **2020**, *11*, 613760. [CrossRef]
39. Su, W.; Ahmad, S.; Ahmad, I.; Han, Q. Nitrogen fertilization affects maize grain yield through regulating nitrogen uptake, radiation and water use efficiency, photosynthesis and root distribution. *Peer J.* **2020**, *8*, e10291. [CrossRef] [PubMed]
40. Piccinnia, G.; Kob, J.; Marekc, T.; Howelld, T. Determination of growth-stage-specific crop coefficients (Kc) of maize and sorghum. *Agric. Water Manag.* **2009**, *96*, 1698–1704. [CrossRef]
41. Skaf, L.; Buonocore, E.; Dumontet, S.; Capone, R.; Franzese, P.P. Food security and sustainable agriculture in Lebanon: An environmental accounting framework. *J. Clean. Prod.* **2019**, *209*, 1025–1032. [CrossRef]
42. Aldababseh, A.; Temimi, M.; Maghelal, P.; Branch, O.; Wulfmeyer, V. Multi-Criteria Evaluation of Irrigated Agriculture Suitability to Achieve Food Security in an Arid Environment. *Sustainability* **2018**, *10*, 803. [CrossRef]
43. Zhang, Y.; Tanga, Y.; Wanga, Z.; Fenga, S.; Wang, F.; Hua, Y. Optimizing nitrogen and irrigation application for drip irrigated sweet potato with plastic film mulching in eastern China. *Agric. Water Manag.* **2024**, *302*, 108997. [CrossRef]
44. Granado-Díaz, R.; Colombo, S.; Romero-Varo, M.; Villanueva, A.J. Farmers' attitudes toward the use of digital technologies in the context of agri-environmental policies. *Agric. Syst.* **2024**, *221*, 104129. [CrossRef]
45. Jaafar, H.H.; Sujud, L.H. High-resolution satellite imagery reveals a recent accelerating rate of increase in land evapotranspiration. *Remote Sens. Environ.* **2024**, *315*, 114489. [CrossRef]

46. Calone, R.; Fiore, A.; Pellis, G.; Cayuela, M.L.; Mongiano, G.; Lagomarsino, A.; Bregaglio, S. A harmonized dataset relating alternative farmer management practices to crop yield, soil organic carbon stock, nitrous oxide emissions, and nitrate leaching generated using IPCC methodologies and meta-analyses. *Data Brief* **2025**, *58*, 111226. [CrossRef] [PubMed]
47. Shaban, A. *Water Resources of Lebanon*; Springer Science Publisher: Cham, Switzerland, 2020; 229p, ISBN 978-3-030-48716-4.
48. Al-Bakri, J.T.; Salahat, M.; Suleiman, A.; Suifan, M.; Hamdan, M.R.; Khresat, S.; Kandakji, T. Impact of climate and land use changes on water and food security in Jordan: Implications for transcending "The Tragedy of the Commons". *Sustainability* **2013**, *5*, 724–748. [CrossRef]
49. Franco-Luesma, S.; Cavero, J.; Álvaro-Fuentes, J. Relevance of the irrigation and soil management system to optimize maize crop production under semiarid Mediterranean conditions. *Agric. Water Manag.* **2025**, *307*, 109272. [CrossRef]
50. Derardja, B.; Khadra, R.; Abdelmoneim, A.A.; El-Shirbeny, M.A.; Valsamidis, T.; De Pasquale, V.; Deflorio, A.M.; Volden, E. Advancements in Remote Sensing for Evapotranspiration Estimation: A Comprehensive Review of Temperature-Based Models. *Remote Sens.* **2024**, *16*, 1927. [CrossRef]
51. FAO. WaPOR Version 3 Data. 2017. Available online: https://www.fao.org/land-water/news-archive/news-detail/en/c/1678391 (accessed on 13 January 2025).
52. AgSAT. Use of Satellite Imagery and Weather Forecasts to Calculate Crop Water Requirements for Irrigation Management Purposes. 2024. Available online: https://agsat.app/ (accessed on 13 January 2025).
53. Karam, F.; Massaad, R.; Skaf, S.; Breidy, J.; Rouphael, Y. Potato response to potassium application rates and timing under semi-arid conditions. *Adv. Hort. Sci.* **2013**, *25*, 265–268. [CrossRef]
54. Karam, F.; Amacha, N.; Fahed, S.; EL Asmar, T.; Domínguez, A. Response of potato to full and deficit irrigation under semiarid climate: Agronomic and economic implications. *Agric. Water Manag.* **2014**, *142*, 144–151. [CrossRef]
55. Donnelly, D.; Leclerc, Y.; Sawwan, J.S.; Hadidi, N.A. Performance and Yield of Potato Grown in the Jordanian Desert. In Proceedings of the 94th Annual International Conference of the American Society for Horticultural Science, Salt Lake City, UT, USA, 23–26 July 1997; Abstracts of Contributed Papers. HortScience: Alexandria, VA, USA, 1997; pp. 475–476.
56. Mohammad, M.J.; Zuraiqi, S.; Quasmeh, W.; Papadopoulos, I. Yield response and nitrogen utilization efficiency by drip irrigated potato. *Nutr. Cycl. Agroecosyst.* **1999**, *54*, 243–249. [CrossRef]
57. Darouich, H.; Karfoul, R.; Eid, H.; Ramos, T.B.; Baddour, N.; Moustafa, A.; Assaad, M.I. Modeling Zucchini squash irrigation requirements in the Syrian Akkar region using the FAO56 dual-Kc approach. *Agric. Water Manag.* **2020**, *229*, 105927. [CrossRef]
58. Abdelmoneim, A.A.; Khadra, R.; Elkamouh, A.; Derardja, B.; Dragonetti, G. Towards Affordable Precision Irrigation: An Experimental Comparison of Weather-Based and Soil Water Potential-Based Irrigation Using Low-Cost IoT-Tensiometers on Drip Irrigated Lettuce. *Sustainability* **2024**, *16*, 306. [CrossRef]
59. Yang, S.; Wu, Q.; Chi, D.; Chen, H.; Zhu, S.; Liu, Q. Water–saving irrigation combined with N–loaded clinoptilolite enhances nutrient yield, and water productivity by improving rice root characteristics: A combined PCA–SEM analysis. *Agric. Water Manag.* **2025**, *307*, 109203. [CrossRef]
60. Zapata-García, S.; Temnani, A.; Berríos, P.; Marín-Durán, L.; Espinosa, P.J.; Monllor, C.; Pérez-Pastor, A. Combined Effects of Deficit Irrigation and Biostimulation on Water Productivity in Table Grapes. *Plants* **2024**, *13*, 3424. [CrossRef] [PubMed]
61. Kundu, A.; Waris, S.; Sanam, S.; Aslam, H.; Khan, B.; Bogati, S.K.; Rizwan, M.; Jamshed, M.; Majeed, S.M.; Ashfaq, M. Precision Irrigation Systems for Sustainable Water Management in Maize Cultivation: Impact on Yield and Water Use Efficiency. *Ind. J. Biosci. Res.* **2024**, *3*, 85–94. [CrossRef]

sustainability

MDPI

Article

Soil Organic Carbon Modelling with Different Input Variables: The Case of the Western Lowlands of Eritrea

Tumuzghi Tesfay [1,2,*], Elsayed Said Mohamed [1,3], Igor Yu. Savin [1,4], Dmitry E. Kucher [1], Nazih Y. Rebouh [1] and Woldeselassie Ogbazghi [2]

1 Department of Environmental Management, Institute of Environmental Engineering, RUDN University, 6 Miklukho-Maklaya St., 117198 Moscow, Russia
2 Department of Land Resources and Environment, NHERI, Hamelmalo Agricultural College, Keren P.O. Box 397, Eritrea
3 National Authority for Remote Sensing and Space Sciences, Cairo 1564, Egypt
4 V.V. Dokuchaev Soil Science Institute, Pyzhevsky per. 7, Building 2, 119017 Moscow, Russia
* Correspondence: tumuzghitesfay7@mail.com

Abstract: In Eritrea, efforts are being made to tackle the widespread land degradation and promote natural resources and the agricultural sector. However, these efforts lack digital resources assessment, mapping, planning and monitoring. Thus, we developed soil organic carbon (SOC) predictor models for the Western Lowlands of the country, employing 6 machine learning models with different input variables (36, 27, 15, and 08) obtained following these variables selection strategies: (1) all proposed SOC predictor variables; (2) very high multicollinearity (≥0.900 **) reduction; (3) high multicollinearity (≥0.700 **) reduction; (4) the Boruta feature selection algorithm. The results revealed that SOC levels were generally low (mean = 0.43%). Grazing lands, rainfed croplands, and irrigated farmlands all exhibited similarly low SOC values, attributed to unsustainable land management practices that deplete soil nutrients. In contrast, natural forestlands exhibited significantly higher SOC concentrations, highlighting their potential for soil carbon sequestration. Among the tested models, the XGBoost algorithm using 27 covariates achieved the highest predictive performance (RMSE = 0.118, R^2 = 0.758, RPD = 2.252), whereas the multiple linear regression (MLR) model with 8 variables yielded the lowest performance (RMSE = 0.141, R^2 = 0.742, RPD = 1.883). Compared to the Boruta-based feature selection, the MLR, PLS, XGBoost, Cubist, and GB models showed performance improvements of 10.41%, 10.06%, 6.72%, 6.50%, and 3.15%, respectively. Rainfall emerged as the most influential predictor of SOC spatial variability in the study area. Other important predictors included temperature, soil taxonomy, SWIR2 and NIR bands from Landsat 8 imagery, as well as sand and clay contents. We conclude that reducing very high multicollinearity is essential for improving model performance across all tested algorithms, while reducing moderate multicollinearity is not consistently necessary. The developed SOC prediction models demonstrate robust predictive capabilities and can serve as effective tools for supporting soil fertility management, land restoration planning, and climate change mitigation strategies in the Western Lowlands of Eritrea.

Keywords: SOC modelling; multicollinearity reduction; model discrepancy; land use; rainfall

1. Introduction

Soil organic carbon (SOC) plays a pivotal role in the global carbon (C) cycle and holds significant potential for addressing global environmental challenges [1]. Enhancing SOC levels contributes to improved soil health and fertility, increased microbial diversity and activity, more efficient nutrient cycling, greater water retention, restoration of degraded ecosystems, and mitigation of climate change [1,2]. Effective management of soil resources can contribute to the achievement of several Sustainable Development Goals (SDGs), including No Poverty (Goal 1), Zero Hunger (Goal 2), Climate Action (Goal 13), and Life on Land (Goal 15) [3]. Soil represents the largest terrestrial carbon reservoir, storing approximately 2500 Gt of carbon comprising about 1550 Gt as organic carbon and 950 Gt as inorganic carbon within the top one meter of the soil profile [4]. However, soils have been extensively and often unsustainably exploited. Agricultural expansion, in particular, has led to the widespread conversion of natural forestlands, resulting in significant losses of SOC stocks and increased atmospheric CO_2 concentrations [5–7]. These land use change-related emissions contribute approximately 25–30% of the total global greenhouse gas emissions [8]. Regrettably, soils have lost approximately 116 Gt of carbon (C) as a result of land cultivation [9]. Conventional and chemical fertilizer-based intensive farming practices have contributed to a range of environmental and socio-economic challenges, including food and nutrition insecurity, deforestation, land degradation, drought, desertification, and climate change. Addressing these issues requires a transition to sustainable agricultural systems, which have the potential to eradicate hunger and poverty, conserve natural resources, mitigate climate change, and promote sustainable development.

Likewise, Eritrea faces widespread environmental challenges, including land degradation, soil erosion, declining soil fertility, food and nutrition insecurity, deforestation, desertification, drought, and the impacts of climate change [10–16]. Eritrea is situated within the Sahel region, where rainfed agriculture presents significant challenges. Approximately 72% of the country's land area is classified as semi-desert or arid. Nevertheless, over 75% of the population relies on mixed subsistence farming—combining crop and livestock production—dependent primarily on rainfed systems for their livelihoods. Crop productivity remains low, averaging less than 0.7 t ha^{-1} [17], and total crop failures and mass livestock deaths are common due to recurrent droughts after each 3–4 years. Several remote sensing and GIS-based spatio-temporal studies have identified deforestation, drought, desertification, and climate change as major environmental challenges in the country. Ref. [18] reported that the country has experienced moderate to severe droughts since the 1960s. Moreover, Ref. [19] notified very high deforestation rate, 62 km^2 y^{-1}, from 1972 to 2014, and as a result, strongly desertified land increased from 38.5% to 66.9%, and 85% of the country was under serious desertification with high temperature and limited rainfall. Ref. [20] also confirmed that the country has been subjected to recurrent droughts ranging from moderate to extreme conditions between 2000 and 2014, accompanied by increasing temperatures, decreasing rainfall, and declining NDVI values. The mean multi-annual and multi-growing season MODIS NDVI for the entire country were 0.158 and 0.184, respectively [20].

To address the aforementioned challenges and promote ecosystem resilience and sustainable agriculture for sustainable development, the country has implemented extensive soil and water conservation (SWC) campaigns through mass mobilization under the self-reliance policy. Consequently, significant SWC works, including soil and stone bunding, hillside terracing, check dams, micro- and macro-dam construction, tree planting, and the establishment of enclosures have been initiated [14]. In recent years, numerous micro- and macro-dams have been constructed across various regions of the country, resulting in significant water conservation within these structures. Notable macro-dams include Kerkebet, Gahtelai, Misilam, Logo, Gerset, Fanco-Rawi, Bademit, and Fanco-Tsimu, with storage capacities of 330, 50, 35, 31, 20, 20, 17, and 14 million cubic meters of water, respectively.

Successively, the extent of irrigated farmland downstream of these dams has increased over time.

Although moisture stress has been partially alleviated, challenges related to soil management and fertility remain largely unaddressed. Soil management practices are generally inadequate, and soil data and research within the country are limited. Furthermore, existing soil studies primarily rely on traditional soil sampling and laboratory analyses, which are destructive to soil structure, labour-intensive, time-consuming, costly, environmentally unsustainable, and fail to provide continuous and comprehensive information across the entire target area [21,22]. Therefore, there is an urgent need for digital assessment, modeling, and mapping of soil resources in the country to enhance food production, support ecosystem restoration, and enhance climate change mitigation strategies and monitoring.

Advancements in remote sensing, geographic information systems (GIS), and machine learning (ML) enable the processing, assessment, modeling, and interpretation of large environmental datasets, thereby improving our understanding and facilitating the development of effective resource management and monitoring plans [23]. Countries with limited resources and capacity must leverage these technological advances to achieve efficient resource management and sustainable development. Although Eritrea currently lacks digital soil mapping (DSM), complex soil-landscape relationships and soil properties, including SOC, can be modelled and predicted with high accuracy using environmental data and machine learning techniques, often at minimal or no cost [24].

The accuracy of digital soil mapping (DSM) depends on various factors, including the selection of models and predictor variables, topographic complexity, as well as the quantity and quality of input data. Several studies have demonstrated that different machine learning models can predict SOC with high accuracy. Ref. [25] reported that extreme gradient boosting (XGBoost) model with a coefficient of determination (R^2) of 0.870, and root mean squared error (RMSE) of 1.818 t C ha^{-1} outperformed random forest (RF) and support vector machine (SVM) models. Ref. [24] also found that XGBoost model with R^2 = 0.7548, RMSE = 7.6792 g kg^{-1}, and ratio of performance to deviation (RPD) = 1.1311 gave the highest SOC prediction accuracy. Ref. [26] revealed that RF (R^2 = 0.79, RMSE = 1.2) and Cubist (R^2 = 0.77, RMSE = 1.2) recorded better accuracies than SVM and gradient boosting (GB) models. Moreover, Ref. [27] testified that Cubist model (R^2 = 0.64, RMSE = 1.95) gave the most precise spatial prediction of soil organic matter (SOM), and [28,29] reported that multiple linear regression (MLR) and GB models performed best in their respective studies. These findings underscore the importance of model selection and highlight the advantage of evaluating multiple models to enable comparison and identify alternative approaches.

Variables selection is critical in SOC modeling, where soil-forming factors such as climate, soil properties, topography, parent material, vegetation, and human activities are commonly used as predictors. Several studies have reported that climate exerts the strongest control on SOC variability [30–35]. However, other studies have identified different primary drivers of SOC, including vegetation [36], climate and altitude [37], altitude and clay content [24], and use/land cover, soil properties, and bioclimatic variables [26], rainfall and vegetation [38,39], as well as geological units, soil taxonomic units, climatic, and topographic factors [40]. Moreover, Ref. [27] noted that the red and near-infrared (NIR) spectral bands were the most important predictors of SOM. These results suggest that including all potential SOC predictor variables enables the identification of key drivers through feature importance ranking. Decision tree-based machine learning algorithms, such as RF, GB, and XGBoost, are particularly robust for this purpose.

This study aimed to develop SOC prediction models and generate SOC map for the Western Lowlands of Eritrea using machine learning approaches employing multiple SOC predictor variables, including soil characteristics, land use, climate, topography, geology, soil taxonomy, Landsat 8 spectral bands, and various vegetation and bare soil spectral indices. Multiple machine learning algorithms—MLR, PLS, Cubist, RF, GB, and XGBoost—were employed to compare model performance and provide alternative SOC prediction approaches.

The study area encompasses the majority of the country's agricultural activities and represents a key target for modern agriculture due to its expansive, and nearly flat terrain with deep lowland soils. Considerable efforts have been made to promote natural resource conservation and to transform from the rain-dependent subsistence farming to sustainable agricultural practices. Accordingly, this study provides valuable insights into the status and drivers of SOC in the region, supporting farmers, policymakers, and land-use planners in developing effective short-, medium-, and long-term management and monitoring strategies to improve soil health, enhance food production, restore ecosystems, and mitigate climate change. Furthermore, the findings offer a valuable resource for researchers working in data-scarce arid environments.

2. Materials and Methods

2.1. Study Area

The study was conducted in the Western Lowlands of Eritrea, located within the Gash Barka administrative zone. Gash Barka is the largest zone in the country, covering approximately 37% of the national territory. It is subdivided into 16 subzones: Sel'a, Kerkebet, Forto, Dghe, Akordet, Mogolo, Mensura, Logo Anseba, Molqi, Shambko, Barentu, Gogne, Haykota, Teseney, Goluj, and Laelay Gash (Figure 1b). The study area encompasses most of the subzones within the Gash Barka, excluding Sel'a in the north, and Mensura, Logo Anseba, Molqi, and parts of Shambko in the east. The Sel'a subzone was excluded due to lack of soil data, while the others were omitted because they are located in the transition zone between the Highlands and Lowlands and exhibit distinct topographic characteristics compared to the main Lowlands.

The climate of the study area is predominantly arid [41] with mean monthly temperatures ranging from 23.7 to 29.4 °C and annual rainfall varying between 184 and 680 mm [42]. Temperature increases from South to North, while rainfall exhibits the opposite trend, decreasing from South to North (Figure 1d,e). The topography predominantly consists of flat lowland plains with deep soils. Elevation generally decreases from South to North and from East to West (Figure 1c). The crop growing season is very short where most of the time rain starts late and ends early; the rainy season expands from early July to early September.

The study area predominantly practices extensive subsistence rainfed agriculture, characterized by mixed crop-livestock and agro-pastoral systems in most parts, with pastoralism prevailing primarily in the northern regions. Thirty years of data (1992–2021) from the Ministry of Agriculture (MoA) indicate that sorghum is the dominant crop, occupying 72.26% of the cultivated area, followed by pearl millet (12.52%), sesame (8.33%), and finger millet (1.68%), among others. Crop productivity remains low, with average yields (t ha^{-1}) of rainfed sorghum at 0.52, pearl millet at 0.32, and sesame at 0.33. Livestock rearing is also widespread, with herds of camels, cattle, sheep, and goats commonly observed. These animals often overgraze the land, contributing to land degradation, as they freely move in search of pasture and water.

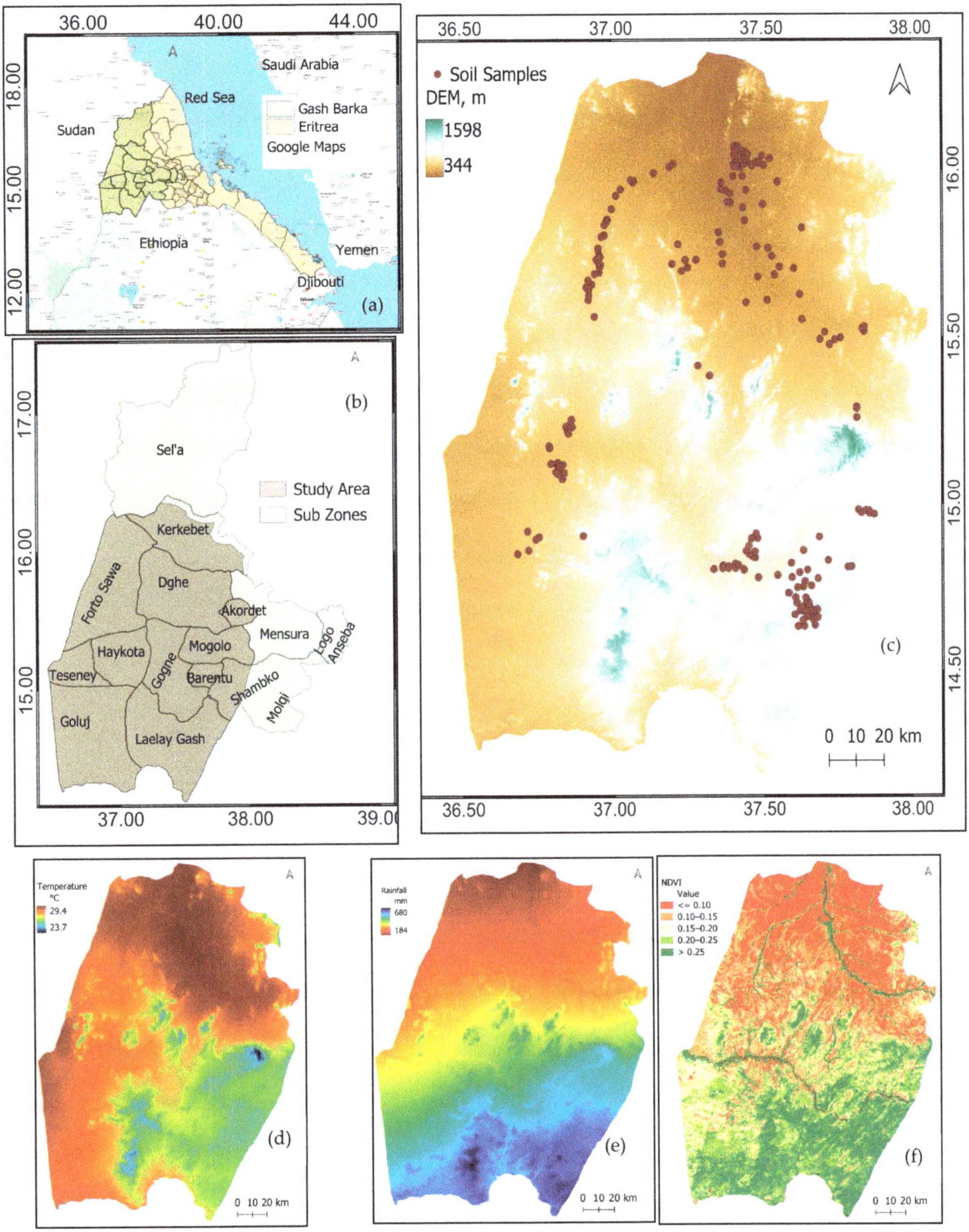

Figure 1. Location map of Eritrea and Gash Barka administration zone (**a**), Gash Barka subzones and the study area (**b**), soil samples and digital elevation model (DEM) imposed on the study area (**c**), mean monthly temperature (**d**), annual rainfall (**e**), NDVI (**f**) of the study area.

The construction of numerous micro- and macro-dams in the study area has promoted irrigated agriculture, enabling the cultivation of various vegetables, fruits, and crops such as banana, tomato, onion, hot pepper, cotton, teff, leafy vegetables, and others. Irrigation using groundwater along the banks of the Barka, Sawa, and Gash Rivers, as well as their tributaries, is also practiced. However, crop productivity remains low; for example, average yields for key irrigated crops—onion, banana, and tomato—are 20, 26, and 25 t ha^{-1}, respectively (MoA records).

The area hosts a diverse range of plant and animal species [16,43]. However, biodiversity is progressively declining due to both anthropogenic and natural disturbances, with agricultural expansion being a primary driver [16]. Most of the study area is characterized by bare soils or scattered Acacia trees and shrubs, particularly in the western, central, and northern regions. The soils predominantly consist of Fluvisols, Leptosols, Lixisols, Vertisols, and Cambisols, which have been conventionally cultivated and grazed for centuries. Cultivation occurs during the rainy season, followed by heavy grazing after harvest, leaving the soils bare for the remainder of the year. Deforestation rates are high in the study area, with several studies reporting hotspots of declining annual and seasonal NDVI trends [20], land productivity decline [10], and serious desertification [12]. The NDVI map of the area (Figure 1f) indicates that vegetation cover is sparse across most of the study area. However, patches of Acacia trees and shrubs are commonly found in the southern regions, along with riverine forests lining the riverbanks.

2.2. Soil, Taxonomic, Geologic, and Land Use Data

In this study, we proposed a range of soil, land use, taxonomic, geological, topographic, climatic, Landsat 8 spectral bands, and spectral index variables as predictors of SOC in the study area, which are discussed below.

Three soil datasets were utilized in this study: (1) soil samples collected during the current study, (2) soil samples obtained from the Regulatory Service of the Ministry of Agriculture (MoA), and (3) legacy soil data from the soil laboratory of the National Agricultural Research Institute (NARI).

A total of 92 georeferenced composite surface soil samples (0–30 cm) were collected between September and November 2023 from various land uses and topographic settings using a 30 cm soil auger, following a stratified sampling technique [15,16]. Each composite sample was formed by collecting five subsamples within a 10 m radius, which were thoroughly mixed, packed, and labeled. The geographic coordinates (latitude and longitude) were recorded at the center of each sampling point.

The soil samples were air-dried, ground, and sieved according to standard procedures at the Soil Laboratory of Hamelmalo Agricultural College, Hamelmalo, Eritrea. They were analyzed for soil pH (using a pH meter) [44], electrical conductivity (using an EC meter) [45], and particle size distribution (via the hydrometer method) [46]. Soil organic carbon (SOC) was analyzed using the wet oxidation method [47] as modified by [48] at the Soil Laboratory of the National Agricultural Research Institute (NARI), Halhale, Eritrea. Additionally, key informant interviews were conducted during the soil sampling campaign to gain a deeper understanding of resource management practices, particularly those related to soils, within the study area.

Georeferenced surface soil samples (0–30 cm; n = 31) were also collected by the Regulatory Service of the Ministry of Agriculture (MoA) from small-scale farmers' irrigated fields along the banks of the Sawa River. These samples were primarily intended to assess general soil conditions, with particular emphasis on salinity. Sampling, preparation, and laboratory analysis followed the same procedures described earlier. All laboratory analyses were conducted at the Soil Laboratory of the National Agricultural Research Institute

(NARI), Halhale. During sampling, interviews were also conducted with farmers to gain insights into soil and water management practices.

In addition, legacy soil data were obtained from the NARI Soil Laboratory. These samples were collected mainly from government-managed large-scale irrigated farms, including Kerkebet, Afhimbol, Fanco, Adi Omer, and Gerset—as well as surrounding grazing lands and rainfed farming systems. From this dataset, only georeferenced surface samples (0–30 cm) with recorded SOC values were extracted, as the dataset also included subsurface samples and entries with missing SOC values or geographic coordinates.

The three datasets were compiled and subjected to Z-score analysis to identify and exclude outliers exceeding ±2 standard deviations. The final dataset consisted of 204 SOC data points, each with associated values for clay, silt, sand, soil texture, pH, electrical conductivity (EC), and geographic coordinates (latitude and longitude).

Parent material has great influence on soil formation and development, and thus, different studies employed soil taxonomy and geology in SOC studies [37,40]. Thereby, geology and soil taxonomy were included in the study. Soil taxonomy data were extracted from the Harmonized World Soil Database (HWSD) v2.0 [49]; Eritrea currently lacks a national soil map and an established soil classification system. In this study, five soil taxonomic units were identified—Cambisols, Lixisols, Leptosols, Fluvisols, and Vertisols—which were numerically coded as 1, 2, 3, 4, and 5, respectively, to convert the categorical data into a quantitative format suitable for modeling. Additionally, geological information was extracted from the national geological map of Eritrea developed by the Ministry of Energy and Mines (MoEM). The identified geological units included gneiss-based formations, alluvial deposits, intrusive rocks-based, meta-volcanic rocks-based, and late basin development-based sediments. These were also numerically coded as 1, 2, 3, 4, and 5, respectively.

Land use/land cover is an important factor that affects soil, and is commonly incorporated in SOC studies [26,37,40,50,51]. Accordingly, land use and land use history data were recorded during soil sampling for most locations. For samples with missing land use information, land use types were identified based on prior knowledge of the study area and verified using Google Earth Pro. Four primary land use types were identified: communal grazing (CG), rainfed farming (RF), irrigated farming (IF), and natural forest (NF), and were numerically coded as 1, 2, 3, and 4, respectively.

In this context, communal grazing (CG) refers to areas used for year-round livestock grazing, excluding regions with significant natural forest cover. Natural forest includes areas with a well-developed forest canopy, even if they fall within communal grazing zones. Rainfed farming fields are also subject to communal grazing during the dry season. Figure 2 presents photographs of different land use types, taken during the soil sampling.

2.3. Topographic, and Climatic Data

Topography plays a critical role in soil formation and distribution; consequently, many studies have incorporated topographic variables into their SOC prediction models [52–55]. The dominant topography of the study area consists of flat lowland plains; therefore, topographic influence on soil properties is expected to be limited. Nonetheless, to assess the potential impact of topography on SOC variability, several topographic variables were included in the analysis. The elevation within the study area ranges from 1598 to 344 m above sea level and exhibits variation in slope, aspect, and other terrain characteristics. Topographic variables were derived from the ASTER Digital Elevation Model (DEM) of Eritrea and included: altitude, slope, aspect, topographic position index (TPI), LS factor (slope length and steepness), terrain roughness index (TRI), surface roughness, hill shade, sediment balance, upslope slope, negative openness, and positive openness.

Climate variables, particularly rainfall and temperature are among the most frequently used predictors in SOC studies, as these factors exert dominant influence on soil formation, organic matter decomposition, and carbon dynamics [30–35]. In this study, long-term historical rainfall and temperature data were extracted from the WorldClim 2.1 dataset (30 arc-seconds resolution), available at https://www.worldclim.org/data/worldclim21.html# (accessed on 3 September 2025) [42]. This approach was necessary due to the scarcity of long-term recorded climate data in Eritrea, and particularly within the study area.

Figure 2. Photos (November 2023) showing grazing land around Mogorayb (**a**), poorly managed drip-irrigated cotton crop at Kerkebet farm (**b**), rainfed field around Kerkebet farm (**c**), and River Barka and its riverine ecosystem around Afhimbol (**d**).

2.4. Landsat 8 and Spectral Indices Data

We downloaded four L8 images with (path, row, dated) of (170, 049, 17 April 2024), (170, 050, 29 February 2024), (171, 049, 8 April 2024), and (171, 050, 8 April 2024) from the USGS archives https://earthexplorer.usgs.gov/ accessed on 24 May 2024. We chose images with <10% cloud cover in the dry months of February to April to avoid the interferences of clouds and seasonal vegetation. We computed different spectral indices using their respective formula (Table 2) using the raster calculator in QGIS 3.38. These spectral indices were normalized difference vegetation index (NDVI), infrared percentage vegetation index (IPVI), optimized soil adjusted vegetation index (OSAVI), modified SAVI2 (MSAVI2), normalized difference water index (NDWI), normalized difference soil index (NDSI), bare soil index (BSI), hyperspectral BSI (HBSI), soil organic carbon index (SOCI), normalized difference moisture index (NDMI), burn ration (BR), BR2, colouration index (CI), hue index (HI), and brightness index 2 (BI2). Different studies employed these spectral indices in different combinations [52–56].

L8 bands B1, Blue, Green, Red, NIR, SWIR1, SWIR2, and B10 [23,52,57] were also included with the dataset as potential SOC predictors.

The initial dataset consisted of 48 input variables. To eliminate both redundant and non-significant predictors, a preliminary Spearman correlation analysis was conducted, as it accounts for both linear and non-linear relationships. Variables that did not show a statistically significant correlation with SOC ($p > 0.05$) were excluded. Additionally, among pairs of variables that were perfectly correlated, one variable from each group was removed to avoid redundancy. As a result, the number of input variables was reduced to 36 (Table 1), which were retained as potential SOC predictor variables. Further refinement to address multicollinearity is described in Section 2.6.

Table 1. Proposed potential SOC predictor variables after preliminary Spearman correlation analysis.

Variable Type	Variables, and Their Short Form, If Given	Resolution	Source
Soil properties	Clay, Silt, Sand, Texture, pH	10 m	Lab measurement
Land use	Land Use (LU)	10 m	LU history
Soil Taxonomy	Taxonomy (Taxom)	1 km	HWSD v2.0
Geology	Geology (Geogy)	1 km	MoEM
Topography	Altitude (Alt), Slope, LS Factor (LSfctr), TRI, Roughness (Rough), Negative Openness (NvOp), Positive Openness (PvOp), Upslope Slope (UspSp),	30 m	ASTER GDEM
Climate	Temperature (Temp), Rainfall (Rain)	1 km	WorldClim 2.1
L8 bands	B1, Blue, Green, Red, NIR, SWIR1, SWIR2	30 m	L8—OLI sensor
	B10	100 m	L8—TIRS sensor
Spectral Indices	NDVI, OSAVI, MSAVI2, NDWI, NDSI, HBSI, SOCI, NDMI, BR2, BI2	30 m	Computed

Table 2. Proposed potential SOC predictor spectral indices, and their formula.

Spectral Index and Formula	Reference	Formula No.
NDVI = (NIR − Red)/(NIR + Red)	[58]	(1)
OSAVI = (NIR − Red)/(NIR + Red + 0.16)	[59]	(2)
MSAVI2 = $0.5\,[2 \times \mathrm{NIR} + 1 - \sqrt{[(2 \times \mathrm{NIR} + 1)^2 - 8(\mathrm{NIR} - \mathrm{Red})]}]$	[60]	(3)
NDWI = (Green − NIR)/(Green + NIR)	[55]	(4)
NDSI = (SWIR1 − Green)/(SWIR1 + Green)	[61]	(5)
HBSI = [(SWIR2 + Green) − (NIR + Blue)]/[(SWIR2 + Green) + (NIR + Blue)]	[62]	(6)
SOCI = Blue/(Red × Green)	[57]	(7)
NDMI = (NIR − SWIR1)/(NIR + SWIR1)	[63]	(8)
BR2 = (SWIR1 − SWIR2)/(SWIR1 + SWIR2)	[64]	(9)
BI2 = $\sqrt{[(\mathrm{Red}^2 + \mathrm{Green}^2 + \mathrm{NIR}^2)/2]}$	[65]	(10)

2.5. Selection of Machine Learning Algorithms

Modelling the relationship between continuous independent variables and a continuous dependent variable typically involves regression fitting, which may take various functional forms such as linear, logarithmic, exponential, or power functions depending on the nature of the relationships. To accommodate these diverse relationships, we employed both linear regression-based and decision tree-based machine learning models. Specifically, we evaluated the performance of six algorithms: Multiple Linear Regression (MLR), Partial Least Squares (PLS), Cubist, Random Forest (RF), Gradient Boosting (GB), and Extreme Gradient Boosting (XGBoost). This approach enabled a comprehensive comparison of model performances and facilitated the identification of robust alternative models for SOC prediction.

MLR model is based on fitting many independent variables linearly to a dependent variable. The model has broad applicability in SOC prediction [28,57,66,67], though it is not robust with multicollinearity and outliers. PLS, a linear relationship-based model, focuses on maximizing covariance between predictors and responses, and seems robust with multicollinearity. This model is commonly used in SOC prediction modeling, and its performance is satisfactory [54,64,68,69].

Decision tree ensemble machine learning algorithms are generally more robust to multicollinearity and outliers compared to traditional regression methods. The Cubist model, a rule-based algorithm, combines decision trees with multiple linear regression models [70]. For each rule, a regression model is fitted based on the subset of data defined by that rule. These rules are hierarchically organized, where recursive partitioning of the response variable is performed to minimize the standard deviation across all potential splits [71]. The Cubist model is widely applied in SOC modeling and has consistently demonstrated strong predictive performance [27,54,72,73].

Random Forest (RF), introduced by [74], constructs multiple decision trees by bootstrapping samples from the training dataset and subsequently aggregates their outputs to form a robust predictive model [71]. This ensemble approach reduces overfitting, improves model accuracy, and decreases computational time compared to individual trees. RF is widely applied across various data mining fields, including SOC modeling, due to its strong predictive performance and robustness [26,38,54,69,72,73,75].

GB is a powerful ensemble learning model that builds models in a sequential manner where each new model corrects the errors made by the previous ones. It combines multiple weak learners to create a strong learner through boosting. It uses the concept of gradient descent to minimize the loss function. This model is also commonly used in SOC prediction studies [32,76,77].

XGBoost is also a decision tree-based ensemble model [78], advantaged by boosting algorithm. It tries to punish the weak learners and develop strong learners. Through parallel learning, computing speed is improved and over-fitting is prevented effectively. XGBoost model is gaining popularity in DSM [24,25,72].

2.6. Selection of Input Variables

Utilizing a large number of input variables in predictive modeling may result in inaccurate conclusions due to multicollinearity among certain independent variables. Therefore, the selection of appropriate input variables is essential for developing robust predictive models [79,80]. This enhances model performance by mitigating multicollinearity and reducing model complexity, thereby facilitating faster model calibration [81]. In DSM, two approaches are commonly employed for covariate selection: (1) Selecting key covariates prior to modeling using correlation analysis or principal component analysis [82,83], which are based on linear relationships and may overlook covariates with nonlinear associations with the target variable [84,85]; (2) Feature importance ranking based on the model during calibration [85–87], which is rooted in nonlinear relationships, but may introduce multicollinearity effects. While most studies have employed one of these approaches, few have compared their impacts on model performance, despite findings suggesting that both approaches are effective for variable selection [29,80].

In this study, we employed both approaches to compare their performance: (1) multicollinearity reduction using Spearman rank correlation, and (2) feature selection based on the Boruta algorithm. Multicollinearity between input variables was progressively reduced based on their correlations with each other and the target variable, as outlined below.

1. In the first step, we executed the SOC modeling with all proposed input variables (36 variables) (Table 1).
2. Based on Spearman correlation, we identified each set of variables that were very highly correlated ($\geq$0.900 **) with each other, and we omitted the variable with weaker correlation with SOC by keeping the one with higher correlation with SOC [56]. For example, the correlation SWIR2 vs. SWIR1 = 0.975 ** (very high collinearity), and SWIR2 vs. SOC = $-$0.534 **, and SWIR1 vs. SOC = $-$0.505 **. Thus, SWIR2 is retained for the fact that it has higher correlation with SOC than SWIR1 ($-$0.534 ** vs. $-$0.505 **) but SWIR1 is removed.
3. From each set of input variables that had high correlation ($\geq$0.700 ** but <0.900 **) with each other, we omitted the variable with weaker correlation with SOC by keeping the one with the higher correlation with SOC.

4. The Boruta feature selection wrapper algorithm was employed to select from the 36 proposed input variables. The algorithm is among the commonly used feature selection techniques in SOC prediction studies [85,88].

2.7. Calibration, Validation and Evaluation of Models

Figure 3 presents the conceptual flowchart of the methodologies employed. The models were trained, calibrated, and validated based on the extraction of independent SOC predictor variables to estimate target SOC values. The SOC modeling process was executed using Python 3.11.7 in the Jupyter Notebook environment. The dataset was randomly partitioned into training (80%) and testing (20%) subsets. We tested multiple algorithms, including MLR, PLS, Cubist, RF, GB, and XGBoost. Hyperparameter tuning and refinement were performed for each algorithm, with the exception of MLR, to optimize model performance, with particular emphasis on minimizing model discrepancies. Models were fitted, trained, calibrated, and validated, and their performance was assessed using commonly established metrics: RMSE, R^2, RPD [24,69,88,89], and model discrepancy (MD), as defined by Equations (11), (12), (13), and (14), respectively.

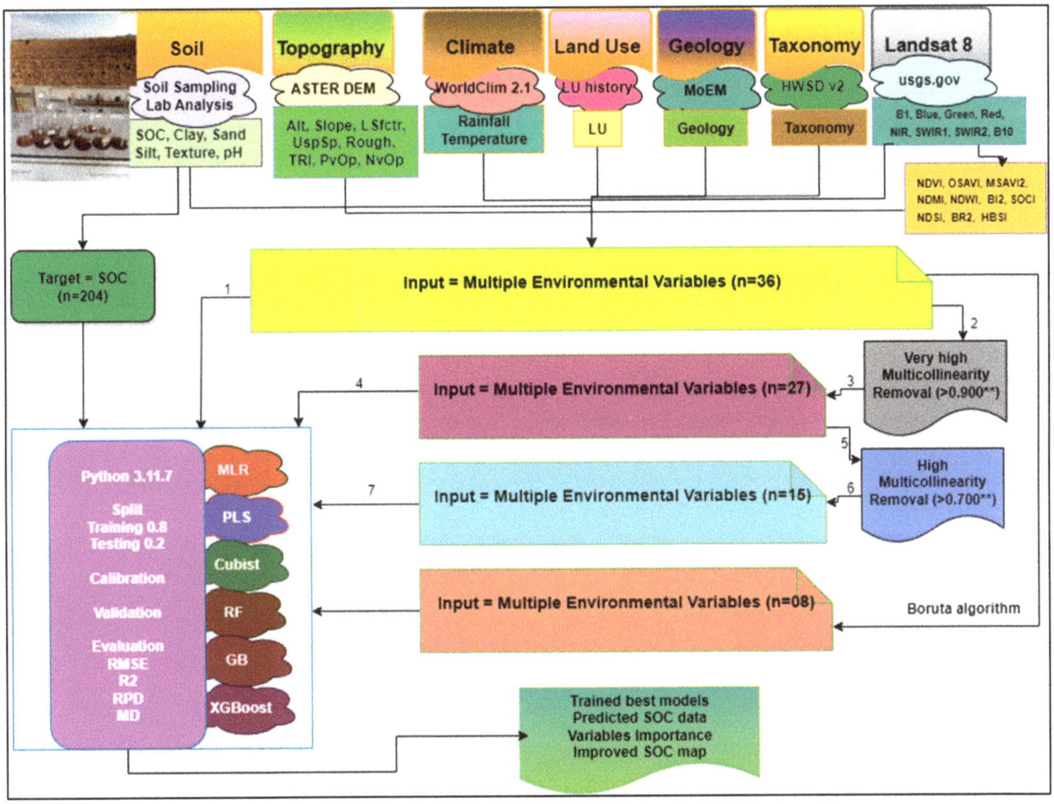

Figure 3. Conceptual flowchart of the methodologies used; ** indicates significant at 0.01.

$$\text{RMSE} = \sqrt{\frac{1}{n}\sum_{i=1}^{n}(\text{Oi} - \text{Pi})^2} \quad (11)$$

$$R^2 = 1 - \frac{\sum_{i=1}^{n}(Oi - Pi)^2}{\sum_{i=1}^{n}\left(Oi - \hat{O}\right)^2} \quad (12)$$

$$RPD = \frac{\sqrt{\sum_{i=1}^{n}\left(Oi - \hat{O}\right)^2}}{\sqrt{\sum_{i=1}^{n}(Oi - Pi)^2}} \quad (13)$$

$$MD = \frac{T - t}{T} * 100\% \quad (14)$$

where n = number of SOC observations, Oi = observed SOC values, Pi = predicted SOC values, Ô = mean of observed SOC values, T = RMSE of the training data, and t = RMSE of the testing data

Furthermore, models are categorized as per their RPD values as; very poor (RPD < 1), poor (1–1.4), fair (1.4–1.8), good (1.8–2.0), very good (2.0–2.5), and excellent (>2.5) [89].

3. Results and Discussions

3.1. Statistical Analysis of Observed Soil Properties

The studied soils exhibited moderate clay and silt content and high sand content, with average percentages of 26.28%, 27.89%, and 46.11%, respectively (Table 3), showing significant spatial variability (CV = 53.92%, 45.19%, and 41.08%, respectively). Soil texture ranged from sandy loam to clay, with sandy loams (32.8%) and loams (24.5%) being the most dominant. The dominant soil order was Fluvisols (46.6%), followed by Lixisols (20.6%) and Leptosols (16.2%). The study area is a lowland region that receives substantial runoff and soil sediments from the surrounding highlands, which may influence soil formation processes.

The soils were non-saline but moderately alkaline, with mean EC and pH values of 0.20 dS m^{-1} and 8.45, respectively. A small proportion of soils in areas such as Adi Hakin, Tekombia, Fanco, Mogolo, and Kerkebet exhibited stronger alkalinity. This alkalinity is likely attributed to the calcareous nature of soils in arid regions [90], where calcium carbonate tests can provide valuable insights for informed management strategies. It is also important to note that alkaline soils may exhibit deficiencies in micronutrients such as manganese, copper, zinc, and boron [91,92].

The SOC content of the soils ranged from 0.02% to 1.01%, with a mean of 0.43%, standard deviation (SD) of 0.27%, and coefficient of variation (CV) of 61.65% (Table 3). SOC content decreased from south to north, following a similar pattern to rainfall distribution, but inversely correlated with temperature across the study area. The Laelay Gash subzone (in the south) recorded the highest mean SOC content, while the Kerkebet subzone (in the north) had the lowest mean SOC content.

According to [91–93], the SOC content is considered poor. This finding is consistent with other studies conducted within the country. For instance, Ref. [16] reported low SOC levels at Laelay Gash (part of the study area), and Refs. [13–15] observed similar trends in other regions. Factors such as conventional tillage practices, continuous monocropping, excessive crop residue removal, minimal organic matter addition, overgrazing, and soil erosion, likely contributed to the low SOC content [13–16].

Table 3. Statistical analysis of observed soil properties.

Parameter	Minimum	Maximum	Mean	Status	SD	CV, %	Skewness	Kurtosis
Sand, %	5.80	87.50	46.11	High	18.94	41.08	0.22	−0.77
Clay, %	4.50	62.50	26.28	Moderate	14.17	53.92	0.46	−1.11
Silt, %	2.50	58.00	27.89	Moderate	12.60	45.19	0.00	−0.41
EC, dS m^{-1}	0.03	1.29	0.20	Non-saline	0.24	121.46	3.06	9.77
pH	7.18	9.72	8.45	Moderately alkaline	0.62	7.33	−0.13	−0.60
SOC, %	0.02	1.01	0.43	Poor	0.27	61.65	0.37	−1.13

SD = standard deviation, CV = coefficient of variation, Status is given based on [91–93].

SOC can be improved through effective agronomic practices such as crop rotation, legume–cereal intercropping, crop residue retention, cover cropping, technology-supported agroforestry practices, push-pull technology, farmyard manure application, integrated and diverse organic farming systems, no/minimal tillage, and the incorporation of biochar or other organic amendments.

In regions with limited natural vegetation, such as the study area, organic amendments play a critical role in enhancing soil fertility. Two abundant organic sources—*Prosopis juliflora* and organic municipal waste—are yet underutilized in the study area, and across the country. Integrating these resources into farming systems could yield multiple benefits, including improvements in food and nutrition security, as well as positive ecological, socio-economic, health, and environmental outcomes. Therefore, promoting composting and biochar technologies using these resources, alongside pilot research initiatives, is crucial.

Given that the study area is impacted by overgrazing, managing the appropriate stocking rate, implementing rotational grazing, allowing sufficient rest periods for pastures, establishing enclosures, promoting cut-and-carry methods, reseeding with legumes, and developing and utilizing grazing and SOC dynamics models [5,94] could significantly enhance soil health and fertility.

3.2. SOC Along Land Uses, Soil Taxonomic Units, and Irrigated Farms

The mean SOC values for the observed land use types were ranked as follows: natural forests (NF) > rainfed farming (RF) > irrigated farming (IF) > communal grazing (CG), with respective values of 0.66%, 0.48%, 0.42%, and 0.33% (Figure 4). The ANOVA test revealed that land use had a highly significant effect ($p < 0.001$) on SOC. However, the Tukey HSD post hoc test indicated no statistically significant difference in SOC between IF and RF, as well as between IF and CG land uses. These results suggest that NF has the potential for soil carbon (C) sequestration, whereas the other land uses are more likely to contribute to soil C release. Ref. [16] also emphasized the importance of conserving and enhancing natural forests for C sequestration, while highlighting the risks associated with RF and CG land uses, which may contribute to C release into the atmosphere.

Our results contradict with the findings that reported significantly higher SOC contents in irrigated farming (IF) compared to rainfed farming (RF) and communal grazing (CG) land use types [15,95]. Ref. [15] emphasized the importance of IF for carbon storage in Keren, Eritrea. In general, numerous studies have highlighted the significant effect of land use on SOC levels [13,15,16,96,97].

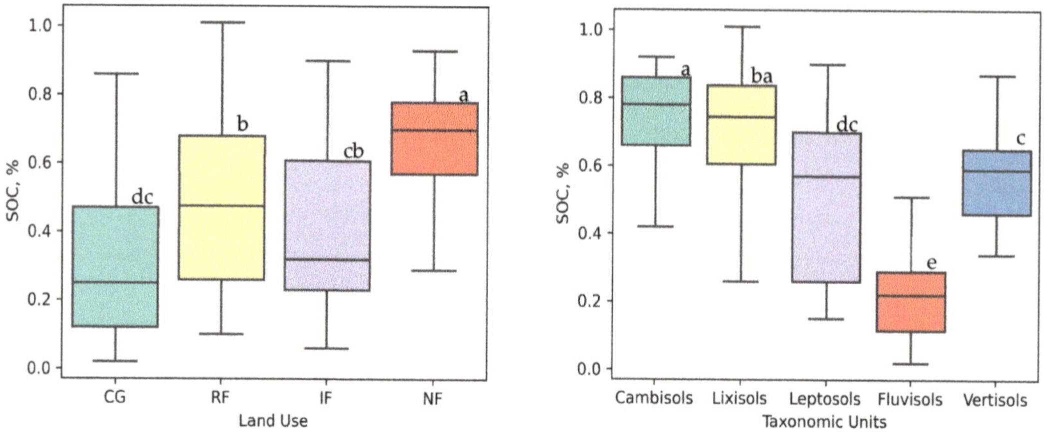

Figure 4. Effect of land use (**left**), and soil taxonomic units (**right**) on SOC; lowercase letters indicate which specific group means are different after conducting ANOVA test.

Based on the interviews with farmers and key informants, along with general observations during soil sampling, it is evident that agronomic and soil management practices in the study area are suboptimal. Soil nutrient depletion is pronounced due to extensive grain and biomass harvesting, coupled with overgrazing and minimal organic matter inputs. Furthermore, irrigated soils are cultivated 2–3 times per year with little to no nutrient replenishment, exacerbating nutrient depletion. Crop biomass is primarily harvested and either fed to livestock or sold for cash. In this arid environment, vegetation is scarce, rangelands and crop fields are overgrazed shortly after the autumn harvest, and the harvested crop biomass constitutes the primary feed for livestock during the winter and spring seasons across much of the study area.

The mean SOC values for the soil taxonomic units were ranked as follows: Cambisols > Lixisols > Vertisols > Leptosols > Fluvisols, with respective values of 0.75%, 0.72%, 0.56%, 0.50%, and 0.22% (Figure 4). The ANOVA test revealed that soil taxonomy had a highly significant effect ($p < 0.001$) on SOC. However, the Tukey HSD post hoc test indicated no significant difference in SOC between Cambisols and Lixisols, nor between Vertisols and Leptosols. Fluvisols exhibited the lowest mean SOC, with most of the sampled Fluvisols located along the banks of the Barka and Sawa Rivers in the northern part of the study area. The low SOC in Fluvisols may be more attributable to the regional climate rather than the inherent properties of the soils, a conclusion that aligns with the findings presented in Section 3.5 on the importance of variables.

Table 4 presents the descriptive statistics of SOC in the irrigated farms. All the irrigated farms, whether government or privately operated, exhibited low SOC content, primarily due to inadequate agronomic and soil fertility management practices. Common issues include mono-cropping, nutrient depletion through frequent grain and crop residue harvesting (2–3 times per year) with minimal nutrient replenishment. Additionally, deep nutrient and water percolation may significantly impact SOC levels, as most farms rely on furrow irrigation systems, despite the soils being sandy and permeable. Furthermore, the practice of deep ploughing with heavy tractors and discs exacerbates nutrient losses. To address these challenges, awareness campaigns and training programs on proper tillage practices, soil nutrient and irrigation water management, crop rotation, legume–cereal intercropping, agroforestry practices, cover cropping, green manuring, composting, and the application of bio-fertilizers and biochar technologies are essential for farmers and land managers in the study area.

Table 4. Descriptive statistics of SOC in irrigated farms.

Subzone	Farm/s	Minimum	Mean	Maximum	SD	CV, %	Skewness	Kurtosis
Kerkebet	Kerkebet *	0.06	0.24	0.42	0.13	53.95	−0.15	−1.50
Dghe	Afhimbol *	0.05	0.14	0.23	0.07	52.77	0.29	−1.70
Forto	Forto Sawa **	0.11	0.26	0.48	0.08	29.33	0.71	1.42
Teseney	Adi Omer *	0.34	0.45	0.68	0.12	26.73	1.73	3.29
Haykota	Fanco *	0.42	0.58	0.87	0.14	23.41	1.08	1.52
Goluj	Gerset *	0.44	0.63	0.86	0.15	24.17	0.37	−0.65
Laelay Gash	Tekombia **	0.56	0.71	0.88	0.12	17.16	0.29	−1.15
Shambko	Shambko **	0.57	0.73	0.90	0.13	18.37	0.05	−2.03

* Government-led large-scale farm, ** farmers' small-scale number of farms.

Farms in the northern part of the study area (Kerkebet, Afhimbol, and Forto Sawa) exhibited relatively lower mean SOC contents compared to farms in other regions. The highest and lowest mean SOC values were recorded at Shambko farms (in the south-east) and Afhimbol farm (in the north), with values of 0.73% and 0.14%, respectively. These differences are likely attributed to climatic variations rather than differences in farm management practices.

3.3. Correlation Analysis, and Multicollinearity Reduction

Spearman correlation analysis revealed that most of the input variables showed strong correlations with SOC (Figure 5). Rainfall (0.746 **) exhibited the highest correlation with SOC, followed by temperature (−0.732 **), altitude (0.729 **), NIR (−0.640 **), Green (−0.632 **), SOCI (0.612 **), BI2 (−0.609 **), Red (−0.591 **), Blue (−0.589 **), and Sand (−0.573), among others. These results suggest that climate factors, specifically rainfall and temperature, may have the greatest impact on SOC and may offer strong predictive capacity for SOC in the study area. Altitude also serves as a surrogate for climate, as higher altitudes generally correspond to cooler temperatures and vice versa.

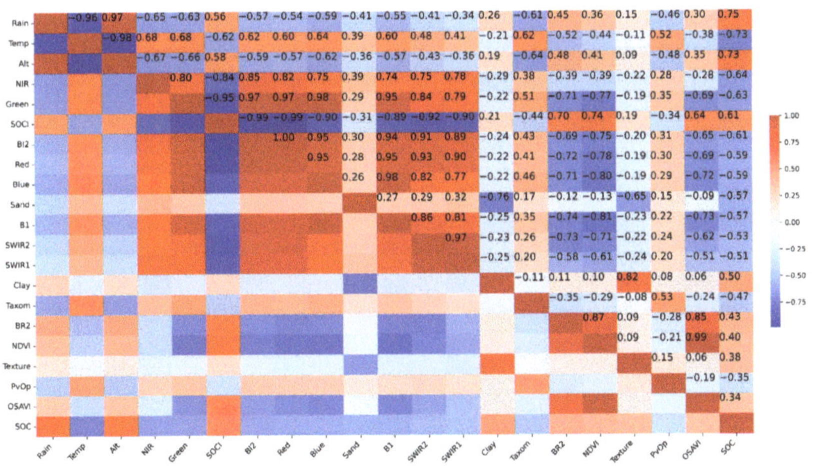

Figure 5. Correlation heatmap: only 20 input variables are displayed (due to limited display space), arranged in decreasing magnitude of their correlations with SOC.

Following the climate variables, L8 bands and bare soil indices demonstrated strong potential for SOC prediction. Soil attributes like sand and clay content, as well as soil taxonomy, also exhibited good correlations with SOC and may contribute significantly to

SOC prediction. However, land use, geology, and topographic variables showed relatively lower correlations with SOC and likely exert less influence on it.

Many of the input variables exhibited very high correlations (≥0.900 **) with each other, necessitating multicollinearity reduction. For example, rainfall, temperature, and altitude were highly correlated, and only rainfall was retained due to its stronger correlation with SOC compared to the other two. Similarly, very high correlations were observed among Green, B1, Blue, Red, BI2, and SOCI, and only the Green band was retained. As a result, nine variables—temperature, altitude, B1, Blue, Red, SWIR1, BI2, SOCI, and OSAVI—were removed due to high multicollinearity with other input variables. Consequently, the number of input variables was reduced from 36 to 27.

High correlations (≥0.700 **) were also observed between different groups of the independent variables. Accordingly, 12 variables were omitted at this step; Green, SWIR2, clay, NDVI, MSAVI2, HBSI, NDWI, NDSI, slope, LS factor, TRI, and roughness, and the input variables were reduced to 15.

According to the Boruta feature selection algorithm, eight variables were selected from the 36 proposed input variables: rainfall, sand, taxonomy, clay, SWIR2, SOCI, Green, and temperature. SOC modeling was then conducted using four categories of input variables: 36, 27, 15, and 8. Based on the Boruta algorithm, the importance score for rainfall was dominant (>0.6), while none of the other variables had an importance score greater than 0.1 (Figure 6), highlighting the significant influence of rainfall on SOC in the study area.

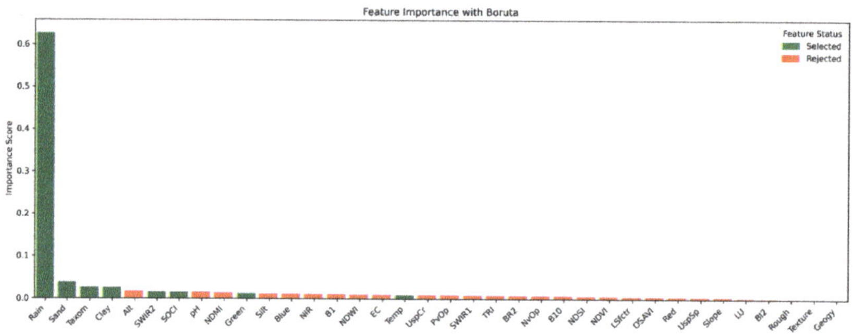

Figure 6. Feature importance score and selection using the Boruta algorithm: selected (green), rejected (red).

3.4. *Performance of Models*

The observed SOC ranged, in %, 0.02–1.01 with a mean = 0.43, SD = 0.27, and CV = 61.65%, and that of the training and testing SOC varied 0.02–1.01, and 0.04–0.85, with means of 0.44, and 0.39, respectively (Table 5). Generally, the soils had low SOC, and showed high spatial variation (CV > 60%), which high spatial variation may challenge SOC modeling.

Table 5. Statistics of total (observed), training and testing SOC (%).

Dataset	Count	Minimum	Maximum	Mean	SD	CV, %	Skewness	Kurtosis
Total SOC	204	0.02	1.01	0.43	0.27	61.65	0.37	−1.13
Training SOC	163	0.02	1.01	0.44	0.27	61.51	0.36	−1.19
Testing SOC	41	0.04	0.85	0.39	0.24	61.82	0.36	−1.01

The performance of the examined models with different input variables to the training and testing data is shown in Table 6. All models fit well to both data types in all the varied number of input variables. For the training data, the highest performance was achieved by the MLR_36 model (MLR model with 36 input variables) with an RMSE of 0.114, R^2 of 0.824, and RPD of 2.342. However, its model discrepancy (MD) was 21.71%, which is high enough to logically question this model. Thus, the highest credit was taken by the XGB_36 model (XGBoost model with 36 input variables) with RMSE = 0.115, R^2 = 0.818, RPD = 2.308, and MD = 3.99%. The lowest model performance was recorded by the PLS_08 model with RMSE = 0.135, R^2 = 0.731, RPD = 1.964, and MD = 3.43% (Table 6, Figure 7a,c). Figure 7 displays the RMSE, RMSE change, and MD of the models with different input variables.

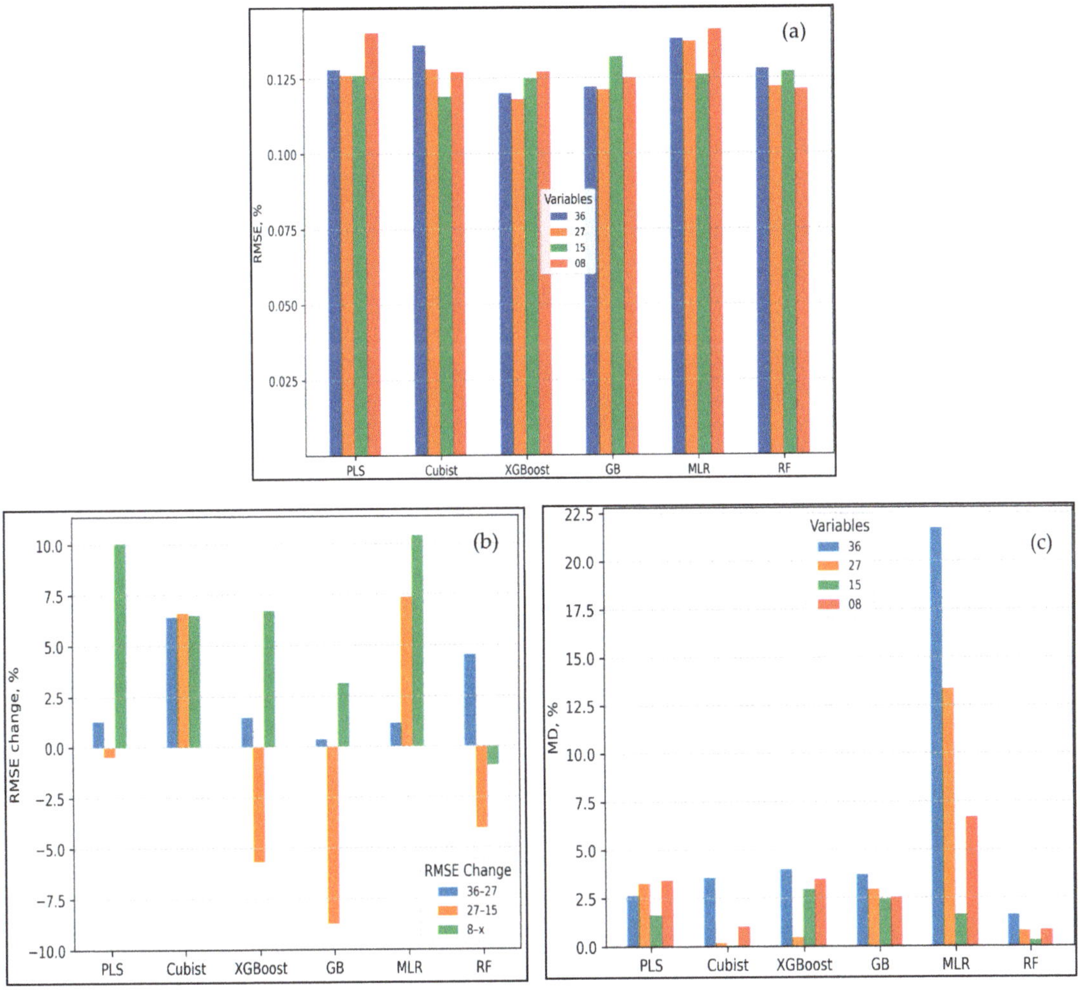

Figure 7. RMSE (**a**), RMSE change (**b**) and MD (**c**) with different input variables: RMSE Change (x stands for the lowest RMSE in each model; positive and negative changes indicate model enhancement and model deterioration, respectively).

Table 6. Performance of models for the training and testing data with different number of input variables.

Model	Dataset	36 Variables RMSE	R^2	RPD	27 Variables RMSE	R^2	RPD	15 Variables RMSE	R^2	RPD	08 Variables RMSE	R^2	RPD
PLS	Training	0.124	0.789	2.139	0.122	0.797	2.181	0.129	0.774	2.067	0.135	0.731	1.964
	Testing	0.128	0.717	2.084	0.126	0.725	2.111	0.126	0.722	2.101	0.140	0.746	1.899
Cubist	Training	0.132	0.763	2.020	0.127	0.778	2.088	0.119	0.806	2.232	0.129	0.773	2.065
	Testing	0.136	0.677	1.951	0.128	0.717	2.084	0.119	0.754	2.232	0.127	0.718	2.086
XGB	Training	0.115	0.818	2.308	0.117	0.811	2.263	0.121	0.799	2.194	0.122	0.795	2.173
	Testing	0.120	0.751	2.219	0.118	0.758	2.252	0.125	0.730	2.131	0.127	0.722	2.101
GB	Training	0.117	0.811	2.263	0.118	0.810	2.255	0.129	0.773	2.064	0.122	0.796	2.175
	Testing	0.122	0.742	2.182	0.121	0.744	2.190	0.132	0.698	2.015	0.125	0.727	2.121
MLR	Training	0.114	0.824	2.342	0.120	0.801	2.207	0.129	0.774	2.067	0.132	0.743	2.009
	Testing	0.138	0.668	1.924	0.137	0.676	1.947	0.126	0.722	2.101	0.141	0.742	1.883
RF	Training	0.126	0.783	2.112	0.121	0.799	2.194	0.127	0.780	2.099	0.122	0.796	2.178
	Testing	0.128	0.716	2.078	0.122	0.741	2.176	0.127	0.720	2.092	0.121	0.746	2.197

For the validation data, the highest model accuracy was recorded by the XGB_27 model with an RMSE of 0.118, R^2 of 0.758, RPD of 2.252, and MD of 0.48%, and on the contrary, the lowest was observed in the MLR_08 model with RMSE = 0.141, R^2 = 0.742, RPD = 1.883, and MD = 6.68 (Table 6, Figure 7a,c). The XGBoost model is a powerful and efficient gradient boosting algorithm that is gaining popularity in many fields including SOC prediction. Different authors also confirmed that the XGBoost model performed the best in their SOC prediction studies [24,25,98]. Generally, the accuracies obtained are on the average of most of DSM and SOC prediction modeling reports [24–27,99].

The highest model discrepancy (MD), 21.71%, was observed in the MLR_36 model, followed by the MLR_27 model (13.38%). This can likely be attributed to the multicollinearity effect, as the MLR model does not address multicollinearity. The lowest MD, 0.02%, was observed in the Cubist_08 model. Overall, the Random Forest (RF) model exhibited the lowest MD, followed by the Cubist model. All models, except the MLR model, maintained an MD of less than 5% across all stages with different input variables (Figure 7c). MD is the difference in performance of a machine learning model when applied to training versus testing data, which is crucial for assessing how well a model generalizes to unseen data. The acceptable level of MD tolerance depends on the accuracy requirements of the target application.

According to their RPD values, most of the models were categorized as 'very good' for both datasets across all different input variable sets. The PLS and MLR models with 8 input variables showed relatively weaker performance but still fell within the 'good' model category, with an RPD > 1.80. RPD represents the factor by which prediction accuracy increases compared to using the mean of the original data. This indicates that all trained, calibrated, and validated models—except for the MLR_36 and MLR_27 models due to their high MD—demonstrated good SOC predictive accuracy. These models can be used for planning and monitoring soil fertility improvements, food production, carbon sequestration, and climate change mitigation.

All models showed improvements when the number of input variables was reduced from 36 to 27 (following the removal of very high multicollinearity), with the greatest improvement recorded by the Cubist model (6.42%), followed by the RF model (4.53%) (Figure 7b). This may indicate that removal of very high multicollinearity (≥0.900 **) is important for better model performance. However, when the number of input variables was reduced from 27 to 15 (after high multicollinearity removal), performance of most of the models deteriorated (GB, XGBoost, RF, and PLS), and only the MLR and Cubist models continued improving (Figure 7b). This may imply that high multicollinearity (≥0.700 ** but <0.900 **) removal may not be necessary especially for the models that deal well with multicollinearity.

In comparison to the Boruta algorithm-based selection of variables (08 input variables), all the models, except the RF model, benefited from the multicollinearity reductions. The highest model performance improvement was obtained by the MLR (10.41%) model followed by the PLS (10.06%), and the XGBoost (6.72%), respectively, and the lowest improvement was in the GB model (3.15%) (Figure 7b). This indicates the potential of Spearman rank correlation-based multicollinearity reduction to enhance performance of models.

During the calibration, and validation of models, the following hyperparameters settings were used: Cubist (n_committees = 100, n_rules = 2), RF (max_depth = 3, min_samples_split = 2, min_samples_leaf = 2), GB (learning_rate = 0.07, max_depth = 1, min_samples_split = 7), XGBoost (colsample_bytree = 0.9, max_depth = 1, learning_rate = 0.07, subsample = 0.7), PLS (n_components = 26, 16, 15, 5, respectively), and no hyperparameter was used with the MLR model.

3.5. Importance of Variables

Variable importance rankings from the XGBoost, Random Forest (RF), and Gradient Boosting (GB) models revealed that, across all categories of input variables, rainfall was the most important variable for SOC prediction in the study area. In some cases, rainfall alone explained more than 60% of the SOC variance, particularly with the GB model.

According to the XGBoost_36 model, the top five most important variables for SOC prediction were rainfall, temperature, taxonomy, altitude, and SWIR2, which together accounted for 73.61% of the SOC spatial variation in the study area. Notably, rainfall alone explained 32.62% of the SOC variance (Figure 8a). Similarly, the RF_36 model identified rainfall, temperature, altitude, taxonomy, and NIR as the top five variables, accounting for 66.32% of SOC variability, with rainfall explaining 20.48% of the variance (Figure 8b). In the GB_36 model, the top five variables were rainfall, altitude, sand, SWIR2, and taxonomy, collectively explaining 92.17% of the variance in SOC, with rainfall contributing the largest share, 58.28% (Figure 8c).

Climate through rainfall (mainly), temperature, and altitude was identified as the first to have the highest impact on SOC in the study area. Altitude is a surrogate variable for climate. From the Spearman correlation analysis, we also realized that SOC had the highest correlation with rainfall followed by temperature and altitude. Moreover, these three variables had very high correlation (>0.900 **) with each other, which shows that they can substitute each other. Different studies in different regions also reported that climate had the highest influence on SOC [30–35].

Similar trends were observed with the 27, 15 and 08 input variables that rainfall occupied the first place in all with the XGBoost, RF, and GB models. Moreover, the importance of rainfall increased as the number of input variables decreased. For instance, rainfall addressed 33.41, 28.51, and 62.50% of the SOC variation according to the XGBoost, RF, and GB models with 08 input variables, respectively (Figure 8d–f). Thus, rainfall has the greatest SOC predictive capacity in the study area.

Figure 9 summarizes the average variables importance for the XGBoost, RF and GB models with 36 variables. It also distinguishes different variable types through different colours. The first most important variable was rainfall and addressed 37.13% of the SOC variance. The second most important SOC predictor variable was altitude (12.00%) followed by temperature (11.36%), taxonomy (7.96%), SWIR2 (5.73%), sand (5.65%), clay (2.87%), NIR (2.09%), pH (1.25%), and NDVI (1.09%), respectively, among others.

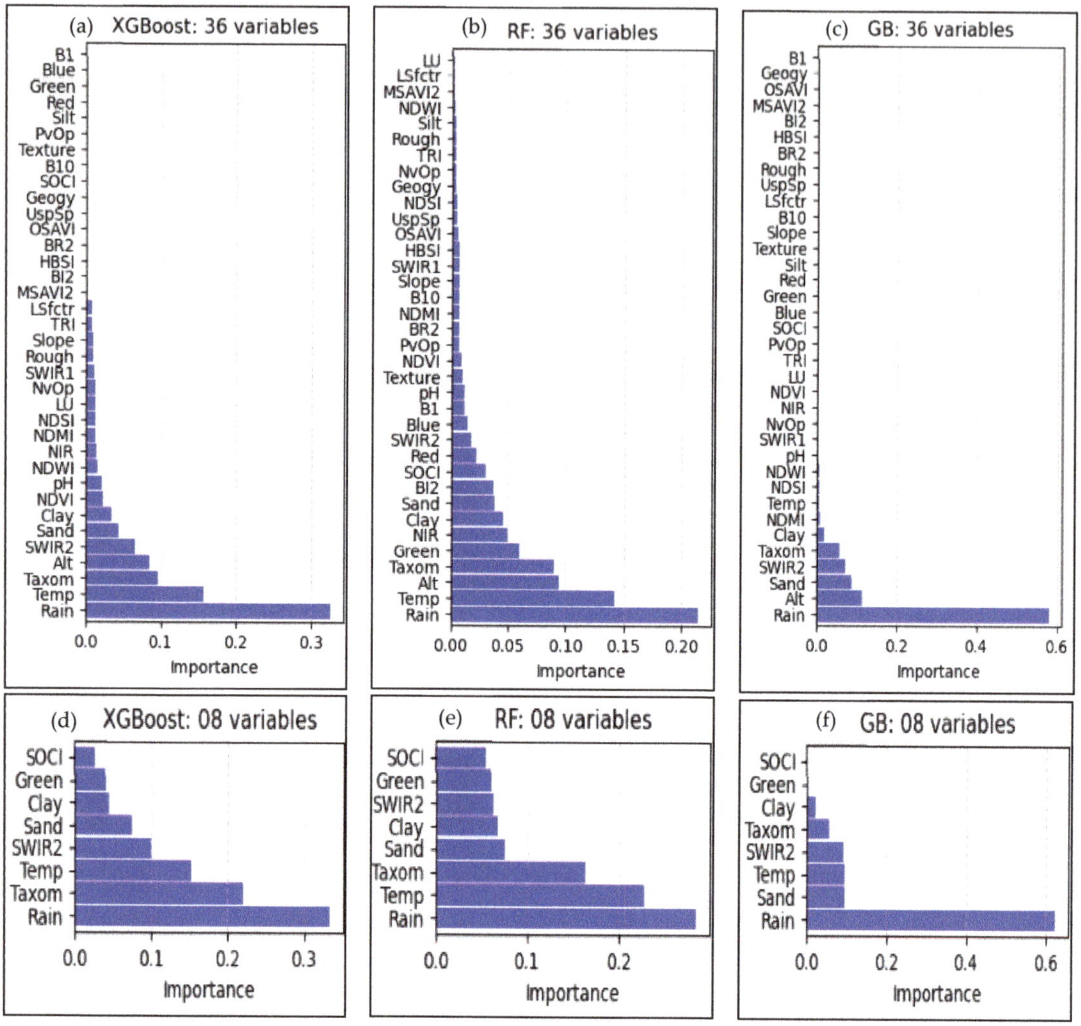

Figure 8. Variables importance for SOC prediction: XGBoost, RF, and GB models with 36 and 08 variables.

Climate, mainly through rainfall, and secondly through temperature, and altitude, holds the highest influence (60.48%) on SOC in the study area. This finding agrees with the findings of [20,100] who reported that rainfall dictates the vegetation dynamics in the country, and [101] also stated that rainfall is crucial variable for deforestation and desertification studies in the country. Since the 1960s, in the country, the frequency of moderate and extreme droughts has increased [18] and temperature has risen in the past 6 decades by about 1.7 °C, which has led to climate change with great loss impacts on agricultural production, biodiversity and overall ecosystem resilience [102]. A comprehensive analysis on East African agriculture and climate change by [103] also presented decreasing rainfed sorghum yield due to climate change in the study area. Thus, adoption of climate-smart agriculture that enhances soil moisture and minimizes water loss could be effective in the study area. In this essence, the country's widespread campaigns on soil and water conservation are encouraged. However, monitoring and evaluation is needed for more effective and efficient actions.

Soil nature (taxonomy, sand, and clay) and L8 spectral bands (SWIR2, NIR) had also good capacity of SOC prediction in the study area. Sand is mostly correlated to SOC negatively as coarse texture is the sign of removal of fine particles by soil deflation, and clay holds the opposite. Thus, these two particles are important indicators of SOC status. Moreover, SWIR2 and NIR are sensitive to soil moisture and soil organic matter [104] and are important SOC predictors. Land use, geology, vegetation indices, bare soil indices, and topographic variables showed weak SOC predictive capacity.

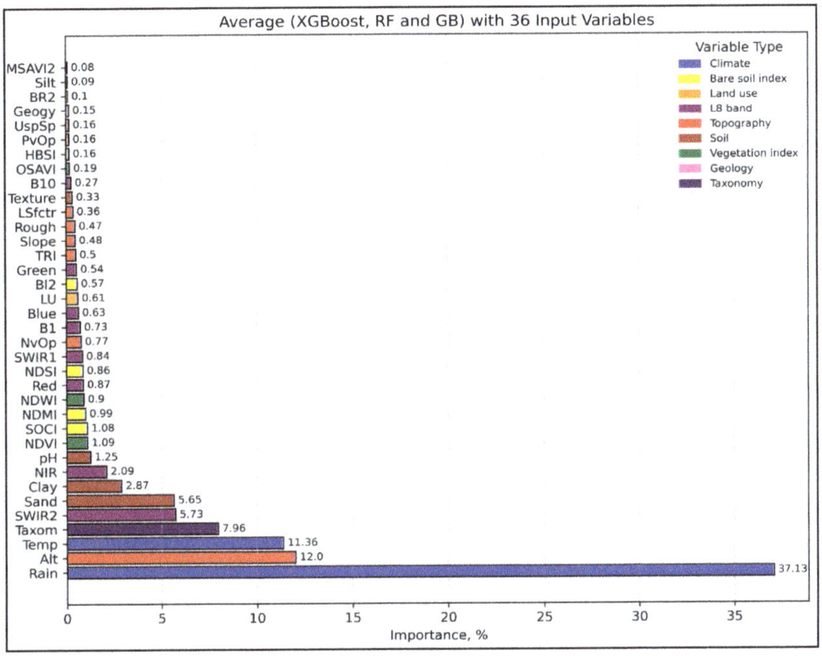

Figure 9. Average variables importance for SOC prediction of the XGBoost, RF, and GB models with 36 variables.

3.6. SOC Mapping

Here, we selected the best one from each model and presented their predicted SOC (Table 7). In all models the SD, CV and skewness of the predicted SOC were reduced as compared to the observed SOC. The lowest CV (50.00%) was recorded by the RF_08 model and the highest CV (56.39%) by the PLS_27 as opposed to 61.65% for the observed SOC. The XGBoost, GB, and RF models gave relatively lower ranges between the minimum and maximum SOC values as compared to the MLR, PLS and Cubist models. Almost all the models fell short of predicting the minimum and maximum values, and the predicted SOC values were shrunk towards the mean. However, the PLS_27 model recorded a minimum SOC value of −0.02%, as opposed to 0.02% for the observed SOC.

Table 7. Descriptive statistics of observed and predicted SOC.

SOC, %	Range	Min	Max	Mean	Median	25%	50%	75%	SD	CV, %	Skewness	Kurtosis
Obs_SOC	0.99	0.02	1.01	0.43	0.35	0.21	0.35	0.68	0.27	61.65	0.37	−1.13
MLR_15	0.86	0.05	0.91	0.44	0.43	0.22	0.43	0.63	0.24	55.24	0.12	−1.33
Cubist_15	0.85	0.06	0.91	0.44	0.43	0.23	0.43	0.67	0.24	54.24	0.18	−1.35
PLS_27	0.94	−0.02	0.92	0.43	0.41	0.21	0.41	0.66	0.24	56.39	0.12	−1.29
RF_08	0.66	0.13	0.79	0.44	0.44	0.24	0.44	0.63	0.22	50.00	0.10	−1.45
GB_27	0.68	0.12	0.80	0.43	0.42	0.25	0.42	0.63	0.22	50.27	0.11	−1.50
XGB_27	0.69	0.12	0.81	0.43	0.40	0.24	0.40	0.64	0.22	51.22	0.11	−1.51

Obs_SOC = observed SOC, Min = minimum, Max = maximum, SD = standard deviation, CV = coefficient of variation.

Rainfall was found to be the main SOC deriver in the study area. Thus, we plotted scatter diagrams of rainfall versus predicted SOC and executed equations and R^2 for each model (Figure 10). The R^2 was the highest in polynomial relations in each model, and degree 2 quadratic equations were developed for each model. Predicted SOC in all models showed high correlation with rainfall; the highest R^2 be 0.8485 for the MLR_15 and the lowest, 0.8172, for the PLS_27. These high correlations indicate the high possibility of calculating and mapping SOC as a function of rainfall using the equations developed in each model for the study area. Figure 10 displays scatter plots of rainfall versus predicted SOC from the selected models and their respective R^2 and quadratic equations.

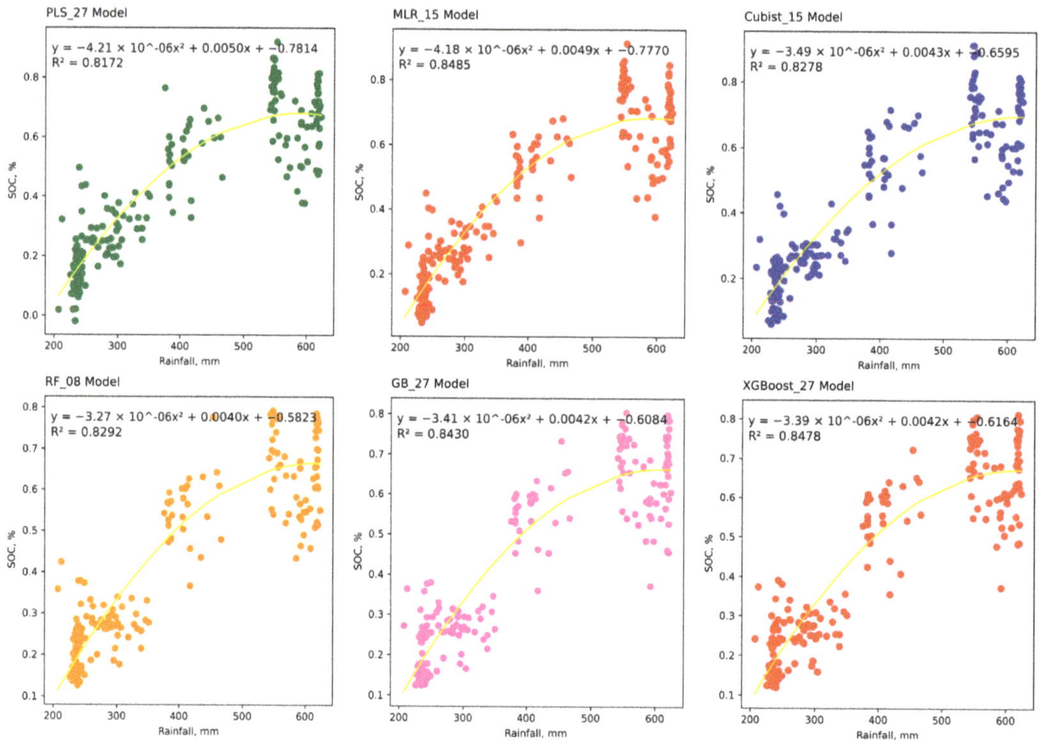

Figure 10. Predicted SOC versus rainfall scatter plots, R^2 and quadratic equations for the different models.

We generated continuous SOC maps using the equations we developed from the rainfall versus predicted SOC scatter plots of each model. These were done using the raster calculator in QGIS 3.38 having raster rainfall map of the study area and executing the equations. All the models showed similar pattern that predicted SOC increased from the North to the South in the study area (Figure 11) in the same pattern as the rainfall and altitude do, and vice versa as the temperature does. The MLR_15 and PLS_27 models showed relatively denser SOC values for the high SOC areas (southern part), and relatively lower SOC values for the low SOC areas (northern part) as compared to the other models. Generally, all the models revealed that the southern and southeastern parts had the highest and northern and northwestern parts had the lowest predicted SOC.

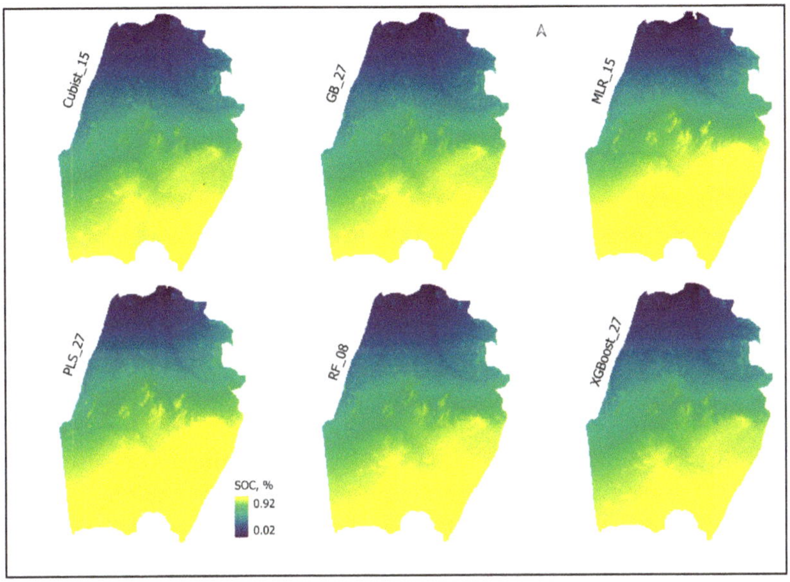

Figure 11. SOC maps developed using the equations from the rainfall versus predicted SOC scatter plots in each model.

3.7. Comparison with the SoilGrids250m Product

Eritrea doesn't have its own soil database, and it adopts the SoilGrids250m [105] product [10,11], which is used for different purposes; for instance, national Land Degradation Neutrality (LDN) assessments and reports [10,11]. Global digital soil maps (DSM) such as the SoilGrids250m are useful assets. However, these may underestimate or overestimate targeted soil properties in regions where there were no/limited soil data for training such as most of the African countries including Eritrea. Thus, we compared our model (Predicted_RF), and SoilGrids250m SOC against the observed SOC. The mean SOC of the observed, predicted (RF model), and SoilGrids250m were 0.43, 0.44, and 0.99%, respectively (Figure 12). The mean observed and predicted SOC were statistically equal. However, the SoilGrids250m highly significantly overestimated the mean SOC. Moreover, the range between the minimum (0.00%) and maximum (2.06%) SOC for the SoilGrids250m was large. Similar studies also reported that SoilGrids250m product overestimated SOC [106–109]. Thus, we have developed workable local SOC predictor models for the Western Lowlands of Eritrea, and the necessity for modeling SOC in other parts of the country seems urgent.

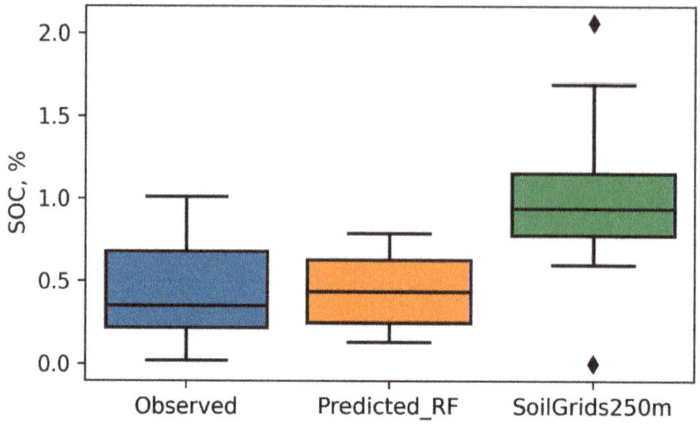

Figure 12. Mean SOC according to Observed, Predicted_RF and SoilGrids250m models; extreme out layers (black diamond shapes) are shown within the SoilGrids250m indicating its high uncertainty for the study area.

3.8. Limitations of the Study

The study focused solely on the spatial distribution of SOC in surface soils (0–30 cm) using only one-time Landsat 8 images, along with other environmental covariates. This limited the analysis of subsurface soil layers and temporal dynamics of SOC. Additionally, the developed models may exhibit deficiencies due to the small amount of training data, variations in timing, and uneven soil sampling frames. Thus, to enhance the models further, it is essential to incorporate spatio-temporal dynamics of both surface and subsurface soils, utilizing multispectral and hyperspectral satellite images, along with other environmental covariates. This should include studies of climate scenarios with a larger and more evenly distributed set of soil samples in the study area, particularly in the context of a changing climate.

4. Conclusions

The soils of the Western Lowlands of Eritrea were non-saline with moderately alkaline pH, characterized by high sand content, moderate silt and clay, and low SOC (0.43%) content with high spatial variability. The low SOC levels require urgent attention due to their critical role in soil fertility, food production, ecosystem restoration, and climate change mitigation. Furthermore, SOC levels in grazing, rainfed, and irrigated farming land uses were poor and statistically similar, primarily due to suboptimal management practices that deplete soil nutrients. In contrast, natural forestlands exhibited significantly higher SOC, highlighting their potential for soil carbon sequestration. To address these challenges, it is essential to implement awareness-raising campaigns and training programs for farmers and land managers in the study area, focusing on soil nutrient and irrigation water management, crop rotation, legume–cereal intercropping, agroforestry practices, cover cropping, green manuring, composting, and the application of bio-fertilizers and biochar technologies.

The spatial distribution of SOC in the study area was modeled with high accuracy using MLR, PLS, Cubist, RF, GB, and XGBoost models. The XGBoost model with 27 input variables achieved the highest SOC prediction accuracy, with RMSE = 0.118%, R^2 = 0.758, RPD = 2.252, and MD = 0.48%. In contrast, the MLR model with 8 input variables showed the lowest accuracy, with RMSE = 0.141, R^2 = 0.742, RPD = 1.883, and MD = 6.68%. Rainfall emerged as the most significant predictor of SOC, explaining over 60% of the variance, while other important predictors included temperature, soil taxonomy, SWIR2, NIR, sand, and clay.

Reducing very high multicollinearity (>0.900**) proved beneficial for all models. However, a reduction in high multicollinearity (>0.700**) did not lead to improvements for the PLS, RF, GB, and XGBoost models. When compared to the Boruta algorithm-based variable selection, the multicollinearity reduction strategy resulted in notable model improvements for the MLR, PLS, XGBoost, Cubist, and GB models, with performance gains of 10.41%, 10.06%, 6.72%, 6.50%, and 3.15%, respectively.

We conclude that all the developed models, except for MLR_36 and MLR_27, demonstrated good accuracy and can be effectively utilized for the planning and monitoring of soil fertility and productivity improvements, ecosystem restoration, and climate change mitigation efforts. For more robust results, we recommend conducting additional SOC modeling studies with larger soil sample sizes from multiple areas and comparing a wider range of models, both in the study area and in other regions of the country.

Author Contributions: Conceptualization, T.T., E.S.M. and W.O.; methodology, T.T.; software, T.T.; validation, I.Y.S., T.T., E.S.M., D.E.K., N.Y.R. and W.O.; formal analysis, T.T.; investigation, T.T.; data curation, T.T.; writing—original draft preparation, T.T.; writing—review and editing, I.Y.S., E.S.M., D.E.K., N.Y.R. and W.O.; visualization, T.T.; supervision, E.S.M.; project administration, E.S.M., D.E.K. and N.Y.R. All authors have read and agreed to the published version of the manuscript.

Funding: This publication has been supported by the RUDN University Scientific Projects Grant System, project no. 202786-2-000.

Institutional Review Board Statement: Ethical review and approval were waived for this study by the Ministry of Health. Ethical clearance is only required for studies involving sensitive demographic, socio-economic, health, or cultural data, which was not applicable to this research. The study involved key informant interviews limited to general agricultural practices and natural resource management.

Informed Consent Statement: Informed consent was obtained from all subjects involved in the study.

Data Availability Statement: Data is available from the corresponding author upon formal and reasonable request.

Acknowledgments: The authors deeply acknowledge RUDN University; Ministry of Agriculture (MoA), Eritrea; National Agricultural Research Institute (NARI); Hamelmalo Agricultural College (HAC); and Eritrean Crops and Livestock Corporation (ECLC). Our great acknowledgements also go to the anonymous peer and editorial reviewers for their constructive comments.

Conflicts of Interest: The authors declare no conflicts of interest.

References

1. FAO. *Soil Organic Carbon: The Hidden Potential*; Food and Agriculture Organization of the United Nations: Rome, Italy, 2017. Available online: https://openknowledge.fao.org/server/api/core/bitstreams/b382a255-5bd5-4656-a8cd-e30fff1a8bfe/content (accessed on 3 September 2025).
2. Page, K.L.; Dang, Y.P.; Dalal, R.C. The ability of conservation agriculture to conserve soil organic carbon and the subsequent impact on soil physical, chemical, and biological properties and yield. *Front. Sustain. Food Syst.* **2020**, *4*, 31. [CrossRef]
3. Kherif, O.; Keskes, M.I.; Pansu, M.; Ouaret, W.; Rebouh, Y.N.; Dokukin, P.; Kucher, D.; Latati, M. Agroecological modelling of nitrogen and carbon transfers between decomposer micro-organisms, plant symbionts, soil and atmosphere in an intercropping system. *Ecol. Model.* **2021**, *440*, 109390. [CrossRef]
4. Tan, Z.X.; Lal, R.; Smeck, N.E.; Calhoun, F.G. Relationships between soil organic carbon pool and site variables in Ohio. *Geoderma* **2004**, *121*, 187–195. [CrossRef]
5. Pringle, M.J.; Allen, D.E.; Phelps, D.G.; Bray, S.G.; Orton, T.G.; Dalal, R.C. The effect of pasture utilization rate on stocks of soil organic carbon and total nitrogen in a semi-arid tropical grassland. *Agric. Ecosyst. Environ.* **2014**, *195*, 83–90. [CrossRef]
6. Schulz, K.; Voigt, K.; Beusch, C.; Almeida-Cortez, J.S.; Kowarik, I.; Walz, A.; Cierjacks, A. Grazing deteriorates the soil carbon stocks of Caatinga forest ecosystems in Brazil. *For. Ecol. Manag.* **2016**, *367*, 62–70. [CrossRef]

7. Tolimir, M.; Kresović, B.; Životić, L.; Dragović, S.; Dragović, R.; Sredojević, Z.; Gajić, B. The conversion of forestland into agricultural land without appropriate measures to conserve SOM leads to the degradation of physical and rheological soil properties. *Sci. Rep.* **2020**, *10*, 13668. [CrossRef]
8. Don, A.; Schumacher, J.; Freibauer, A. Impact of tropical land-use change on soil organic carbon stocks–A meta-analysis. *Glob. Change Biol.* **2011**, *17*, 1658–1670. [CrossRef]
9. Sanderman, J.; Hengl, T.; Fiske, G.J. Soil carbon debt of 12000 years of human land use. *Proc. Natl. Acad. Sci. USA* **2017**, *114*, 9575–9580. [CrossRef]
10. Berhe, S.M. Final Country Report of the Land Degradation Neutrality Target Setting Programme in Eritrea. In *The State of Eritrea*; UNCCD National Focal Point: Bonn, Germany; LDN National Working Group, Ministry of Agriculture: Asmara, Eritrea, 2018. Available online: https://www.unccd.int/sites/default/files/ldn_targets/Eritrea%20LDN%20TSP%20Country%20Report.pdf (accessed on 3 September 2025).
11. MoA. *National Land Degradation Neutrality Targets*; Ministry of Agriculture: Asmara, Eritrea, 2018. Available online: https://www.unccd.int/sites/default/files/ldn_targets/Eritrea%20LDN%20Country%20Commitments.pdf (accessed on 26 October 2022).
12. Ghebrezgabher, M.G.; Taibao, Y.; Xuemei, Y.; Congqiang, W. Assessment of desertification in Eritrea: Degradation based on Landsat images. *J. Arid. Land* **2019**, *11*, 319–331. [CrossRef]
13. Nuguse, M.T.; Singh, B.; Ogbazghi, W. Studies on soil organic carbon and some physico-chemical properties as affected by different land uses in Eritrea. *J. Soil Water Cons.* **2019**, *18*, 213–222. [CrossRef]
14. Tesfay, T.; Ogbazghi, W.; Singh, B. Effects of soil and water conservation interventions on some physico-chemical properties of soil in Hamelmalo and Serejeka Sub-zones of Eritrea. *J. Soil Water Cons.* **2020**, *19*, 229–234. [CrossRef]
15. Tesfay, T.; Mohamed, E.S.; Ghebretnsae, T.W.; Ghebremariam, S.B.; Mehrteab, M. Soil organic carbon stock assessment for soil fertility improvement, ecosystem restoration and climate-change mitigation. In Proceedings of the RIEEM 2024, E3S Web of Conferences, Moscow, Russia, 26–28 April 2024; Volume 555. [CrossRef]
16. Tesfay, T.; Mohamed, E.S.; Medhanie, M.; Ghebretnsae, T.W.; Sereke, T.E. Soil organic carbon losses following conversion of natural forests into agriculture. *Dokuchaev Soil Bull.* **2025**, *123*, 100–115. [CrossRef]
17. Tesfay, T.; Ogbazghi, W.; Singh, B.; Tsegai, T. Factors Influencing Soil and Water Conservation Adoption in Basheri, Gheshnashm and Shmangus Laelai, Eritrea. *IRA Int. J. Appl. Sci.* **2018**, *12*, 7–14. [CrossRef]
18. Ghebrezgabher, M.G.; Yang, X. Spatio-temporal assessment of climate change in Eritrea based on precipitation and temperature variables. *World Wide J. Multidiscip. Res. Dev.* **2018**, *4*, 1–10.
19. Ghebrezgabher, M.G.; Yang, T.; Yang, X.; Wang, X. Extracting and analyzing forest and woodland cover change in Eritrea based on Landsat data using supervised classification. *Egypt J. Remote Sens. Space Sci.* **2016**, *19*, 37–47. [CrossRef]
20. Measho, S.; Chen, B.; Trisurat, Y.; Pellikka, P.; Guo, L. Spatio-Temporal Analysis of Vegetation Dynamics as a Response to Climate Variability and Drought Patterns in the Semiarid Region, Eritrea. *Remote Sens.* **2019**, *11*, 724. [CrossRef]
21. Adhikari, K.; Hartemink, A.E.; Minasny, B.; Kheir, R.B.; Greve, M.B.; Greve, M.H. Digital mapping of soil organic carbon contents and stocks in Denmark. *PLoS ONE* **2014**, *9*, e105519. [CrossRef]
22. Mohamed, E.S.; Saleh, A.M.; Belal, A.B.; Gad, A.A. Application of near infrared reflectance for quantitative assessment of soil properties. *Egypt. J. Remote Sens. Space Sci.* **2018**, *21*, 1–14. [CrossRef]
23. Gouda, M.; Abu-hashim, M.; Nassrallah, A.; Khalil, M.N.; Hendawy, E.; Benhasher, F.F.; Shokr, M.S.; Elshewy, M.A.; Mohamed, E.S. Integration of remote sensing and artificial neural networks for prediction of soil organic carbon in arid zones. *Front. Environ. Sci* **2024**, *12*, 1448601. [CrossRef]
24. Chen, Q.; Wang, Y.; Zhu, X. Soil organic carbon estimation using remote sensing data-driven machine learning. *PeerJ* **2024**, *12*, e17836. [CrossRef]
25. Nguyen, T.T.; Pham, T.D.; Nguyen, C.T.; Delfos, J.; Archibald, R.; Dang, K.B.; Hoang, N.B.; Guo, W.; Ngo, H.H. A novel intelligence approach based active and ensemble learning for agricultural soil organic carbon prediction using multispectral and SAR data fusion. *Sci. Total Environ.* **2022**, *804*, 150187. [CrossRef] [PubMed]
26. Meliho, M.; Boulmane, M.; Khattabi, A.; Dansou, C.E.; Orlando, C.A.; Mhammdi, N.; Noumonvi, K.D. Spatial prediction of soil organic carbon stock in the Moroccan high atlas using machine learning. *Remote Sens.* **2023**, *15*, 2494. [CrossRef]
27. Suleymanov, A.; Tuktarova, I.; Belan, L.; Suleymanov, R.; Gabbasova, I.; Araslanova, L. Spatial prediction of soil properties using random forest, k-nearest neighbors and cubist approaches in the foothills of the Ural Mountains, Russia. *Model. Earth Syst. Environ.* **2023**, *9*, 3461–3471. [CrossRef]
28. Devine, S.; O'Geen, A.; Liu, H.; Jin, Y.; Dahlke, H.; Larsen, R.; Dahlgren, R. Terrain attributes and forage productivity predict catchment-scale soil organic carbon stocks. *Geoderma* **2020**, *368*, 114286. [CrossRef]
29. Zhou, T.; Geng, Y.; Chen, J.; Pan, J.; Haase, D.; Lausch, A. High-resolution digital mapping of soil organic carbon and soil total nitrogen using DEM derivatives, Sentinel-1 and Sentinel-2 data based on machine learning algorithms. *Sci. Total Environ.* **2020**, *729*, 138244. [CrossRef]

30. Zhao, F.; Wu, Y.; Hui, J.; Sivakumar, B.; Meng, X.; Liu, S. Projected soil organic carbon loss in response to climate warming and soil water content in a loess watershed. *Carbon Balance Manag.* **2021**, *16*, 24. [CrossRef]
31. Shen, C.; Xiao, W.; Chen, J.; Hua, L.; Huang, Z. Climate-Sensitive Spatial Variability of Soil Organic Carbon in Multiple Forests, Central China. *Glob. Eco. Cons* **2023**, *46*, e02555. [CrossRef]
32. Zhou, Y.; Chartin, C.; Van Oost, K.; van Wesemael, B. High-resolution soil organic carbon mapping at the field scale in Southern Belgium (Wallonia). *Geoderma* **2022**, *422*, 115929. [CrossRef]
33. Galluzzi, G.; Plaza, C.; Priori, S.; Giannetta, B.; Zaccone, C. Soil organic matter dynamics and stability: Climate vs. time. *Sci. Total Environ.* **2024**, *929*, 172441. [CrossRef]
34. Naderi, M.; Saatsaz, M.; Behrouj Peely, A. Extreme climate events under global warming in Iran. *Hydrol. Sci. J.* **2024**, *69*, 337–364. [CrossRef]
35. von Fromm, S.F.; Doetterl, S.; Butler, B.M.; Aynekulu, E.; Berhe, A.A.; Haefele, S.M.; McGrath, S.P.; Shepherd, K.D.; Six, J.; Tamene, L.; et al. Controls on timescales of soil organic carbon persistence across sub-Saharan Africa. *Glob. Change Biol.* **2024**, *30*, e17089. [CrossRef] [PubMed]
36. Ayala Izurieta, J.E.; Jara Santillán, C.A.; Márquez, C.O.; García, V.J.; Rivera-Caicedo, J.P.; Van Wittenberghe, S.; Delegido, J.; Verrelst, J. Improving the remote estimation of soil organic carbon in complex ecosystems with Sentinel-2 and GIS using Gaussian processes regression. *Plant Soil* **2022**, *479*, 159–183. [CrossRef] [PubMed]
37. Negassa, M.K.; Haile, M.; Feyisa, G.L.; Wogi, L.; Liben, F.M. Soil Organic Carbon Stock Prediction: Fate under 2050 Climate Scenarios, the Case of Eastern Ethiopia. *Sustainability* **2023**, *15*, 6495. [CrossRef]
38. Emadi, M.; Taghizadeh-Mehrjardi, R.; Cherati, A.; Danesh, M.; Mosavi, A.; Scholten, T. Predicting and mapping of soil organic carbon using machine learning algorithms in Northern Iran. *Remote Sens.* **2020**, *12*, 2234. [CrossRef]
39. Baul, T.K.; Charkraborty, A.; Peuly, T.A.; Karmakar, S.; Nandi, R.; Kilpeläinen, A. Effects of varying forest management on soil carbon and nutrients in hill and coastal home gardens in Bangladesh. *J. Soil Sci. Plant Nutr.* **2022**, *22*, 719–731. [CrossRef]
40. Ayala Izurieta, J.E.; Márquez, C.O.; García, V.J.; Jara Santillán, C.A.; Sisti, J.M.; Pasqualotto, N.; Van Wittenberghe, S.; Delegido, J. Multi-predictor mapping of soil organic carbon in the alpine tundra: A case study for the central Ecuadorian páramo. *Carbon Balance Manag.* **2021**, *16*, 32. [CrossRef]
41. *Pre-Investment Study on Forestry and Wildlife Sub-Sector of Eritrea*; FAO: Rome, Italy, 1997.
42. Fick, S.E.; Hijmans, R.J. WorldClim 2: New 1 km spatial resolution climate surfaces for global land areas. *Int. J. Clim* **2017**, *37*, 4302–4315. [CrossRef]
43. Naty, A. Environment, Society and the State in Western Eritrea. *Africa* **2002**, *72*, 569–597. [CrossRef]
44. Thomas, G.W. Soil pH and Soil Acidity. In *Method of Soil Analysis, Part 3: Chemical Methods*; SSSA Inc.: Madison, WI, USA; ASA Inc.: Madison, WI, USA, 1996; pp. 475–490.
45. Rhoades, J.D. Salinity: Electrical conductivity and total dissolved solids. In *Methods of Soil Analysis: Part 3*; SSSA Book Series No. 5; SSSA Inc.: Madison, WI, USA; ASA Inc.: Madison, WI, USA, 1996; pp. 417–435.
46. Lavkulich, L.M. *Methods Manual: Pedology Laboratory*; University of British Columbia, Department of Soil Science: Vancouver, BC, Canada, 1981.
47. Walkley, A.J.; Black, I.A. Estimation of soil organic carbon by the chromic acid titration method. *Soil Sci.* **1934**, *37*, 29–38. [CrossRef]
48. FAO. *Standard Operating Procedure for Soil Organic Carbon Walkley-Black Method (Titration and Colorimetric Method)*; Global Soil Laboratory Network GLOSOLAN: Rome, Italy, 2019. Available online: https://www.fao.org/3/ca7471en/ca7471en.pdf (accessed on 3 September 2025).
49. *Harmonized World Soil Database Version 2.0*; FAO: Rome, Italy; IIASA: Laxenburg, Austria, 2023. [CrossRef]
50. Ostle, N.J.; Levy, P.E.; Evans, C.D.; Smith, P. UK land use and soil carbon sequestration. *Land. Use Policy* **2009**, *26S*, S274–S283. [CrossRef]
51. Wiesmeier, M.; Spörlein, P.; Geuß, U.; Hangen, E.; Haug, S.; Reischl, A.; Schilling, B.; von Lützow, M.; Kögel-Knabner, I. Soil organic carbon stocks in southeast Germany (Bavaria) as affected by land use, soil type and sampling depth. *Glob. Change Biol.* **2012**, *18*, 2233–2245. [CrossRef]
52. Sodango, T.H.; Sha, J.; Li, X.; Noszczyk, T.; Shang, J.; Aneseyee, A.B.; Bao, Z. Modelling the Spatial Dynamics of Soil Organic Carbon Using Remotely-Sensed Predictors in Fuzhou City, China. *Remote Sens.* **2021**, *13*, 1682. [CrossRef]
53. Liu, F.; Wu, H.; Zhao, Y.; Li, D.; Yang, J.; Song, X.; Shi, Z.; Zhu, A.; Zhang, G. Mapping high-resolution national soil information grids of China. *Sci. Bull.* **2022**, *67*, 328–340. [CrossRef] [PubMed]
54. Zeraatpisheh, M.; Garosi, Y.; Reza Owliaie, H.; Ayoubi, S.; Taghizadeh-Mehrjardi, R.; Scholten, T.; Xu, M. Improving the spatial prediction of soil organic carbon using environmental covariates selection: A comparison of a group of environmental covariates. *Catena* **2022**, *208*, 105723. [CrossRef]
55. Yami, B.; Singh, N.J.; Handique, B.K.; Swami, S. Mapping and monitoring of soil organic carbon using regression analysis of spectral indices. *Curr. Sci* **2023**, *124*, 1431–1444. [CrossRef]

56. Hosseinpour-Zarnaq, M.; Moshiri, F.; Jamshidi, M.; Taghizadeh-Mehrjardi, R.; Tehrani, M.M.; Meymand, F.E. Monitoring changes in soil organic carbon using satellite based variables and machine learning algorithms in arid and semi-arid regions. *Environ. Earth Sci* **2024**, *83*, 582. [CrossRef]
57. Zhang, Y.; Wang, Y.; Bai, Y.; Zhang, R.; Liu, X.; Ma, X. Prediction of Spatial Distribution of Soil Organic Carbon in Helan Farmland Based on Different Prediction Models. *Land* **2023**, *12*, 1984. [CrossRef]
58. Rouse , J.W., Jr.; Haas, R.H.; Schell, J.A.; Deering, D.W. *Monitoring Vegetation Systems in the Great Plains with ERTS*; NASA Special Publication; NASA Goddard Space Flight Center: Greenbelt, MD, USA, 1974; Volume 351, p. 309.
59. Mokarram, M.; Roshan, G.; Negahban, S. Landform classification using topography position index (case study: Salt dome of Korsia-Darab plain, Iran. *Model. Earth Syst Environ.* **2015**, *1*, 1–7. [CrossRef]
60. Qi, J.; Chehbouni, A.; Huete, A.R.; Kerr, Y.H.; Sorooshian, S. A modified soil adjusted vegetation index. *Remote Sens. Environ.* **1994**, *48*, 119–126. [CrossRef]
61. Deng, Y.; Wu, C.; Li, M.; Chen, R. RNDSI: A ratio normalized difference soil index for remote sensing of urban/suburban environments. *Int. J. Appl. Earth Obs. Geoinf.* **2015**, *39*, 40–48. [CrossRef]
62. Salas, E.A.L.; Kumaran, S.S. Hyperspectral Bare Soil Index (HBSI): Mapping Soil Using an Ensemble of Spectral Indices in Machine Learning Environment. *Land* **2023**, *12*, 1375. [CrossRef]
63. Skakun, R.S.; Wulder, M.A.; Franklin, S.E. Sensitivity of the thematic mapper enhanced wetness difference index to detect mountain pine beetle red-attack damage. *Remote Sens. Environ.* **2003**, *86*, 433–443. [CrossRef]
64. Dvorakova, K.; Heiden, U.; Pepers, K.; Staats, G.; van Os, G.; van Wesemael, B. Improving soil organic carbon predictions from a Sentinel–2 soil composite by assessing surface conditions and uncertainties. *Geoderma* **2023**, *429*, 116128. [CrossRef]
65. Carmona, J.Á.S.; Quirós, E.; Mayoral, V.; Charro, C. Assessing the potential of multispectral and thermal UAV imagery from archaeological sites: A case study from the Iron Age hillfort of Villasviejas del Tamuja (Cáceres, Spain). *J. Archaeol. Sci. Rep.* **2020**, *31*, 102312. [CrossRef]
66. Wehrhan, M.; Sommer, M. A parsimonious approach to estimate soil organic carbon applying unmanned aerial system (UAS) multispectral imagery and the topographic position index in a heterogeneous soil landscape. *Remote Sens.* **2021**, *13*, 3557. [CrossRef]
67. Deng, Y.; Zhao, X.; Tian, Y.; Zhang, X.; Cao, J.; Yin, L.; Zhang, B. Impact of different environmental covariate selection strategies on mapping accuracy of soil organic carbon in salt-affected coastal farmland. *Ecol. Indic.* **2025**, *178*, 113956. [CrossRef]
68. Zhu, J.; Jin, Y.; Zhu, W.; Kun Lee, D. VIS-NIR spectroscopy and environmental factors coupled with PLSR models to predict soil organic carbon and nitrogen. *Int. Soil Water Conserv. Res.* **2024**, *12*, 844–854. [CrossRef]
69. Wang, S.; Zhou, M.; Zhuang, Q.; Guo, L. Prediction Potential of Remote Sensing-Related Variables in the Topsoil Organic Carbon Density of Liaohekou Coastal Wetlands, Northeast China. *Remote Sens.* **2021**, *13*, 4106. [CrossRef]
70. Kuhn, M. Classification and Regression with Random Forest. In *R Package Version 4.6–12*; R Foundation for Statistical Computing: Vienna, Austria, 2017. Available online: https://CRAN.R-project.org/package=randomForest (accessed on 3 September 2025).
71. Gomes, L.C.; Faria, R.M.; de Souza, E.; Veloso, G.V.; Schaefer, C.E.G.R.; Filho, E.I.F. Modelling and mapping soil organic carbon stocks in Brazil. *Geoderma* **2019**, *340*, 337–350. [CrossRef]
72. Mahmoudzadeh, H.; Matinfar, H.R.; Taghizadeh-Mehrjardi, R.; Kerry, R. Spatial prediction of soil organic carbon using machine learning techniques in western Iran. *Geoderma Reg.* **2020**, *21*, e00260. [CrossRef]
73. Rostaminia, M.; Rahmani, A.; Mousavi, S.R.; Taghizadeh-Mehrjardi, R.; Maghsodi, Z. Spatial prediction of soil organic carbon stocks in an arid rangeland using machine learning algorithms. *Environ. Monit. Assess.* **2021**, *193*, 815. [CrossRef]
74. Breiman, L. Random forests. *Mach. Learn.* **2001**, *45*, 5–32. [CrossRef]
75. Fathizad, H.; Taghizadeh-Mehrjardi, R.; Hakimzadeh Ardakani, M.A.; Zeraatpisheh, M.; Heung, B.; Scholten, T. Spatio-temporal assessment of soil organic carbon change using machine learning in arid regions. *Agronomy* **2022**, *12*, 628. [CrossRef]
76. Safanelli, J.L.; Chabrillat, S.; Ben-Dor, E.; Demattê, J.A.M. Multispectral models from bare soil composites for mapping topsoil properties over Europe. *Remote Sens.* **2020**, *12*, 1369. [CrossRef]
77. Xie, B.; Ding, J.; Ge, X.; Li, X.; Han, L.; Wang, Z. Estimation of soil organic carbon content in the Ebinur Lake Wetland, Xinjiang, China, based on multisource remote sensing data and ensemble learning algorithms. *Sensors* **2022**, *22*, 2685. [CrossRef]
78. Elith, J.; Leathwick, J.R.; Hastie, T. A working guide to boosted regression trees. *J. Anim. Ecol.* **2008**, *77*, 802–813. [CrossRef]
79. Zebari, R.; Abdulazeez, A.; Zeebaree, D.; Zebari, D.; Saeed, J. A comprehensive review of dimensionality reduction techniques for feature selection and feature extraction. *J. Appl. Sci. Technol. Trends* **2020**, *1*, 56–70. [CrossRef]
80. Wang, T.; Zhou, W.; Xiao, J.; Li, H.; Yao, L.; Xie, L.; Wang, K. Soil organic carbon prediction using Sentinel-2 data and environmental variables in a karst trough valley area of Southwest China. *Remote Sens.* **2023**, *15*, 2118. [CrossRef]
81. Wadoux, A.M.J.C.; Minasny, B.; McBratney, A.B. Machine learning for digital soil mapping: Applications, challenges and suggested solutions. *Earth-Sci. Rev.* **2020**, *210*, 103359. [CrossRef]
82. Hamzehpour, N.; Shafiadeh-Moghadam, H.; Valavi, R. Exploring the driving forces and digital mapping of soil organic carbon using remote sensing and soil texture. *Catena* **2019**, *182*, 104141. [CrossRef]

83. Taghizadeh-Mehrjardi, R.; Hamzehpour, N.; Hassanzadeh, M.; Heung, B.; Ghebleh Goydaragh, M.; Schmidt, K.; Scholten, T. Enhancing the accuracy of machine learning models using the super learner technique in digital soil mapping. *Geoderma* **2021**, *399*. [CrossRef]
84. Zeraatpisheh, M.; Ayoubi, S.; Jafari, A.; Tajik, S.; Finke, P. Digital mapping of soil properties using multiple machine learning in a semi-arid region, central Iran. *Geoderma* **2019**, *338*, 445–452. [CrossRef]
85. Zhang, X.; Chen, S.; Xue, J.; Wang, N.; Xiao, Y.; Chen, Q.; Hong, Y.; Zhou, Y.; Teng, H.; Hu, B.; et al. Improving model parsimony and accuracy by modified greedy feature selection in digital soil mapping. *Geoderma* **2023**, *432*, 116383. [CrossRef]
86. Poggio, L.; de Sousa, L.M.; Batjes, N.H.; Heuvelink, G.B.M.; Kempen, B.; Ribeiro, E.; Rossiter, D. SoilGrids 2.0: Producing soil information for the globe with quantified spatial uncertainty. *Soil* **2021**, *7*, 217–240. [CrossRef]
87. Hu, B.; Xie, M.; Shi, Z.; Li, H.; Chen, S.; Wang, Z.; Zhou, Y.; Ni, H.; Geng, Y.; Zhu, Q.; et al. Fine-resolution mapping of cropland topsoil pH of Southern China and its environmental application. *Geoderma* **2024**, *442*, 116798. [CrossRef]
88. Zeraatpisheh, M.; White, A.; Darby, H.; Neher, D.A.; Faulkner, J.; Hancock, L.; Kretzler, B.; El-Naboulsi, N.; Turner, H.C.; von Wettberg, E.J.B.; et al. Spatial Mapping and Predictive Modeling of Soil Organic Carbon Stocks in Vermont Agricultural Lands Using Machine Learning and Environmental Variables. *Comput. Electron. Agric.* **2025**, *237*, 110727. [CrossRef]
89. Viscarra-Rossel, R.A.; Taylor, H.J.; McBratney, A.B. Multivariate calibration of hyperspectral γ- ray energy spectra for proximal soil sensing. *Eur. J. Soil Sci.* **2007**, *58*, 343–353. [CrossRef]
90. Estefan, G.; Rolf, S.; John, R. *Methods of Soil, Plant, and Water Analysis: A Manual for the West Asia and North Africa region*, 3rd ed.; ICARDA (International Center for Agricultural Research in the Dry Areas): Beirut, Lebanon, 2013.
91. Hazelton, P.; Murphy, B. *Interpreting Soil Test Results What Do All the Numbers Mean?* 2nd ed.; CSIRO Publishing: Collingwood, Australia, 2007.
92. Hazelton, P.; Murphy, B. Interpreting soil test results. In *Interpreting Soil Test Results*, 3rd ed.; CSIRO Publishing: Clayton, Australia, 2016. [CrossRef]
93. Husein, H.H.; Lucke, B.; Bäumler, R.; Sahwan, W. A Contribution to Soil Fertility Assessment for Arid and Semi-Arid Lands. *Soil Syst.* **2021**, *5*, 42. [CrossRef]
94. Ritchie, M.E. Grazing Management, Forage Production and Soil Carbon Dynamics. *Resources* **2020**, *9*, 49. [CrossRef]
95. Weldewahid, Y.; Habtu, S.; Taye, G.; Teka, K.; Gessesse, T.A. Effects of long-term irrigation practice on soil quality, organic carbon and total nitrogen stocks in the drylands of Ethiopia. *J. Arid. Environ.* **2023**, *214*, 104982. [CrossRef]
96. Ngatia, L.W.; Moriasi, D.; Grace, J.M., III; Fu, R.; Gardner, C.S.; Taylor, R.W. Land Use Change Affects Soil Organic Carbon: An Indicator of Soil Health. In *Environmental Health*; IntechOpen: London, UK, 2021. [CrossRef]
97. Ghimire, P.; Lamichhane, U.; Bolakhe, S.; Lee, C.H.J. Impact of Land Use Types on Soil Organic Carbon and Nitrogen Stocks: A Study from the Lal Bakaiya Watershed in Central Nepal. *Int. J. For. Res.* **2023**, *2023*, 9356474. [CrossRef]
98. Ha, N.T.; Manley-Harris, M.; Pham, T.D.; Hawes, I. The use of radar and optical satellite imagery combined with advanced machine learning and metaheuristic optimization techniques to detect and quantify above ground biomass of intertidal seagrass in a New Zealand estuary. *Int. J. Remote Sens.* **2021**, *42*, 4712–4738. [CrossRef]
99. Ye, Z.; Sheng, Z.; Liu, X.; Ma, Y.; Wang, R.; Ding, S.; Liu, M.; Li, Z.; Wang, Q. Using machine learning algorithms based on GF-6 and google earth engine to predict and map the spatial distribution of soil organic matter content. *Sustainability* **2021**, *13*, 14055. [CrossRef]
100. Iyob, B. Mapping Vegetation Using Landsat TM and ETM+ in Eritrea. Master's Thesis, Department of Geosciences, Oregon State University, Corvallis, OR, USA, 2005.
101. Ghebrezgabher, M.G.; Yang, T.; Yang, X. Remote Sensing and GIS Analysis of Deforestation and Desertification in Central Highland and Eastern Region of Eritrea (1972–2014). *Int. J. Sci. Basic Appl. Res.* **2014**, *18*, 161–176.
102. *Eritrea's Intended Nationally Determined Contribution (INDCs) Report*; Ministry of Land, Water and Environment (MoLWE): Asmara, Eritrea, 2015.
103. Waithaka, M.; Nelson, G.C.; Thomas, T.S.; Kyotalimye, M. *East African Agriculture and Climate Change: A Comprehensive Analysis*; International Food Policy Research Institute: Washington, DC, USA, 2013; ISBN 9780896292055. [CrossRef]
104. Zhang, S.; Tian, J.; Lu, X.; Tian, Q. Temporal and spatial dynamics distribution of organic carbon content of surface soil in coastal wetlands of Yancheng, China from 2000 to 2022 based on Landsat images. *Catena* **2023**, *223*, 106961. [CrossRef]
105. Hengl, T.; Mendes de Jesus, J.; Heuvelink, G.B.M.; Ruiperez Gonzalez, M.; Kilibarda, M.; Blagotić, A.; Shangguan, W.; Wright, M.N.; Geng, X.; Bauer-Marschallinger, B.; et al. SoilGrids250m: Global gridded soil information based on machine learning. *PLoS ONE* **2017**, *12*, e0169748. [CrossRef]
106. Mulder, V.L.; Lacoste, M.; Richer-de-Forges, A.C.; Martin, M.P.; Arrouays, D. National versus global modelling the 3D distribution of soil organic carbon in mainland France. *Geoderma* **2016**, *263*, 16–34. [CrossRef]
107. Tifafi, M.; Guenet, B.; Hatté, C. Large differences in global and regional total soil carbon stock estimates based on SoilGrids, HWSD and NCSCD: Intercomparison and evaluation based on field data from USA, England, Wales and France. *Glob. Biogeochem. Cycles* **2018**, *1*, 42–56. [CrossRef]

108. Liang, Z.; Chen, S.; Yang, Y.; Zhou, Y.; Shi, Z. High-resolution three-dimensional mapping of soil organic carbon in China: Effects of SoilGrids products on national modelling. *Sci. Total Environ.* **2019**, *685*, 480–489. [CrossRef]
109. Silatsa, F.B.T.; Yemefack, M.; Tabi, F.O.; Heuvelink, G.B.M.; Leenaars, J.G.B. Assessing countrywide soil organic carbon stock using hybrid machine learning modelling and legacy soil data in Cameroon. *Geoderma* **2020**, *367*, 114260. [CrossRef]

sustainability

MDPI

Article

AI Advice for Amateur Food Production: Assessing Sustainability of LLM Recommendations

Agnieszka Krzyżewska

Department of Hydrology and Climatology, Institute of Earth and Environmental Sciences, Maria Curie Skłodowska University, 20-718 Lublin, Poland; agnieszka.krzyzewska@umcs.pl

Abstract: Large language models (LLMs) are increasingly consulted by amateur gardeners who rely on them for diagnosing plant problems and selecting management strategies. This study evaluates whether such AI systems promote environmentally sustainable or chemically oriented practices. Fifteen real images of edible plants showing typical health issues were collected during 2024–2025, and four major models—ChatGPT 5.0, Gemini 2.5 Pro, Claude Sonnet 4.5, and Perplexity AI (standard version)—were queried in October 2025 using an identical user-style prompt. Each response was coded across four sustainability dimensions (ecological prevention, diagnostic reasoning, nutrient management, and chemical control) and aggregated into a composite Eco-Score (−1 to +1). Across cases, all models prioritized preventive and low-impact advice, emphasizing pruning, hygiene, compost, and organic sprays while recommending synthetic fungicides or pesticides only occasionally. The highest sustainability alignment was achieved by Perplexity AI (Eco-Score = 0.71) and Gemini 2.5 Pro (0.69), followed by ChatGPT 5.0 (0.57) and Claude Sonnet 4.5 (0.41). Although the models frequently converged in general reasoning, no case achieved full agreement in Eco-Score values across systems. These findings demonstrate that current LLMs generally reinforce sustainable reasoning but vary in interpretative reliability. While they can enhance ecological awareness and accessible plant care knowledge, their diagnostic uncertainty underscores the need for human oversight in AI-assisted amateur food production.

Keywords: large language models (LLMs); sustainable agriculture; home gardening; amateur food production; plant disease diagnosis; AI-driven decision support; environmentally friendly practices; ChatGPT; Gemini; Claude; Perplexity

1. Introduction

Home gardens—including balcony, rooftop and backyard plots—are increasingly recognized as an important element of sustainable food production and urban resilience. Even small-scale cultivation provides environmental, social, and economic benefits: enhanced biodiversity, local carbon sequestration, mitigation of urban heat island effects, and improved access to fresh produce [1,2]. In the context of climate change, such decentralized food production systems have become particularly valuable. Changing precipitation regimes, increasing temperatures and disruptions in global food supply highlight the importance of localized, resilient production methods that combine ecological awareness with social engagement [3].

In the Polish context, family allotment gardens (ROD—Rodzinne Ogrody Działkowe) represent a particularly well-developed form of amateur food cultivation. Their origins date back to the eighteenth century, when the first sites were established to provide working-class residents with access to green space and homegrown products [4]. During the socialist period, RODs played a dual role—supporting household subsistence and providing recreation—and today they remain integral to urban green infrastructure. Poland currently hosts over 4667 allotment complexes, divided into almost one million individual plots covering about 40,863 ha [5]. Studies show that Polish allotment gardens deliver significant ecosystem services, support biodiversity and community well-being, and can enable practices such as composting and rainwater harvesting; however, their ecological contributions vary with use type and management intensity [6–8]. At the same time, they face pressures from urban development and exhibit a long-term shift from predominantly productive to more recreational functions [7,9].

Similar forms of allotment and community gardens are widespread across Europe—for example, "Kleingärten" in Germany, statutory allotments in the United Kingdom, "volkstuinen" in the Netherlands, and "kolonihaver" in Denmark and Sweden. These gardens have deep cultural and planning roots and provide multiple ecosystem and social services, including food provision, biodiversity conservation, and community cohesion [4,10]. In the UK, a well-documented historical decline in allotment land since the mid-twentieth century has been followed by renewed interest linked to food security and urban sustainability [11]. In the Nordic region, long-term ecological studies demonstrate that Swedish "koloniträdgårdar" retain high levels of biodiversity and embody "social–ecological memory"—the transmission of adaptive management knowledge across generations [12,13]. In Denmark, "kolonihaver" have a documented history of urban green provisioning and worker-garden movements, and studies highlight their evolution from worker allotments in the nineteenth century to enduring green infrastructure elements in Danish cities [14,15].

Large language models (LLMs) represent a rapidly developing class of AI systems whose increasing deployment spans both everyday decision support and specialized advisory domains—including agriculture and food production. Studies indicate that LLMs may enable improvements in agricultural extension by simplifying scientific knowledge, tailoring recommendations to individual users, and scaling advisory reach [16]. At the same time, a growing body of research emphasizes that LLMs impose substantial environmental burdens: one survey observes that training and deploying LLMs entail high energy consumption, significant carbon emissions, and substantial water usage, pointing to an urgent need for sustainability-conscious design and deployment [17,18]. Moreover, a dedicated review of applications in agriculture highlights both opportunities (e.g., decision-making, image-based diagnostics) and challenges (e.g., modality alignment, data bias, domain adaptation) for LLMs in food production systems [19]. Thus, while LLMs hold promise for supporting amateur food producers, their environmental profile and advisory alignment with sustainability principles require careful scrutiny.

Despite the growing integration of large language models into domains connected with sustainability, there remains a notable research gap: few studies have empirically examined whether the advice generated by LLMs genuinely supports environmentally responsible behavior and sustainable production practices. Existing work on AI and sustainability tends to focus either on the environmental footprint of model training and operation [17] or on high-level conceptual frameworks for "green AI" [18], but not on the substantive content of AI-generated recommendations in everyday contexts.

In the sphere of amateur food production—which directly relates to key United Nations Sustainable Development Goals, including SDG 2 (Zero Hunger), SDG 12 (Responsible Consumption and Production), and SDG 13 (Climate Action) [20]—this absence of evaluation is particularly relevant. Home and allotment gardeners represent a vast, decentralized user base whose decisions cumulatively affect resource use, biodiversity, and local carbon cycles. Understanding whether LLMs reinforce or undermine sustainable cultivation norms therefore carries both environmental and educational significance.

Moreover, although outputs produced by generative AI systems often appear authoritative and may be readily perceived as expert advice, these models still lack the ability to reliably assess the validity, contextual relevance, and environmental consequences of their own recommendations [17]. As a result, there is a growing risk that users may adopt AI-generated guidance uncritically, including advice that could inadvertently promote unsustainable or environmentally harmful practices.

By qualitatively analyzing and descriptively quantifying the responses of four major large language models—ChatGPT (OpenAI [21]), Gemini (Google DeepMind [22]), Claude (Anthropic [23]), and Perplexity AI [24]—this study evaluates whether, and to what extent, their agricultural recommendations can be considered sustainable. Using a coding framework derived from sustainability science (covering reliance on organic inputs, prevention-based management, avoidance of synthetic chemicals, and resource efficiency [25]), the research systematically compares model outputs across real-world plant health scenarios. To complement the qualitative interpretation, a composite *Eco-Score* was developed to provide a simple quantitative indicator of each model's alignment with sustainable cultivation principles. The findings aim to identify which AI systems most consistently produce environmentally responsible guidance and to inform the design of future AI-assisted advisory tools aligned with the United Nations Sustainable Development Goals.

2. Materials and Methods

2.1. Study Design

This study employed a qualitative comparative design to evaluate the sustainability orientation of agricultural advice generated by four large language models. A qualitative content analysis approach was chosen because it enables the systematic interpretation of meaning patterns in AI-generated texts and the transparent construction of coding frames—categories, definitions, and decision rules—rather than testing model accuracy per se [26]. The analysis focused on how model outputs translate common plant health prompts into actionable recommendations and the extent to which these recommendations align with sustainability principles. The study is situated within the emerging research on the role of large language models in agriculture and sustainability, which highlights both potential advisory benefits and environmental risks of such systems, as well as within the broader debate on the limits of language-only architectures for reliable judgment and understanding [16,17,19,27–29].

2.2. Data Collection

Photographic material illustrating typical plant health issues was collected between 2024 and 2025 in home and allotment gardens located in the vicinity of Lublin, eastern Poland. In total, fifteen representative cases were documented, covering commonly cultivated crops such as tomato (*Solanum lycopersicum*), pepper (*Capsicum annuum*), zucchini (*Cucurbita pepo*), pumpkin (*Cucurbita maxima*), grapevine (*Vitis vinifera*), raspberry (*Rubus idaeus*), currant (*Ribes rubrum*), strawberry (*Fragaria* × *ananassa*), and green bean (*Phaseolus vulgaris*). Thirteen photographs were taken by the author, while two (cases 14 and 15) were provided by another gardener with consent for research use. All images were captured under natural lighting using a standard smartphone camera, without professional calibration, ensuring sufficient clarity for visual diagnosis. The setup of consumer-grade smartphones—Samsung Galaxy S22 (cases 1–13) and iPhone 11 (cases 14–15)—was intentionally chosen to reflect the quality and variability typical of images uploaded by amateur gardeners.

Although parameters such as lighting, focus, or background complexity may influence model interpretation, each large language model received exactly the same input images, ensuring full comparability across systems. The selection of cases followed a purposive sampling strategy, aiming to capture a broad range of both biotic and abiotic stresses typically encountered in amateur food production. The dataset included fungal infections, nutrient deficiencies, physiological disorders, and irrigation-related stress. Several healthy plants were also included to serve as control samples for comparative interpretation.

Each case consisted of a high-resolution image accompanied by a concise textual description of visible symptoms. The images were subsequently used as direct inputs to the large language models, each paired with the same natural language prompt: "What happened to my plant and how can I fix it?" This phrasing was deliberately simple and context-free, designed to reflect how a typical non-expert user might seek advice from an AI system. The use of identical prompts ensured comparability across all models (see Section 2.3).

2.3. *Querying the Large Language Models*

Four large language models were queried in October 2025: ChatGPT 5.0 (OpenAI) [21], Gemini 2.5 Pro (Google DeepMind) [22], Claude Sonnet 4.5 (Anthropic) [23], and Perplexity AI [24]. Each model was accessed through its official web interface in separate sessions, ensuring full isolation between queries and avoiding carryover effects from prior conversations.

For each of the fifteen plant cases, the same high-resolution image was uploaded together with a uniform prompt phrased as follows: "What happened to my plant and how can I fix it?" The prompt was intentionally simple and context-free, designed to emulate the behavior of an average amateur gardener seeking advice from an AI assistant. No additional background information—such as plant species, location, or cultivation method—was provided, and only the model's first answer was recorded for analysis, even when follow-up clarification questions were generated by the system.

This procedure yielded a total of sixty outputs (15 cases × 4 models), all written in English. The full text of each response is available in the Supplementary Materials (Tables S1–S15). The responses were subsequently exported in their original form for coding and sustainability assessment, as described in Section 2.4.

2.4. *Coding Framework and Sustainability Criteria*

The analysis applied a mixed qualitative content analysis approach [26], in which textual responses from the models were coded according to four sustainability-related categories derived from sustainability science, agroecology, and international frameworks such as the FAO Sustainability Assessment of Food and Agriculture Systems [30] and the United Nations 2030 Agenda for Sustainable Development [20], and conceptual syntheses of sustainable agrifood systems [25]. The categories captured distinct dimensions of environmentally responsible cultivation and were defined as follows:

1. ECO—Ecological and preventive practices.

   This category encompassed environmentally sound, non-chemical, or preventive management strategies. Typical examples included removing infected leaves, pruning to improve air circulation, watering at the base of plants, crop rotation, compost use, and employing biological or low-impact remedies. In Case 3 (zucchini), several models advised "remove infected leaves" and "water at the base" while suggesting organic sprays such as milk or baking soda solutions. In Case 4 (currant), Claude recommended pruning affected leaves and "encouraging natural predators," explicitly discouraging broad-spectrum insecticides. Similar ecological reasoning appeared in Cases 11 and 14 (cucurbits and beans), emphasizing ventilation, hygiene, and resistant varieties.

2. DIAG—Diagnostic reasoning.

   This code captured explicit reasoning about the causes of symptoms and suggestions for verification, reflecting the "knowledge-based management" principle of sustainable agriculture. For instance, in Case 2 (grapevine), Gemini advised conducting a soil pH or petiole test to distinguish between magnesium deficiency and root rot stress. In Case 1 (pepper), ChatGPT and Claude both discussed pH-induced nutrient lockout and overwatering as alternative explanations.

3. NUTR—Nutrient management.

   This dimension included advice concerning plant nutrition through either organic or mineral inputs, consistent with the FAO SAFA Soil Quality and Nutrient Cycling themes and SDG 2.4 (soil health). Recommendations such as "apply a balanced liquid fertilizer," "use compost tea," or "spray Epsom salts (MgSO4)" were coded as NUTR = 1. In Case 1 (pepper), multiple models advised Epsom salt foliar feeding; in Case 2 (grapevine), ChatGPT suggested a foliar Mg spray (15–20 g L^{-1}); and in Cases 12 and 13, models recommended compost or manure amendments.

4. CHEM—Synthetic chemical treatments.

   This category recorded explicit references to chemical control products such as fungicides or insecticides, which SAFA and SDG 12.4 classify as potential environmental risks. In Case 2 (grapevine), Claude proposed a copper-based fungicide (Bordeaux mixture) and sulfur dust as curative measures. Similar mentions occurred in Cases 9–11 and 14, where several models recommended copper or systemic fungicides against leaf spot and powdery mildew infections.
   For consistency, all chemical treatments—including those based on copper or sulfur that are conditionally accepted in organic farming—were coded under the CHEM category. This decision reflects a precautionary interpretation of sustainability, recognizing that even organically approved products may cause harm when misused or applied excessively [31–33]. Given that amateur gardeners often lack formal agronomic training, the risk of overapplication or incorrect handling justifies their inclusion as environmentally adverse within the Eco-Score framework.

Each response was manually coded using a binary scheme (1 = present, 0 = absent) for the four categories, with each coded feature counted only once per model output. The binary approach ensured consistency across heterogeneous text lengths while enabling subsequent comparisons between models. The coded data were later aggregated into a composite Eco-Score indicator (see Section 2.5).

2.5. Eco-Score Calculation

To obtain a concise, comparable indicator of sustainability orientation for each model output, a compensatory Eco-Score was computed from four binary codes: ECO (ecological/preventive practices), DIAG (diagnostic reasoning), NUTR (nutrient management), and CHEM (synthetic chemical control). The presence of a dimension in the response was coded as 1, and its absence as 0 (each code counted at most once per response). The index penalizes chemical recommendations by subtracting CHEM in the numerator while still including it in the denominator:

$$\text{Eco-Score} = \frac{ECO + DIAG + NUTR - CHEM}{ECO + DIAG + NUTR + CHEM} \tag{1}$$

The score ranges from −1 (purely chemical, no sustainable dimensions) to +1 (fully aligned with sustainable practice without chemicals). In this study, scores were interpreted qualitatively (higher values indicate stronger sustainability alignment) and were averaged across cases to compare models (see Table 1).

The four sustainability dimensions were treated as equally weighted to ensure transparency and comparability across heterogeneous model outputs. Each dimension represents a distinct and complementary component of sustainable cultivation practices: ecological and preventive actions (ECO), diagnostic reasoning (DIAG), and nutrient management (NUTR) were assigned positive values, while synthetic chemical control (CHEM) was given a negative value due to its recognized environmental risks under FAO SAFA guidelines and SDG 12.4 (Responsible Consumption and Production) [20,30]. This approach also reflects the specific user context of the study: amateur gardeners cultivating home or allotment plots typically lack formal agricultural training, and inappropriate use of chemical treatments could therefore result in disproportionate ecological and health risks. Penalizing such recommendations within the Eco-Score framework highlights the importance of promoting safe, low-impact, and knowledge-based management strategies in this group.

Example: Step-by-step calculations: Case 2: grapevine leaves with interveinal chlorosis and reddening (Table 1).

Coding notes from full responses (Supplementary Materials Table S2):

- ChatGPT: Accurately identified magnesium deficiency; suggested confirming the cause through soil nutrient check and pH testing and explicitly asked whether symptoms start from edges or interveins → DIAG = 1; promoted non-chemical correction (Epsom salt, pH balance, avoiding excess K) and irrigation adjustment → ECO = 1; included nutrient-focused advice (Mg/K balance, steady watering) → NUTR = 1; Applied a copper-based fungicide → CHEM = 1.
- Gemini: Correctly recognized magnesium/potassium deficiency; clearly advised soil and petiole analysis to confirm whether deficiency is Mg or K → DIAG = 1; encouraged balanced soil management and moderate fertilization rather than immediate chemical action → ECO = 1; maintained strong nutritional focus (Mg, K, pH optimization, dolomitic lime) → NUTR = 1; no environmentally harmful substances → CHEM = 0.
- Claude AI: Misdiagnosed downy mildew; did not propose any verification steps (no check, test, or confirmatory observation) → DIAG = 0; recommended copper/sulfur fungicides, hence chemical input → CHEM = 1; included some ecological advice (airflow, pruning, and timing of watering) → ECO = 1; no nutritional measures were suggested → NUTR = 0.
- Perplexity AI: Incorrectly attributed symptoms to Grapevine Red Blotch Disease (viral); called for laboratory testing to confirm viral infection, satisfying verification criterion → DIAG = 1; emphasized removal of infected material and monitoring, no toxic inputs → ECO = 1; mentioned maintaining nutrient balance and plant vigor → NUTR = 1; no harmful chemicals were recommended → CHEM = 0.

Table 1. Eco-Score evaluation for Case 2: Grapevine leaf chlorosis and reddening.

Model	ECO	DIAG	NUTR	CHEM	Numerator (ECO + DIAG + NUTR − CHEM)	Denominator (ECO + DIAG + NUTR + CHEM)	Eco-Score
ChatGPT 5.0	1	1	1	1	2	4	0.50
Gemini 2.5 Pro	1	1	1	0	3	3	1.00
Claude Sonnet 4.5	1	0	0	1	1	3	0.00
Perplexity	1	1	1	0	2	4	1.00

Interpretation (qualitative): In this illustrative case, Gemini and Perplexity achieved the maximum Eco-Score (all sustainable dimensions, no chemicals), ChatGPT was positioned mid-range (balanced by a copper option), whereas Claude scored lower due to explicit chemical control despite preventive/diagnostic content.

2.6. GenAI Usage

In this study, four large language models—ChatGPT 5.0 (OpenAI [21]), Gemini 2.5 Pro (Google DeepMind [22]), Claude Sonnet 4.5 (Anthropic [23]), and Perplexity AI [24]—were employed in a controlled and documented manner, as described in Section 2 (Materials and Methods), to generate plant health advice forming the empirical dataset.

ChatGPT 5.0 was additionally used as a methodological assistant to support the manuscript's development. It helped identify potential weaknesses in clarity or structure, provided language-editing suggestions to improve readability and scientific style, and assisted in verifying the internal consistency of the Eco-Score coding scheme. All analytical choices, code assignments, data interpretation, and final revisions were performed solely by the author. No GenAI tools were used to generate or manipulate empirical data, figures, or quantitative results.

3. Results

3.1. Overview of Model-Generated Diagnoses and Recommendations

Table 2 presents a summary of the diagnostic outputs provided by the four large language models (ChatGPT, Gemini, Claude, and Perplexity) for the 15 crop cases included in the dataset. Each case represented a real-world example of visible plant stress (e.g., nutrient deficiency, environmental stress, chlorosis, leaf deformation, fungal infection). The table summarizes the primary diagnosis and key recommendations suggested by each model, reflecting their reasoning patterns, terminology, and emphasis on ecological or chemical interventions. Full LLM outputs can be found in Supplementary Materials (Tables S1–S15).

Table 2. Overview of diagnoses and management advice across 15 plant cases analyzed by large language models.

LLM	ChatGPT 5.0	Gemini 2.5 Pro	Claude Sonnet 4.5	Perplexity
Case 1.  (See full answer in Table S1 in Supplementary Materials)	Diagnosis: Nutrient deficiency (magnesium and potassium), overwatering + poor drainage, possible fungal stress (as a possibility, not the real threat) Solutions: liquid fertilizer with magnesium (Mg), potassium (K), and micronutrients, foliar spray of Epsom salt, remove worst leaves, check for pests	Diagnosis: Nutrient deficiency (magnesium and nitrogen), overwatering, fungal or bacterial infection Solution: fertilizer with an N-P-K ratio like 10-10-10, Epsom Salt for magnesium, for prevention: add compost, adjust watering, prune damaged leaves	(needed information about the plant) Diagnosis: nutrient deficiency (magnesium and potassium) Advice: Add Epsom salt, check watering, use a balanced fertilizer, check soil pH, remove damaged leaves	Diagnosis: Nutrient deficiencies (Lack of magnesium, potassium, sulfur, or other micronutrients), watering issues, fungal, bacterial diseases and insects Advice: watering, supplement nutrients: balanced fertilizer with micronutrients. Epsom salts (magnesium sulfate), inspect for pests, remove affected leaves, practice good sanitation

Table 2. *Cont.*

LLM	ChatGPT 5.0	Gemini 2.5 Pro	Claude Sonnet 4.5	Perplexity
Case 2.  (See full answer in Table S2 in Supplementary Materials)	Diagnosis: Magnesium Deficiency (most probable), Flavescence Dorée/Grapevine Yellows (less likely but serious), Fungal leaf spot (possible secondary issue) Solutions: 1. Foliar Mg spray → Epsom salt (15–20 g/L). 2. Check soil nutrients → Mg/K/Ca balance. 3. Avoid drought stress → steady irrigation. 4. Inspect for leafhoppers or mildew → use copper-based fungicide if present.	Diagnosis: Magnesium (Mg) Deficiency, Potassium (K) Deficiency, Grapevine Leafroll Virus, Fungal Diseases (e.g., Downy Mildew) Solutions: 1. Confirmation: soil test, petiole analysis (leaf stem test) 2. Short-Term Solution (foliar spray) 3. Long-Term Solution (soil amendment) 4. General Care and Prevention (balanced fertilization, monitor soil pH, watering)	Diagnosis: downy mildew (*Plasmopara viticola*), Solutions: 1. Immediate steps: Remove affected leaves, improve air circulation, avoid overhead watering; Water in the morning 2. Treatment options: Apply copper-based fungicides (like Bordeaux mixture), use organic sulfur sprays, consider systemic fungicides for severe infections 3. Prevention for next season: Apply preventive fungicide sprays before wet periods, ensure proper vine spacing for airflow, choose resistant grape varieties if replanting, keep the area around vines clean of fallen leaves	Diagnosis: Grapevine Red Blotch Disease (GRBD), nutrient deficiencies or other stresses. Solutions: 1. Confirm Diagnosis, 2. Remove Infected Vines, 3. Optimize Vineyard Health (adequate nutrition and manage water appropriately) 4. Monitor for Vectors
Case 3.  (See full answer in Table S3 in Supplementary Materials)	Diagnosis: powdery mildew Solutions: 1. Remove infected leaves 2. Natural sprays (baking soda, milk spray, Garlic extract) 3. Chemical Fungicides 4. Prevention advice	Diagnosis: powdery mildew Solutions: 1. Prune infected leaves and dispose them (not compost), sanitize tools 2. organic and DIY sprays (baking soda, potassium bicarbonate, milk spray, neem oil) 3. Commercial fungicides 4. Future prevention	Diagnosis: powdery mildew Solutions: 1. Immediate steps (remove infected leaves, improve circulation, water at the base) 2. Treatment (baking soda spray, milk spray, neem oil, fungicidal spray) 3. Prevention tips	Diagnosis: powdery mildew Solutions: 1. baking soda spray 2. Milk spray 3. Vinegar Spray 4. Neem Oil 5. Hydrogen peroxide 6. Remove infected leaves

Table 2. *Cont.*

LLM	ChatGPT 5.0	Gemini 2.5 Pro	Claude Sonnet 4.5	Perplexity
Case 4.  (See full answer in Table S4 in Supplementary Materials)	Diagnosis: gall mites (Eriophyidae) Solution: 1: Prune affected leaves 2: Encourage plant health 3. Biological control (natural predators, avoid broad-spectrum insecticides) 4. Prevention (horticultural oil, sulfur-based sprays)	Diagnosis: Currant Blister Aphid (*Cryptomyzus ribis*). Solution: 1. Natural and Mechanical Methods (strong water spray, manual removal, natural predators) 2. Organic sprays (Insecticidal soap, neem oil, horticultural oil) 3. Chemical control (use with caution) 4. Prevention tips	Diagnosis: peach leaf curl, a fungal disease caused by *Taphrina deformans* Solution: 1. Remove infected leaves 2. Do not compost them 3. Water and fertilize 4. Monitor new growth 4. Prevention—spray fungicide in late fall	Diagnosis: leaf galls 1. Do not use chemical sprays 2. Prune affected leaves 3. Encourage plant health 4. Future tips: prevent with horticultural oil
Case 5.  (See full answer in Table S5 in Supplementary Materials)	Diagnosis: leaf scorch or sunburn, and possibly some stress-related damage (e.g., watering or transplant shock) Solutions: 1. Remove damaged leaves 2. Provide shade temporarily 3. Add water and mulch 4. Avoid fertilizers now	Diagnosis: sunscald Solutions: 1. Provide shade 2. Do not remove damaged leaves yet 3. Watering	Diagnosis: powdery mildew 1. Immediate steps (remove infected leaves, improve circulation, water at the base) 2. Treatment (baking soda spray, milk spray, neem oil, fungicidal spray) 3. Prevention tips	Diagnosis: environmental stress, sunscald, or potential disease issues Solutions: 1. Remove affected leaves 2. Check sun exposure 3. Water consistently 4. Assess soil and fertilizer 5. Improve airflow 6. Companion Shade or Row Cove 7. Monitor for pests and diseases

Table 2. *Cont.*

LLM	ChatGPT 5.0	Gemini 2.5 Pro	Claude Sonnet 4.5	Perplexity
Case 6.  (See full answer in Table S6 in Supplementary Materials)	Diagnosis: generally healthy, but there were a few issues: Nutrient deficiency (especially magnesium or nitrogen). Fungal diseases, such as downy mildew (*Plasmopara viticola*) or powdery mildew (Erysiphe necator). Water stress. Botrytis (gray mold) or sunburn. Solutions: 1. Check and Improve Airflow 2. Inspect for Disease 3. Fungicide or Organic Treatments —Use copper-based fungicide (like Bordeaux mixture) or sulfur spray for organic control —Use a baking soda spray 4. Nutrition and Water (balanced fertilizer (e.g., 10-10-10 NPK), magnesium supplement 5. Harvest Timing	Diagnosis: woody canes (stems) appear to be dead. The healthy-looking green leaves and grape clusters are growing from other parts of the vine that are still alive. Grapevine Trunk Diseases (GTDs): Eutypa dieback and Phomopsis, Winter Injury (Freeze Damage) Solutions: 1. Prune in dormant season 2. Identify and Remove All Dead Wood 3. Practice Strict Sanitation (tools, destroy pruned wood) 4. Plan for Vine Renewal 5. Prevent Future Infections	Diagnosis: frost damage, cold injury Solutions: 1. Now: Harvest the grapes, do not prune yet 2. During dormant season: Remove all dead and damaged wood, prune to healthy buds on viable canes 3. For next season: Protect from late frosts, choose cold-hardy varieties, proper fall preparation, reduce watering in late summer, mulch	Diagnosis: appears healthy, but there are signs of possible powdery mildew, downy mildew, or botrytis (gray mold), common vineyard pests, nutrient deficiency or environmental stress. Solutions: 1. Remove any moldy, shriveled, or affected grapes and leaves. 2. Prune the plant to improve air circulation 3. In case of humid weather, consider a fungicide specifically for grapes 4. Consistent watering 5. Inspect the undersides of leaves for insect pests like aphids or mites—if you spot these, insecticidal soap can help
Case 7.  (See full answer in Table S7 in Supplementary Materials)	Diagnosis: pest damage and nutrient deficiency (possibly nitrogen or magnesium). Solution: 1. Control pests (inspect plant for insects, eggs and slugs) 2. For flea beetles or caterpillars, use neem oil spray, pyrethrin-based insecticide, or biological control (e.g., Bacillus thuringiensis for caterpillars) in the evening 3. Control nutrition (Apply a balanced fertilizer (like 10-10-10 NPK) or a nitrogen-rich source (e.g., diluted compost tea, blood meal, fish emulsion), spray with Epsom salt solution	Diagnosis: pest infestation, a fungal disease, and potentially a nutrient deficiency Solutions: 1: Prune and remove damaged leaves 2. For pests: handpicking, insecticidal soap, neem oil 3. For powdery mildew	Diagnosis: powdery mildew Solution: 1. Remove affected leaves 2. Improve air circulation. 3. Reduce humidity 4. Fungicide treatment options: a. Organic: baking soda + dish soap b. Natural: Neem oil c. Commercial: Sulfur-based or potassium bicarbonate fungicides Preventive care: Ensure plants receive adequate sunlight Avoid overfertilizing with nitrogen Clean up plant debris around the base	Diagnosis: nutrient deficiency, watering problems, pest damage, or disease Solutions: 1. Check for pests beneath leaves; remove insects by hand or use organic treatments like neem oil, insecticidal soap, or Bacillus thuringiensis if pest pressure is high 2. Ensure proper watering 3. Test and adjust soil pH 4. Apply a balanced fertilizer if nutrient deficiency is suspected 5. Remove severely affected leaves

Table 2. *Cont.*

LLM	ChatGPT 5.0	Gemini 2.5 Pro	Claude Sonnet 4.5	Perplexity
Case 8.  (See full answer in Table S8 in Supplementary Materials)	Diagnosis: Nutrient Deficiency (most likely magnesium or manganese) or Fungal/Bacterial Leaf Spot, Environmental Stress (drought + heat) Solutions: 1. Remove the most affected leaves 2. Check soil pH (target 6.0–6.5 for most vegetables). 3. Apply Epsom salt as a foliar spray 4. If no improvement or spots continue spreading, switch to fungicidal treatment (copper-based). 5. Avoid watering leaves directly and ensure good air circulation.	Diagnosis: Interveinal Chlorosis, Necrotic Spots Solutions 1: Immediate Care and Inspection (inspect the underside, remove damaged leaves, improve air circulation, water properly) 2: Address Potential Nutrient Deficiency (check soil pH, apply a corrective treatment: For Suspected Magnesium Deficiency: Epsom salts For Suspected Manganese Deficiency: Apply a foliar feed containing manganese sulfate. For General Health: Use a balanced, all-purpose fertilizer that contains a full range of micronutrients. Treat for Potential Fungal Disease—apply a fungicide. Organic Option: A copper-based fungicide or a product containing *Bacillus subtilis* Neem Oil:	Diagnosis: Pest infestation, Yellowing/chlorosis: Nutrient deficiency (especially nitrogen or iron), Pest damage, Overwatering or poor drainage, environmental stress Solutions: 1. Remove heavily damaged leaves 2. Treat the pests (Spray with insecticidal soap or neem oil on both sides of leaves) 3. Check watering 4. Fertilize—Apply a balanced liquid fertilizer if you have not recently 5. Isolate from uninfected plants	Diagnosis: 1. Nutrient Deficiency: (interveinal chlorosis). 2. Fungal or Bacterial Infection 3. Environmental Stress 4. Pest Damage Solutions: 1. Check and improve soil drainage. 2. Test soil nutrient levels, especially iron and magnesium, and amend with appropriate fertilizers. 3. Remove and dispose of affected leaves to prevent spread. 4. Consider applying a fungicide if infection is suspected. 5. Avoid overwatering and ensure the plant has proper air circulation.
Case 9.  (See full answer in Table S9 in Supplementary Materials)	Diagnosis: leaf spot disease, most likely *Mycosphaerella fragariae* (also known as *Ramularia tulasnei*) Solutions: 1. Remove infected leaves 2. Improve air circulation 3. Avoid overhead watering 4. Use Fungicides (if infection is severe): a. Organic options: Copper-based fungicides (e.g., copper hydroxide, copper oxychloride). b. Conventional options: Captan, Dithane (mancozeb), or chlorothalonil 5. Crop rotation, hygiene and mulching	Diagnosis: Common Leaf Spot (*Mycosphaerella fragariae*) Solutions: 1. Remove Infected Leaves, Improve Air Circulation, Adjust Your Watering Technique, mulch 2. Treatment: a. Organic/home remedies (neem oil, copper-based fungicides, baking soda spray) b. Chemical fungicides (myclobutanil or chlorothalonil for strawberries) c. Long-Term Prevention (plant resistant varieties, crop rotation, good garden hygiene)	Diagnosis: leaf spot disease, caused by fungal pathogens like *Mycosphaerella fragariae* (common leaf spot) or *Diplocarpon earlianum* (leaf scorch). Solutions: 1. Immediate actions (Remove infected leaves, Improve air circulation, Water properly, Clean up debris 2. Ongoing management: Mulch—Spacing Fungicide (if severe)—Use copper-based fungicides or organic options like neem oil, following label instructions	Diagnosis: Strawberry Leaf Spot (caused by *Mycosphaerella fragariae*) or Leaf Scorch (caused by *Diplocarpon earliana*). Solutions: 1. Remove Affected Leaves: 2. Increase Air Circulation: 3. Avoid Overhead Watering: 4. Use Mulch: 5. Fungicide: If the outbreak is severe, consider a fungicide labeled for strawberries. Options include copper-based fungicides or potassium bicarbonate for organic management.

Table 2. *Cont.*

LLM	ChatGPT 5.0	Gemini 2.5 Pro	Claude Sonnet 4.5	Perplexity
Case 10.  (See full answer in Table S10 in Supplementary Materials)	Diagnosis: leaf scorch (*Diplocarpon earlianum*) or leaf spot (*Mycosphaerella fragariae*), Dry, cracked soil, possible phosphorus deficiency, Solutions: 1. Remove affected leaves 2. Improve soil and watering 3. Mulch 4. Fungicide (if infection spreads) Apply a biological fungicide (like *Bacillus subtilis*-based products, e.g., Serenade) or a copper-based spray after removing infected leaves. 5. Apply a balanced fertilizer (like 10-10-10 or compost tea)	Diagnosis: nutrient deficiency (Phosphorus Deficiency) and a possible fungal disease, Environmental Stress, Natural Aging Solutions: 1. Immediate Care—Prune Damaged Leaves, Water Correctly - Address 2. Nutrient Needs (Fertilize: Apply a balanced fertilizer that is rich in phosphorus. Look for a fertilizer where the middle number (P) in the N-P-K ratio is equal to or higher than the first number (N). Organic options like bone meal are excellent sources of phosphorus —Check Soil pH 3. Prevent and Control Disease (Improve Air Circulation, Mulch, Consider a Fungicide (If Necessary): Look for one containing copper or potassium bicarbonate for strawberries	Diagnosis: Frost damage, Water stress, Nutrient deficiency, Disease or root damage Solutions: 1. Assess the damage, 2. Remove dead foliage 3. Check watering 4. Protect from frost (Mulching around the base, Covering plants overnight, Moving potted plants to a protected location) 5. Wait and observe Asks what plant it is	Diagnosis: Leaf Scorch, Common Leaf Spot, Nutrient Deficiency, Environmental Stress Solutions: 1. Remove Damaged Leaves 2. Improve Airflow 3. Water Properly 4. Add Mulch and Fertilizer (phosphorus-rich fertilizer) 5. Fungicide Treatment (if severe): use organic fungicides like neem oil 6. Soil Care: Ensure soil pH is between 5.5 and 6.9
Case 11.  (See full answer in Table S11 in Supplementary Materials)	Diagnosis: powdery mildew Solutions: 1. Remove Infected Leaves 2. Improve Air Circulation 3. Treat the Infection —Organic options: Baking soda spray, Milk spray, Neem oil —Chemical options (if infection is severe) Fungicides containing myclobutanil, sulfur, or trifloxystrobin 4. Prevention for Next Season	Diagnosis: Powdery Mildew Solutions: 1. Immediate Damage Control (prune, do not compost, improve airflow) 2. Apply a Fungicide Spray—Home Remedy: Baking Soda, Milk spray—Organic Commercial option: Neem Oil, Potassium Bicarbonate 3. Prevention for the Future	Diagnosis: powdery mildew Solutions: 1. Remove affected leaves 2. Improve air circulation 3. Water properly 4. Apply treatments: Baking soda spray, Milk spray, Neem oil, Sulfur or copper fungicides 5. Prevention	Diagnosis: powdery mildew Solutions: 1. Remove Infected Leaves 2. Improve Airflow. 3. Spray with a Remedy: Hydrogen peroxide spray, milk solution, baking soda mix 4. Commercial Fungicides: sulfur, potassium bicarbonate, or horticultural oils 5. Prevent Spreading

Table 2. *Cont.*

LLM	ChatGPT 5.0	Gemini 2.5 Pro	Claude Sonnet 4.5	Perplexity
Case 12.  (See full answer in Table S12 in Supplementary Materials)	Diagnosis: nutrient deficiency (especially nitrogen or magnesium), powdery mildew onset, or aging leaves, fungal infection, Weed and moisture competition Solutions: 1. Remove affected leaves 2. Check watering practices 3. Improve air circulation 4. Feed the plant (balanced fertilizer or compost tea rich in nitrogen and potassium, alternatively liquid fertilizer for cucurbits) 5. Prevent fungal spread (spray with a mild baking soda solution, use a biological fungicide containing *Bacillus subtilis* or Trichoderma)	Diagnosis: Nutrient Deficiency (Magnesium and/or Potassium), Possible Fungal Disease: Downy Mildew Solutions; 1. Add nutrition: Foliar Spray Epsom salt) or Soil Amendment, all-purpose vegetable fertilizer that contains magnesium and potassium 2. (Check the Underside of Leaves, improve Air Circulation, Watering Technique, Fungicide Application: For organic gardening, copper-based fungicides or products containing chlorothalonil can be effective)	Diagnosis: Powdery Mildew or Downy Mildew, Nutrient Deficiency, Natural Senescence Solutions: 1. Remove affected leaves 2. Water at the base 3. Improve air circulation 4. Fertilize (balanced fertilizer with nitrogen) 5. Fungicide treatment—Use organic options like neem oil, sulfur spray, or a baking soda solution 6. Mulch 7. Morning watering	Diagnosis: Nutrient Deficiency, Watering Issues, Pest Damage, Fungal or Viral Diseases Solutions: 1. Fertilize with a balanced or nitrogen-dominant product 2. Inspect for bugs and treat with insecticidal soap, neem oil, or by removing infested leaves 3. If disease is suspected (extensive spots, powdery residue, or severe distortion), prune affected leaves and increase air circulation; 4. Adjust watering 5. Use crop rotation and disease-resistant varieties
Case 13.  (See full answer in Table S13 in Supplementary Materials)	Diagnosis: drought stress, nutrient imbalance, and possibly pest or fungal damage or dust/dry soil splash Solutions: 1. Improve watering 2. Feed the plant (balanced fertilizer (e.g., NPK 10-10-10) or a berry fertilizer rich in potassium and magnesium. Add compost or well-rotted manure around the base). 3. Check for pests (if present, spray with soapy water or use neem oil or an insecticidal soap. 4. Manage disease (If white patches persist and are not dust, treat with sulfur-based fungicide or baking soda spray), improve circulation 5. Soil care	Diagnosis: severe water stress (drought—most probable), Fungal Disease (e.g., Gray Mold/Botrytis), Pests (e.g., Spider Mites) Solutions: 1. Water Deeply, Immediately 2. Prune Damaged Fruit 3. Apply Mulch 4. Future prevention (improve watering, air circulation and check for pests)	Diagnosis: Fungal disease (Gray mold: Botrytis, Anthracnose, Cane blight) and environmental stress (water). Solutions: 1. Immediate: Remove all infected fruit and damaged canes, dispose of infected material away from the garden, Water deeply and regularly 2. Ongoing care (Improve air circulation, Mulch around the base, Water at the base of the plant, consider applying a fungicide labeled for raspberries if the problem persists 3. For next season (prune properly in late winter/early spring, space plants adequately for good air flow, ensure consistent watering throughout the growing season, clean up all fallen debris around plants	Diagnosis: Nutrient Deficiency, Incorrect soil pH, water stress, fungal or viral diseases Solutions: Test Soil pH and Amend, Feed with Balanced Fertilizer, Correct Nutrient Deficiencies, Water Properly, Mulch and Improve Soil Texture, Inspect for Disease Extra tips: —Prune damaged or diseased canes at ground level. —Avoid high-chloride fertilizers. —Always test before amending soil or using chemical treatments

Table 2. *Cont.*

LLM	ChatGPT 5.0	Gemini 2.5 Pro	Claude Sonnet 4.5	Perplexity
Case 14.  (See full answer in Table S14 in Supplementary Materials)	Diagnosis: powdery mildew Solutions: 1. Prune for airflow 2. Natural treatments (Baking soda spray, Milk spray) 3. Fungicides (if severe) Use sulfur-based or potassium bicarbonate fungicides labeled for powdery mildew. 4. Prevention for next season (spacing, water, clean up, choose resistant varieties)	Diagnosis: Powdery Mildew Solutions: 1. Immediate Actions: Prune and Remove Infected Parts, Improve Air Circulation, Apply a Fungicide Spray: Baking Soda Spray (DIY), Neem Oil, Commercial Fungicides 2. Prevention for Next Season: (Fall/Winter Cleanup, Dormant Spray, Proper Watering, Ensure Good Location, Resistant Varieties)	Diagnosis: powdery mildew Solutions: 1. Remove affected leaves 2. Improve conditions (air circulation, Water at the base, Ensure adequate sunlight) 3. Treatment options (Baking soda spray, Neem oil, Milk solution, Fungicide: For severe cases, use a commercial fungicide labeled for powdery mildew 4. Prevention:	Diagnosis: powdery mildew Solutions: 1. Remove and dispose of affected leaves 2. Improve air circulation 3. Avoid overhead watering; 4. Treat the plant with fungicides like neem oil, sulfur sprays, or commercial fungicides labeled for powdery mildew. 5. Maintain good garden hygiene and monitor regularly for early signs.
Case 15.  (See full answer in Table S15 in Supplementary Materials)	Diagnosis: sooty mold on honeydew of aphids, whiteflies, or scale insects Solutions: 1. Inspect for pests 2. Control the pests —For mild infestations: Wash leaves gently with water —For heavier infestations: Use insecticidal soap or neem oil spray, introduce beneficial insects 3. Remove heavily affected leaves 4. Improve air circulation 5. Rinse the leaves	Diagnosis: sooty mold on honeydew of aphids, whiteflies, mealybugs Solutions: 1. Get Rid of the Sooty Mold (Water Spray, Soapy Water Solution) 2. Identify and Control the Pest (Manual Removal, Insecticidal Soap, Neem Oil, Encourage Natural Predators) 3. Prevent Future Problems (Regular Inspections, Good Air Circulation, Proper Watering and Fertilizing, Control Ants)	Diagnosis: sooty mold on honeydew of aphids, whiteflies, mealybugs Solutions: 1. Address the pest problem first: Inspect plant, Spray with insecticidal soap or neem oil; for heavy infestations, use a systemic insecticide or horticultural oil 2. Remove the sooty mold: wipe leaves with a damp cloth or sponge with mild soapy water, rinse with clean water 3. Prevent recurrence: Monitor for pests Improve air circulation, avoid overfertilizing	Diagnosis: raspberry leaf rust or fungus *Phragmidium rubi-idaei* Solutions: 1. Remove affected leaves 2. Clear leaf litter 3. Thin canes for airflow 4. Water wisely 5. Fungicide option: If the disease is severe, copper-based fungicides can help limit spread, applied in spring as a preventive measure. 6. Fertilize appropriately

Cases 1–13: photographs by Dr. Agnieszka Krzyżewska; cases 14–15: photographs by Dr. Marta Ziółek.

Across most cases, large language models demonstrated a generally high capacity to recognize plant health problems and provide plausible diagnostic reasoning. However, several misclassifications occurred, most frequently in the outputs of Claude Sonnet 4.5 (five cases: 2, 3, 5, 7, and 13) and sporadically in Gemini 2.5 Pro (case 6) and Perplexity AI (case 15). Notably, even in situations where plants were relatively healthy (cases 6, 7, and 12), all models tended to overinterpret minor discoloration or structural irregularities as signs of disease. This pattern suggests a cautious or "problem-seeking" diagnostic bias rather than an underestimation of plant condition.

Despite these discrepancies, the structure and tone of recommendations were strikingly consistent across models. In nearly all responses, LLMs began their advice with basic preventive and environmentally responsible measures—such as pruning infected leaves, ensuring adequate watering and nutrient balance, or improving air circulation—before suggesting any chemical intervention. Subsequent steps typically included low-impact or homemade treatments, including baking soda, milk, or neem oil sprays, reflecting alignment with ecological or "green" gardening principles. Only when fungal infections or pest damage were explicitly recognized did the models mention commercial fungicides or insecticides, usually framed as a last resort. Such an ordering of advice indicates an inherent sustainability bias in LLM-generated guidance, prioritizing non-chemical and resource-efficient management approaches.

3.2. Eco-Score Results and Sustainability Orientation

Quantitative analysis of coded responses revealed clear differences between the four large language models in terms of their sustainability orientation (Table 3), although a Friedman test revealed no statistically significant differences in Eco-Score among models ($\chi^2(3) = 3.37$, $p = 0.34$, W = 0.075); therefore, no post hoc tests were conducted. The Eco-Score, ranging from −1 (fully chemical) to +1 (fully sustainable), averaged between 0.41 and 0.71 across models. The highest alignment with sustainable practices (Table 3) was achieved by Perplexity AI (Eco-Score = 0.71), followed closely by Gemini 2.5 Pro (0.69). Both models consistently emphasized ecological management, preventive reasoning, and low reliance on synthetic chemical inputs. ChatGPT 5.0 obtained a moderate Eco-Score (0.57), reflecting a balanced pattern of advice that combined strong ecological reasoning with relatively frequent, though often optional, references to chemical control. By contrast, Claude Sonnet 4.5 showed the lowest sustainability alignment (0.41), largely due to its recurrent inclusion of copper- or sulfur-based fungicides in at least ten of the fifteen analyzed cases.

Detailed analysis showed that across all 15 cases, ecological prevention (ECO) recommendations appeared consistently in all model outputs (100%). However, clear variation was observed in the remaining categories. ChatGPT 5.0 and Claude Sonnet 4.5 produced the highest proportion of chemical control advice (67%), while Gemini 2.5 Pro (47%) and Perplexity AI (40%) showed a more conservative pattern. Requests for further diagnostic checks were frequent in ChatGPT 5.0 and Gemini 2.5 Pro (80%), but notably lower in Claude Sonnet 4.5 (47%). Nutrient management advice was most common in ChatGPT 5.0 and Perplexity AI (67%) and least represented in Claude Sonnet 4.5 (27%). Although no statistically significant differences were detected, the consistent directional variation in category frequencies (Table 3) indicates systematic, model-specific tendencies rather than random variation.

Table 3. Summary of sustainability-related coding results and average Eco-Score values for the large language models (full calculations are in Supplementary Materials Table S16).

LLM	ECO (sum)	ECO (%)	DIAG (sum)	DIAG (%)	NUTR (sum)	NUTR (%)	CHEM (sum)	CHEM (%)	Eco-Score
ChatGPT 5.0	15	100%	12	80%	10	67%	10	67%	0.5667
Gemini 2.5 Pro	15	100%	12	80%	8	53%	7	47%	0.6889
Claude Sonnet 4.5	15	100%	7	47%	4	27%	10	67%	0.4111
Perplexity AI	15	100%	9	60%	10	67%	6	40%	0.7111

All four systems demonstrated full coverage of the ECO dimension (environmentally responsible actions), indicating that preventive and non-chemical strategies were universally present. Differences were instead driven by the DIAG and NUTR categories—representing diagnostic reasoning and nutrient management—where Gemini and ChatGPT performed more consistently. The CHEM dimension (synthetic chemical control) appeared most frequently in Claude's and ChatGPT's outputs, confirming their greater readiness to recommend commercial pesticides or fungicides. Overall, the Eco-Score analysis supports the qualitative observation that large language models tend to prioritize sustainable and preventive cultivation methods, though their precision and restraint in chemical recommendations remain model-dependent.

Figure 1 provides a comparative visualization of Eco-Score values across all fifteen analyzed plant health cases. The graph highlights clear, case-specific variability in the sustainability orientation of recommendations generated by different LLMs. Perplexity AI and Gemini 2.5 Pro consistently achieved higher Eco-Scores, maintaining stable alignment with ecological and preventive practices throughout most scenarios. ChatGPT 5.0 showed greater internal variability—alternating between strongly sustainable and more chemically oriented responses—while Claude Sonnet 4.5 produced the lowest and most uneven scores, reflecting its recurrent inclusion of fungicidal treatments.

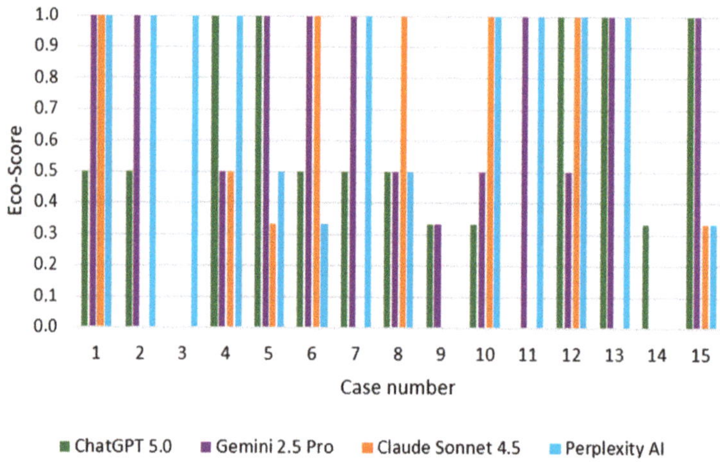

Figure 1. Comparative Eco-Score values across fifteen plant cases and four large language models.

The graphical comparison revealed considerable variation in the Eco-Score values between models across individual plant health cases. The greatest divergence occurred in Case 11, where the models generated the most polarized results—Eco-Score values ranged from 0 (fully chemical) to 1 (fully sustainable), reflecting fundamentally different diagnostic and management interpretations of the same image. In contrast, the highest level of agreement was observed for Case 14 (three models scoring 0, one 0.33) and Case 9 (two models scoring 0, two 0.33), both representing situations where the symptoms were more clearly identifiable and prompted similar management reasoning.

These patterns emphasize that while overall sustainability orientation remained positive, model-specific biases and interpretative uncertainty can lead to substantial inconsistencies in recommendations for the same plant case. Such variability illustrates the need for careful validation of AI-generated advice before its practical application in sustainable horticulture.

4. Discussion

4.1. Sustainable Reasoning Patterns and Implications for Amateur Food Production

The present analysis demonstrates that large language models (LLMs) can generate predominantly sustainability-oriented advice in the context of amateur food production. Across all fifteen plant health scenarios, the four examined systems consistently emphasized preventive and low-impact management practices, with limited reliance on chemical control. These findings are consistent with earlier research suggesting that LLMs tend to promote environmentally responsible decision pathways and democratize access to sustainability knowledge [16,19,27]. In particular, the ranking obtained through the Eco-Score metric indicates that Perplexity AI and Gemini 2.5 Pro provided the most environmentally aligned guidance, while Claude Sonnet 4.5 showed a stronger inclination toward curative, fungicide-based management.

The prominence of ecological and diagnostic reasoning in LLM-generated advice suggests that such systems can reinforce preventive thinking and potentially enhance user awareness of plant health processes. This feature may be particularly valuable in the context of home and allotment gardening, where non-expert users often lack access to expert consultation. However, the results also highlight the potential for overconfidence and misinterpretation—both by the models and by human users. For instance, in Case 7 (zucchini), several models misclassified natural whitish leaf patterns as powdery mildew, subsequently recommending fungicide treatments. Similar misdiagnoses have been observed in other AI-based plant detection systems, underscoring that language models do not possess genuine visual understanding but instead rely on probabilistic pattern recognition [28,29]. Consequently, users may attribute excessive authority to LLM-generated diagnoses, overlooking the need for critical verification or expert confirmation.

4.2. Limitations and Methodological Considerations

While the Eco-Score proved useful for systematically comparing model outputs, it should be interpreted with caution. The index simplifies complex sustainability dimensions into binary variables and may overpenalize models that mention chemical treatments even when these are consistent with organic standards. For example, copper-based fungicides or sulfur dust—although classified here as CHEM—are permissible within certified organic farming systems [34]. Likewise, some "home remedies" proposed by the models (e.g., milk, baking soda or vinegar sprays) may be environmentally benign but agronomically ineffective, raising questions about the practical soundness of LLM advice [35,36]. Therefore, the Eco-Score should not be viewed as an absolute measure of sustainability performance but rather as a comparative indicator of model tendencies.

Furthermore, even organically permitted substances such as copper- or sulfur-based fungicides can pose environmental risks when applied excessively or without professional supervision. Long-term use of copper products has been shown to lead to soil accumulation, reduced microbial activity, and toxicity to beneficial organisms, while overapplication of sulfur-based agents may cause soil acidification and shifts in microbial community structure [31–34]. These effects underline that the environmental safety of such inputs strongly depends on user knowledge and precision in dosage and timing. In the context of this study, the target group—amateur gardeners cultivating home or allotment plots—typically lacks formal agricultural training and may unintentionally misuse even permitted products, increasing the potential for ecological harm. Therefore, all chemical treatments, including copper and sulfur compounds, were conservatively coded as environmentally adverse within the Eco-Score framework. This precautionary approach better reflects the real-world risks associated with nonprofessional plant care practices and emphasizes the need for education and responsible input management among amateur food producers.

Image-related variables such as resolution, lighting, and background complexity can influence the diagnostic reasoning of multimodal LLMs. In this study, all models were exposed to identical, unedited smartphone photographs, which mitigates comparative bias but does not eliminate potential effects of image quality on absolute performance. Future research could incorporate standardized image quality metrics or controlled visual datasets to quantify the sensitivity of LLMs to these visual factors.

Moreover, the scope of this study was limited to a small, image-based sample and a single type of user prompt, which may not reflect real-world diversity in language, context, or crop conditions. Future work should expand the dataset, include multilingual prompts, and test the reproducibility of results under different conversational settings. Additionally, longitudinal research could explore whether model updates or fine-tuning toward agroecological data sources could improve the precision and environmental alignment of generated recommendations.

The performance differences observed among models likely stem from a combination of structural and training-related factors rather than random variability. Although the exact composition of training and fine-tuning strategies is not publicly disclosed, previous research suggests that differences in data diversity, inclusion of sustainability-related sources, and multimodal alignment procedures can influence advisory tone and domain reasoning [16,19,27]. Additionally, models with more restrictive safety-alignment mechanisms may exhibit a stronger bias toward low-risk, preventive advice, while others trained with broader general-domain data may display greater variability and occasional chemical recommendations. Given the proprietary nature of current LLM architectures, these influences cannot be directly verified, and the models in this study were therefore treated as "black boxes". Nevertheless, considering these plausible factors provides context for the observed variation and underscores the need for greater transparency in model training and safety-tuning processes.

In addition to methodological constraints, several limitations are inherent to the models themselves. Because their training data, update cycles, and alignment strategies are proprietary, reproducibility and reliability remain limited. Model outputs can vary across sessions and prompts, leading to inconsistent advice. For amateur gardeners, such variability carries practical implications: model-generated guidance should be interpreted as indicative rather than prescriptive and ideally cross-verified with local horticultural or extension resources. Although the models in this study exhibited a generally consistent sustainability orientation, they occasionally produced incomplete or context-insensitive recommendations. Future research should therefore assess the stability and reliability of model responses across time and conditions and explore how prompt design or fine-tuning could improve their safety and practical usefulness for nonprofessional users.

Overall, while LLMs exhibit promising potential as sustainability support tools for amateur food producers, their use should complement—not replace—expert agronomic judgment. Educating users about the probabilistic and non-expert nature of LLM outputs is essential to ensure that AI-assisted plant care advice remains both environmentally responsible and contextually appropriate.

5. Conclusions

This study provides one of the first systematic evaluations of the sustainability orientation of large language models (LLMs) in plant health diagnostics. Despite differences among systems, all four models demonstrated an inherent preference for preventive, low-impact, and resource-efficient cultivation strategies. This pattern indicates that LLMs have internalized sustainability-aligned reasoning, likely as a reflection of their training data and conversational safety filters.

However, diagnostic reliability remains inconsistent. The observed tendency to over-diagnose problems in relatively healthy plants, coupled with the occasional recommendation of unnecessary chemical treatments, highlights the limitations of LLMs as autonomous advisors. The Eco-Score index proposed in this study effectively captured relative sustainability tendencies across models but should be interpreted with caution—it simplifies complex agronomic realities and does not fully account for context-specific appropriateness of chemical or organic interventions.

From a practical perspective, the findings suggest that AI systems should remain supportive, interpretive tools rather than sources of prescriptive agricultural advice. Amateur gardeners are encouraged to cross-check model recommendations with verified local resources, such as agricultural extension services, organic certification guidelines, or official pest management databases, before applying any chemical or treatment-based intervention.

For developers, future improvements in sustainability alignment could be achieved by expanding fine-tuning datasets with sustainability-oriented, context-specific agricultural literature and by designing prompt structures that prioritize ecological prevention, diagnostic reasoning, and nutrient management before recommending chemical control. Embedding such principles into model alignment and user interfaces would enhance both the environmental responsibility and practical reliability of AI advisory tools. Overall, the findings suggest that while current LLMs can reinforce ecological awareness and promote sustainable plant care behaviors, they should be regarded as supportive tools rather than expert systems. Their integration into citizen science or home gardening applications could enhance environmental literacy, provided that users remain aware of the models' interpretative nature and the need for critical human oversight.

Supplementary Materials: The following supporting information can be downloaded at https://www.mdpi.com/article/10.3390/su172310466/s1. Table S1. Full LLMs answers to prompt: What happened to my plant and how can I fix it? (case 1). Table S2. Full LLMs answers to prompt: What happened to my plant and how can I fix it? (case 2). Table S3. Full LLMs answers to prompt: What happened to my plant and how can I fix it? (case 3). Table S4. Full LLMs answers to prompt: What happened to my plant and how can I fix it? (case 4). Table S5. Full LLMs answers to prompt: What happened to my plant and how can I fix it? (case 5). Table S6. Full LLMs answers to prompt: What happened to my plant and how can I fix it? (case 6). Table S7. Full LLMs answers to prompt: What happened to my plant and how can I fix it? (case 7). Table S8. Full LLMs answers to prompt: What happened to my plant and how can I fix it? (case 8). Table S9. Full LLMs answers to prompt: What happened to my plant and how can I fix it? (case 9). Table S10. Full LLMs answers to prompt: What happened to my plant and how can I fix it? (case 10). Table S11. Full LLMs answers to prompt: What happened to my plant and how can I fix it? (case 11). Table S12. Full LLMs answers to prompt: What happened to my plant and how can I fix it? (case 12). Table S13. Full LLMs answers to prompt: What happened to my plant and how can I fix it? (case 13). Table S14. Full LLMs answers to prompt: What happened to my plant and how can I fix it? (case 14). Table S15. Full LLMs answers to prompt: What happened to my plant and how can I fix it? (case 15). Table S16. The results of calculating eco-score.

Funding: This research received no external funding.

Institutional Review Board Statement: Not applicable.

Informed Consent Statement: Not applicable.

Data Availability Statement: Data is contained within the article or Supplementary Materials.

Acknowledgments: The author wishes to thank Marta Ziółek from the Department of Hydrology and Climatology, Institute of Earth and Environmental Sciences, Maria Curie Skłodowska University for kindly providing two plant images used in the analysis of cases 14 and 15. The study employed four large language models—ChatGPT 5.0 (OpenAI), Gemini 2.5 Pro (Google DeepMind), Claude Sonnet 4.5 (Anthropic), and Perplexity AI—as described in the Section 2. The author has reviewed and edited all model-generated outputs and takes full responsibility for the final content of this publication.

Conflicts of Interest: The author declares no conflicts of interest.

References

1. Osei Boateng, B.; Mensah, H.; Accomford, O. The Contribution of Home Gardens to Sustainable Community Development from the Perspectives of Residents in Northern Ghana. *Discov. Cities* **2025**, *2*, 39. [CrossRef]
2. Kanosvamhira, T.P. Sustainable Urban Agriculture: Unlocking the Potential of Home Gardens in Low-Income Communities. *Prof. Geogr.* **2024**, *76*, 587–596. [CrossRef]
3. Hansen, L.S.; Kihagi, G.W.; Agure, E.; Muok, E.M.O.; Mank, I.; Danquah, I.; Sorgho, R. Sustainable Home Gardens in Western Kenya: A Qualitative Study for Co-Designing Nutrition-Sensitive Interventions. *J. Rural Stud.* **2023**, *103*, 103132. [CrossRef]

4. Poniży, L.; Latkowska, M.J.; Breuste, J.; Hursthouse, A.; Joimel, S.; Külvik, M.; Leitão, T.E.; Mizgajski, A.; Voigt, A.; Kacprzak, E.; et al. The Rich Diversity of Urban Allotment Gardens in Europe: Contemporary Trends in the Context of Historical, Socio-Economic and Legal Conditions. *Sustainability* **2021**, *13*, 11076. [CrossRef]
5. Trembecka, A.; Kwartnik-Pruc, A. An Analysis of the Changes in the Structure of Allotment Gardens in Poland and of the Process of Regulating Legal Status. *Sustainability* **2018**, *10*, 3829. [CrossRef]
6. Speak, A.F.; Mizgajski, A.; Borysiak, J. Allotment Gardens and Parks: Provision of Ecosystem Services with an Emphasis on Biodiversity. *Urban For. Urban Green.* **2015**, *14*, 772–781. [CrossRef]
7. Dymek, D.; Wilkaniec, A.; Bednorz, L.; Szczepańska, M. Significance of Allotment Gardens in Urban Green Space Systems and Their Classification for Spatial Planning Purposes: A Case Study of Poznań, Poland. *Sustainability* **2021**, *13*, 11044. [CrossRef]
8. Melon, M.; Dzieduszyński, T.; Gawryszewska, B.; Lasocki, M.; Hoppa, A.; Przybysz, A.; Sikorski, P. Urban Allotment Gardens with Turf Reduce Biodiversity and Provide Limited Regulatory Ecosystem Services. *Sustainability* **2025**, *17*, 6216. [CrossRef]
9. Moskalonek, Ż.; Połom, M.; Puzdrakiewicz, K. Changes in the Function of Allotment Gardens in an Attractive Location Based on the Example of Tri-City in Poland. *Land* **2020**, *9*, 464. [CrossRef]
10. Haase, D.; Gaeva, D. Allotments for All? Social–Environmental Values of Urban Gardens for Gardeners and the Public in Cities: The Example of Berlin, Germany. *People Nat.* **2023**, *5*, 1207–1219. [CrossRef]
11. Dobson, M.C.; Edmondson, J.L.; Warren, P.H. Urban Food Cultivation in the United Kingdom: Quantifying Loss of Allotment Land and Identifying Potential for Restoration. *Landsc. Urban Plan.* **2020**, *199*, 103803. [CrossRef]
12. Barthel, S.; Folke, C.; Colding, J. Social-Ecological Memory in Urban Gardens-Retaining the Capacity for Management of Ecosystem Services. *Glob. Environ. Change* **2010**, *20*, 255–265. [CrossRef]
13. Andersson, E.; Barthel, S.; Ahrné, K. Measuring Social-Ecological Dynamics behind the Generation of Ecosystem Services. *Ecol. Appl.* **2007**, *17*, 1267–1278. [CrossRef]
14. Damin, A.; Palmer, J.F. Rural Life in the City: The Chalet Garden in Denmark. In *Proceedings of the 2002 Northeastern Recreation Research Symposium*; Schuster, R., Ed.; Gen. Tech. Rep. NE-302; US Department of Agriculture, Forest Service, Northeastern Research Station: Newtown Square, PA, USA, 2003; pp. 59–64.
15. Kwartnik-Pruc, A.; Droj, G. The Role of Allotments and Community Gardens and the Challenges Facing Their Development in Urban Environments—A Literature Review. *Land* **2023**, *12*, 325. [CrossRef]
16. Tzachor, A.; Devare, M.; Richards, C.; Pypers, P.; Ghosh, A.; Koo, J.; Johal, S.; King, B. Large Language Models and Agricultural Extension Services. *Nat. Food* **2023**, *4*, 941–948. [CrossRef] [PubMed]
17. Rillig, M.C.; Ågerstrand, M.; Bi, M.; Gould, K.A.; Sauerland, U. Risks and Benefits of Large Language Models for the Environment. *Environ. Sci. Technol.* **2023**, *57*, 3464–3466. [CrossRef] [PubMed]
18. Jegham, N.; Abdelatti, M.; Elmoubarki, L.; Hendawi, A. How Hungry Is AI? Benchmarking Energy, Water, and Carbon Footprint of LLM Inference. *arXiv*, 2025; *preprint*. [CrossRef]
19. Li, H.; Wu, H.; Li, Q.; Zhao, C. A Review on Enhancing Agricultural Intelligence with Large Language Models. *Artif. Intell. Agric.* **2025**, *15*, 671–685. [CrossRef]
20. United Nations. *Transforming Our World: The 2030 Agenda for Sustainable Development*; United Nations: New York, NY, USA, 2015; Volume 1.
21. OpenAI. ChatGPT. Available online: https://chatgpt.com/ (accessed on 10 September 2024).
22. Google. Gemini Advanced. Available online: https://gemini.google.com/app (accessed on 10 September 2024).
23. Anthropic. Available online: https://claude.ai/ (accessed on 18 February 2025).
24. PerplexityAI. Available online: https://www.perplexity.ai/ (accessed on 10 September 2024).
25. Zhang, Q.F. From Sustainable Agriculture to Sustainable Agrifood Systems: A Comparative Review of Alternative Models. *Sustainability* **2024**, *16*, 9675. [CrossRef]
26. Mayring, P. *Qualitative Content Analysis: Theoretical Foundation, Basic Procedures and Software Solution*; SSOAR: Klagenfurt, Austria, 2014.
27. De Clercq, D.; Nehring, E.; Mayne, H.; Mahdi, A. Large Language Models Can Help Boost Food Production, but Be Mindful of Their Risks. *Front. Artif. Intell.* **2024**, *7*, 1–17. [CrossRef] [PubMed]
28. Khanifar, J. Evaluating AI-Generated Responses from Different Chatbots to Soil Science-Related Questions. *Soil Adv.* **2025**, *3*, 1–5. [CrossRef]
29. Bender, E.M.; Koller, A. Climbing towards NLU: On Meaning, Form, and Understanding in the Age of Data. In Proceedings of the 58th Annual Meeting of the Association for Computational Linguistics, Online, 5–10 July 2020; pp. 5185–5198. [CrossRef]
30. FAO. *Sustainability Assessment of Food and Agricultural System: Guidelines*; FAO: Rome, Italy, 2014; ISBN 9789251084854.
31. Fan, J.; He, Z.; Ma, L.Q.; Stoffella, P.J. Accumulation and Availability of Copper in Citrus Grove Soils as Affected by Fungicide Application. *J. Soils Sediments* **2011**, *11*, 639–648. [CrossRef]
32. Wightwick, A.M.; Salzman, S.A.; Reichman, S.M.; Allinson, G.; Menzies, N.W. Effects of Copper Fungicide Residues on the Microbial Function of Vineyard Soils. *Environ. Sci. Pollut. Res.* **2013**, *20*, 1574–1585. [CrossRef]

33. Liang, X.; Zhang, Y.; Wan, S.; Xia, L.; Li, P.; Wang, S.; Yang, Y.; Zhu, W. Impacts of Sulfur Application on Microbial Communities and Functional Attributes in Rubber Plantation Soil. *BMC Microbiol.* **2025**, *25*, 1–12. [CrossRef] [PubMed]
34. Tamm, L.; Thuerig, B.; Apostolov, S.; Blogg, H.; Borgo, E.; Corneo, P.E.; Fittje, S.; de Palma, M.; Donko, A.; Experton, C.; et al. Use of Copper-Based Fungicides in Organic Agriculture in Twelve European Countries. *Agronomy* **2022**, *12*, 673. [CrossRef]
35. Bettiol, W. Effectiveness of Cow's Milk against Zucchini Squash Powdery Mildew (Sphaerotheca Fuliginea) in Greenhouse Conditions. *Crop Prot.* **1999**, *18*, 489–492. [CrossRef]
36. Mostafa, Y.S.; Hashem, M.; Alshehri, A.M.; Alamri, S.; Eid, E.M.; Ziedan, E.S.H.E.; Alrumman, S.A. Effective Management of Cucumber Powdery Mildew with Essential Oils. *Agriculture* **2021**, *11*, 1177. [CrossRef]

sustainability

MDPI

Article

Production Costs of Grass-Fed Organic Milk in the Northeastern United States: Empirical Results from Survey Data and Implications for Sustainable Development

Qingbin Wang [1,*], Sara Ziegler [2], Sarah Flack [3], Hakan Unveren [1], Avery Anderson [1] and Heather Darby [2]

[1] Department of Community Development and Applied Economics, University of Vermont, Burlington, VT 05405, USA; hakan.unveren@uvm.edu (H.U.); avery.anderson@uvm.edu (A.A.)
[2] University of Vermont Extension, Burlington, VT 05405, USA; sara.ziegler@uvm.edu (S.Z.); heather.darby@uvm.edu (H.D.)
[3] Sarah Flack Consulting, Enosburg Falls, VT 05450, USA; sarahflackconsulting@gmail.com
* Correspondence: qwang@uvm.edu

Abstract: While there is very limited information on the cost of production (COP) for the emerging 100% grass-fed organic dairy sector, this study (1) estimates the COP using primary data collected from on-farm surveys, (2) assesses the correlation between COP and key production and management factors, (3) examines how land, feed and labor efficiency, and production scale affect the COP, and (4) derives recommendations for enhancing the economic efficiency of grass-fed organic dairy farms. Data collected via annual surveys in the Northeastern United States from 2019 to 2022 were analyzed through descriptive statistics, correlation analysis, hypothesis tests, and regression analysis. At an average cost of USD 45.91 per hundredweight equivalent of milk, the marginal impacts of the cows managed per full time equivalent labor and milk sold per cow on the COP were −USD 0.166 and −USD 0.003, respectively. Conversely, the COP increased by USD 1.44 when the crop acres per cow increased by one unit, and the COP of small farms with less than 45 cows was USD 6.20 higher than other farms. As farms are significantly different in resource endowment and other factors, the strategies for reducing the COP and improving the economic returns should be identified for individual farms. However, our analyses highlight the importance of enhancing labor efficiency in forage production, land management, milking and feeding, improving herd management and optimizing nutrition and dry matter intake to support high milk productivity. This study may help existing grass-fed dairy farms improve their farm management and reduce COP and help prospective farms assess their suitability for transitioning to grass-fed operation.

Keywords: grass-fed dairy; cost of production (COP); Northeastern United States; New York; Vermont; correlation analysis; hypothesis tests; regression analysis; survey

1. Introduction

While dairy has been one of the largest agricultural sectors in the United States, many American dairy farms, especially small and medium-sized farms, are struggling to stay in business due to increasing production costs and a downward trend in farmgate milk prices in real terms (i.e., after inflation is adjusted) [1–3]. For example, while the average farmgate price per hundredweight (cwt) for Class I milk increased moderately from USD 13.77 in 1980 to USD 17.00 in 2023 [4], the real price in 1980 dollars dropped remarkably from USD 13.77 to USD 4.60 over the period (calculated from the nominal price using the consumer price index data of the U.S. Bureau of Labor Statistics [5]).

To overcome these challenges and move towards more sustainable development, the dairy industry is actively seeking innovative strategies in accordance with rapidly changing consumer preferences and market conditions. Many dairy farms have made great efforts to change their production practices to differentiate their milk and meet consumer demand for healthier and more environmentally friendly dairy products such as organic milk and yogurt [6–10]. For example, the total retail sales of organic milk and cream in the United States increased from about USD 1.1 billion in 2005 to USD 4.2 billion in 2023 [11,12]. The rapid growth in the organic dairy markets in the United States is also reflected in the remarkable increase in the number of organic milk cows, from 2265 in 1992 to 352,289 in 2021 [4]. Organic dairy markets have also allowed farmers to capture price premiums for their products. For example, organic milk price received by farmers in New York in 2023 averaged USD 37.73 per cwt, while the average price received by dairy farmers for all milk in the state that year was only USD 21.66 per cwt [13,14]. In addition to organic operations, an increasing number of dairy farms have switched to 100% grass-fed organic operations [15]. In 100% grass-fed organic dairy production systems, cows on certified organic farms are not fed any grain or grain byproducts, and the nutrient needs of the animals are met with grazed and stored forages, minerals, and limited amounts of permitted supplemental non-grain energy sources such as molasses [16]. Note that a 100% grass-based diet is not necessary to be considered organic in the United States. In the rest of this paper, grass-fed refers to 100% grass-fed organic operations as defined above.

Research has shown that grass-fed milk has a significantly different nutritional profile compared to conventional milk [17–23] which may contribute to many human health benefits, including protecting against oxidative stress, cell damage, inflammation, and promoting cardiovascular health and immune support [24–28]. Consumer demand for grass-fed milk has increased significantly in the past decade due to these perceived health benefits as well as concerns over environmental impacts and sustainability in the industrial food production. For example, the number of grass-fed dairy farms in the United States increased about 400% from 2016 to 2019 [29]. As of August 2025, Organic Valley, the largest grass-fed milk supplier in the United States, had 285 grass-fed dairies across the United States—a metric that had doubled since 2024 [30]. According to the Verified Market Research [31], the grass-fed dairy market in the United States was valued at USD 4.64 billion in 2023 and is projected to grow at a compound annual growth rate of roughly 6% to reach USD 6.98 billion by 2031. Globally, grass-fed milk market revenue was valued at USD 12.5 billion in 2024 and is estimated to reach USD 20 billion by 2033, growing at a compound annual growth rate of 6.5% over the period [32].

Despite interest in both consumption and production of grass-fed milk, economics may be a restrictive factor for both consumers and producers limiting future industry growth. Currently retail costs of certified 100% grass-fed milk are between 50% and 100% higher than certified organic milk and over 200% higher than conventional milk. These higher retail prices reflect higher production costs which may be related to differences in milk productivity, feed production costs, land needs, and labor efficiency [33,34].

Understanding the major factors influencing the cost of producing grass-fed milk is crucial not only for producer economic viability but reducing retail prices to expand consumption. This study focuses on the COP rather than the profit due to three considerations: First, while COP is determined by many internal and individual factors and revenue is the result of many external factors, focusing on the COP provides a more direct way for farmers to analyze their cost components as compared to their peers and identify actionable ways to reduce their COP with more predictable impacts. Second, although dairy farms are significantly different in resource endowment and many other factors, they face very competitive markets and the variation in their profit is basically determined by their milk productivity and COP. Third, information on the COP can help consumers understand why grass-fed milk is more expensive than conventional and organic milk and appreciate farmers' efforts.

This study is motivated by the significant development of the grass-fed dairy sector in the United States and many other countries and the lack of information on the COP of grass-fed milk and its contributing factors in the existing literature. Specifically, this study will address the following research questions through empirical analysis of primary data collected from a farm survey: (1) How do farm management and production practices influence the COP for grass-fed milk? (2) What are some potential strategies for grass-fed dairy farms to enhance their production efficiency and economic viability? Answering these questions is expected to help enhance the economic viability of the existing grass-fed dairy farms, informing farmers who are considering starting grass-fed operations, and helping consumers understand and appreciate the reasons behind the higher price of grass-fed milk.

This paper is organized into four sections. Following this Section 1, Section 2 introduces the data and methods, Section 3 reports the empirical results and discusses the limitations, and Section 4 summarizes the major conclusions and recommendations.

2. Data and Methods

In 2019, 2020, and 2022, farm financial data were collected from grass-fed dairy farms in the northeast region, primarily New York and Vermont. Only farms that were 100% grass-fed organic and selling wholesale milk were selected. Data was collected during visits to the farms using the Dairy TRANS 20.20 tool [35]. This tool is an Excel-based workbook developed by Iowa State University Extension that translates farm-level data inputs into summaries including (1) a balance sheet that provides a snapshot of beginning and ending values of the farm on a fair market value and/or cost, (2) farm cash incomes and expenses taken from farmer's financial records which contain more detail than those taken directly from the tax form Schedule F, (3) non-farm income, income taxes paid, capital purchases, and sales, and (4) production data with milk output, number of cows, productive crop/pasture acres, opportunity costs of labor and equity, unpaid labor hours, and fulltime labor equivalents (1 FTE = 3000 h).

To allow for more equal comparisons between farms with different debt loads and dependence on the dairy enterprise to cover living expenses, the tool assumes several standardizations: (1) farms' actual asset (market) value, income, and all expenses other than interest are included, (2) an unpaid labor charge of USD 40,000 per owner operator and per additional FTE unpaid is applied, (3) inventory change adjustments to factor in changes in herd size or equipment inventory and value are included, (4) a 4% charge on the farm's assets is included instead of interest payments, and (5) dairy-related non-milk income (e.g., crop sales, calf sales, etc.) is converted into an equivalent weight of milk which is then added to the total weight of milk sold and is referred to as hundredweight equivalents (cwt eq.). This is performed to account for the expenses associated with producing the dairy-related non-milk incomes that cannot accurately be separated from the expenses to produce milk. This tool was selected due to its simultaneous simplicity and comprehensive assessment of the factors of the total cost of production as well as its suitability for research applications through the standardizations and familiarity among farmers in the region due to prior uses.

Financial data was collected by one individual to decrease variability. Farms were contacted primarily by mail and phone due to the high proportion of Plain community farmer participants who do not use computers. Data was collected primarily via in-person interviews; however, some were collected over the phone once a farm had participated for several years and was familiar with the process. Data were input directly into the Dairy -Trans tool as they were collected. Data collected includes the following: (1) a modified balance sheet for market value of farm assets, feed and cattle inventory, and farm acreage, (2) quantity of milk, livestock, and crops sold and their respective farm incomes, (3) all dairy farm expenses except interest payments, and (4) farm labor hours for both paid and unpaid labor. Farm liability data were not collected.

The production and financial data from the Excel-based DairyTrans 20.20 [35] together with farm demographic data were analyzed in Stata SE 17.0 and SPSS 29 through descriptive statistics, correlation analysis and hypothesis tests, and regression analysis.

Descriptive statistics were generated for the whole sample over three years to summarize and describe the key variables of the dataset (e.g., mean, standard deviation, minimum value, maximum value, frequency, etc.). Such statistics provide a clear and concise overview of the data. For example, the average values of COP by year, region, and farmer groups and their variations provide highly needed information for dairy farms, extension specialists, researchers, policymakers, and other stakeholders of the grass-fed dairy sector. Correlation analysis and hypothesis tests were conducted to identify potential relationships between COP and key farm production and management factors as well as farm demographics. Pearson correlation coefficients were calculated, and alternative hypotheses were tested at the 0.01, 0.05, and 0.10 significance levels to address several research questions. Regression analysis was used to identify the major factors affecting the COP and quantify each factor's marginal impact on the COP when other factors are controlled. Specifically, the following linear regression model was estimated to examine the impacts of four groups of factors: land efficiency, output efficiency, labor efficiency, and farm demographics,

$$Y_{ij} = a + b_1X_{1ij} + b_2X_{2ij} + b_3X_{3ij} + \ldots + b_nX_{nij} + e_{ij} \quad (1)$$

where Y_{ij} is the COP per cwt eq. of farm i in year j, X_{1ij}, X_{2ij}, X_{3ij}, …, X_{ijn} are the values of n independent variables of farm i in year j, a is the intercept, b_1, b_2, b_3, …, b_n are the corresponding coefficients to be estimated, and e_{ij} is the error term. Alternative specifications of this linear regression model were estimated with different combinations of independent variables.

While the overall goodness of fit of each estimated regression model was measured by R-squared and adjusted R-squared and tested via the F-test, each estimated coefficient is tested for its significance level.

3. Empirical Results

This section first reports the descriptive statistics, including the estimated COP, then presents the results of correlation analysis and hypothesis tests, and finally reports the estimated regression models. Note that some interpretation and explanation of the empirical findings are based on the authors' field observations and communication with many grass-fed dairy farms over several years.

3.1. Descriptive Statistics

While the primary data collected from the surveys were recorded in Excel and then processed in STATA for statistical analysis, the descriptive statistics for numerical variables and categorical variables are reported in Tables 1 and 2, respectively.

Table 1. Descriptive statistics of the dataset.

Variable Definition	N	Mean	SD	Min.	Max.
Cost of production (USD/cwt eq.)	64	45.91	13.30	25.50	92.38
Land per cow (acres/cow)	64	5.52	2.45	1.71	11.65
Fuel cost (USD/cwt eq.)	64	1.83	1.21	0.65	8.75
Insurance cost (USD/cwt eq.)	64	0.90	0.95	0.00	3.86
Cost of supplies (USD/cwt eq.)	64	2.64	1.75	0.47	10.95
Repair cost (USD/cwt eq.)	64	2.96	2.08	0.00	10.30
Seed and fertilizer expense (USD/acre)	64	25.80	33.90	0.00	201.28
Cost of molasses (USD/cwt eq.)	64	0.63	1.19	0.00	4.81
Milk sold (cwt/cow)	64	84.65	21.25	44.85	149.91
Cows per FTE (cows/worker)	64	28.24	12.12	9.70	66.10
Milk sold per FTE (cwt/worker)	64	2369.94	1077.24	629.45	5764.49
Labor earnings (USD/h)	64	6.24	9.33	−13.55	34.88
Total cash expense (USD/cow)	64	459.28	773.00	1130.00	4668.06
Total cash expense (USD/cwt eq.)	64	25.95	7.50	11.70	49.50
Average price (USD/cwt eq.)	64	38.80	3.61	27.46	46.17
Yearly milk sales (USD1000)	64	200.71	112.06	54.33	729.60
Number of cows	64	62.00	32.54	23.00	220.00
Number of acres	64	345.65	221.62	69.00	870.00
Yearly milk sales (cwt)	64	5239.40	3029.85	1491.80	19,195.43

Table 2. Farm demographic and management information.

Variable	Frequency	Percentage
Plain vs. English		
Plain	25	39.06
English	39	60.94
Daily milking frequency		
Twice a day	49	79.03
Three times in two days	3	4.84
Once a day	10	16.13
Calf rearing		
Raising extra animals	29	46.03
Matching cull rate	34	53.97
Seasonality		
Year-round	45	70.30
Seasonal	19	29.70

The overall COP per cwt eq. averaged USD 45.91, with a wide range from USD 25.51 to USD 92.38. Farms managed, on average, 346 acres for pasture and stored forage production with an average herd size of 62 cows, resulting in an average of 5.5 acres per mature cow. While the acres per cow varied greatly from 1.71 to 11.65, this large range indicates different production models: some farms were managing very small land bases of primarily pasture and purchasing all their stored forage needs while other farms were managing both pasture and hay land to produce stored forage. Similarly, some farms opted to purchase supplementary energy in the form of molasses, while other farms relied only on forages to meet cow energy requirements. This is also evidenced by the wide range in molasses costs from USD 0 to USD 4.81 per cwt eq.

As Snider et al. [15] indicated, a substantial proportion of grass-fed dairy producers (64%) nationally identify as Plain sect community members (e.g., Amish, Mennonite, etc.). Because these religious beliefs can influence management practices and access to information, this information was collected in the survey. This dataset hosts a smaller proportion (39.06%) of Plain sect community farmers than that of Snider et al. [15]. The term English is used to distinguish farmers who do not identify as Plain sect community members. While most participating farms (70.31%) milked year-round, there were some fully seasonal herds (29.70%). Additionally, most farms (79.03%) milked twice daily while few herds milked once per day (16.13%) or three times every two days (4.84%).

While COP varied significantly among the farms in the dataset, Figure 1 reports COP by five groups defined by their percentiles of COP, from the 15% with the lowest COP to the 15% with the highest COP. While the average COP for the lowest group and the highest group were, respectively, USD 31.70 per cwt eq. and USD 67.30 per cwt eq., the majority (70% or 45 farms) had a COP between USD 38.00 and USD 50.00 per cwt eq.

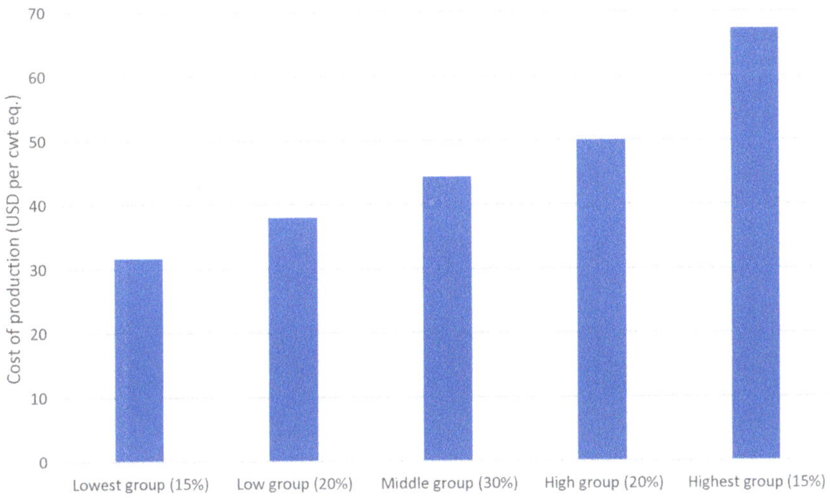

Figure 1. Distribution of farms by their COP.

3.2. Correlation Analysis and Hypothesis Tests

Correlation analysis was conducted to identify potential relationships between COP and key farm production and management factors. This allows for the identification of management or other factors that are highly associated with increasing or decreasing COP. As reported in Table A1 in the Appendix A, positive correlations were identified between COP and land per cow (r = 0.3751, $p = 0.0021$), fuel expense (r = 0.6365, $p < 0.0001$), supply expense (r = 0.3422, $p = 0.0053$), repair expense (r = 0.4642, $p = 0.0001$), and molasses expense (r = 0.6107, $p = 0.0002$). These indicate that increasing acreage and expenses relating to forage production and energy supplementation significantly increase production costs in grass-fed dairy operations.

Conversely, negative correlations were identified between COP and labor earnings (r = −0.6473, $p = 0.0000$), milk sold per cow (r = −0.5545, $p = 0.0000$), herd size (r = −0.2710, $p = 0.0201$), and cows managed per FTE (r = −0.2912, $p = 0.0186$). The inverse relationships between COP, milk sold per cow, and herd size suggest that enhanced milk production efficiency per cow and increased herd size are linked with diminished production costs, thereby emphasizing the benefits of increased productivity per animal and per farm on cost reduction. However, it should be recognized that milk sold does not equate to milk production. A large portion of milk produced may be utilized to raise youngstock on the farm and therefore may contribute to lower milk sales. Additional information on milk diverted to youngstock rearing was not available for all years in this dataset and therefore cannot be fully explored here. However, preliminary data from 2023 show that, on average, grass-fed farms fed 24.07 cwt of milk per calf and raised an average of 17 calves per farm per year with an average milking herd size of 62. Therefore, while milk sold per cow averaged 84.65 cwt, total milk produced per cow averaged 92.50 cwt. While future data collection of this nature will allow for a more robust analysis of these factors, animal production efficiency is clearly a critical factor in lowering farm COP. In addition, the negative relationships between COP, cows managed per FTE, and labor earnings per hour indicate that labor efficiency is a critical determinant of COP and profitability on grass-fed dairy farms.

One-tailed *t*-tests were conducted to investigate the difference in COP and other variables between groups of farms utilizing different management practices in terms of buying forage vs. producing forage, supplementing energy or not, and year-round twice daily milking vs. alternative milking schedules. First, since fuel costs and other cash expenses related to forage production were highly correlated to COP, a comparison was made between farms that were producing most of their stored forages and farms that were buying most of their stored forages. This was determined by the purchased forage expense per cow where farms spending less than USD 30 per cow were assigned to the 'producing forage' group. Upon conducting a one-tailed *t*-test between these groups, the results, reported in Table 3, are statistically significant for all variables included in the table, which are indicative of operational and economic aspects of farming practices. The variables in Table 3 were selected due to their association with producing stored forages and their influence on COP.

Table 3. *t*-test of the mean difference for producing forage vs. buying forage.

Variable	Producing Forage	Buying Forage	Difference	*p*-Value
Cost of production (USD/cwt eq.)	51.57	42.73	8.84	0.0005
Land per cow (acres/cow)	7.05	4.67	2.38	0.0003
Labor earnings (USD/h)	1.66	8.99	−7.33	0.0000
Milk sold (cwt/cow)	77.55	88.63	−11.08	0.0213
Supplies (USD/cwt eq.)	3.30	2.27	1.03	0.0121
Fuel cost (USD/cwt eq.)	2.50	1.45	1.05	0.0000
Repair (USD/cwt eq.)	3.95	2.41	1.54	0.0002

Farms producing most of their stored forage needs managed 2.38 more acres per cow and incurred higher production costs than farms purchasing most of their forage needs. Similarly, individual expense categories related to making forage were higher for the 'producing forage' group including supplies, fuel, and repairs. These indicate substantial production cost differences between these two management paradigms. Like many other agricultural sectors, specialization in forage production may provide economies of scale, especially relative to labor. Reliance on more resource-efficient equipment and specialization of labor could be contributing to generally lower forage prices on the market compared to realized feed production costs on grass-fed dairy farms. Furthermore, differences in quality of the forages made on-farm or purchased off-farm could also impact milk production and thus sales which is highly negatively correlated to COP. In addition, "producing forage" farms earned less per hour influenced by their lower milk productivity. As these differences could reflect variations in labor economics, scales of operation, herd management, feeding practices, forage quality, or genetic selection, more information is needed to fully understand the underlying causes of these differences.

Second, as reported in Table 4, the farms that supplemented energy were significantly different from their counterparties in terms of COP and several other variables. Supplying additional energy to the animals would be expected to increase milk productivity, therefore COP and milk sold per cow were the primary variables investigated in this comparison. Farms feeding supplemental energy had substantially higher production costs with no statistical difference in milk sold per cow. It is possible that differences in the quality of forages fed in the two groups could explain the lack of increase in milk sales. Reliance on poorer quality forages and energy supplementation at the levels used by these farms may represent a smaller increase in milk sales than feeding higher quality forages without supplemental energy. Without information on the quality of forages or quantity of molasses fed, this cannot be elucidated. It is also possible that farms supplementing energy produced more milk, but the additional milk was diverted to feed youngstock and therefore is not reflected in milk sales. Information on youngstock rearing would be needed to further explore this possibility. Regardless, energy supplementation appears to be a significant driver of COP for grass-fed dairy farms, and more information on the economics of feeding supplemental energy is needed to aid producer decision-making.

Table 4. *t*-test of the mean difference for supplementing energy vs. not supplementing energy.

Variable	Supplementing Energy	Not Supplementing Energy	Difference	*p*-Value
Cost of production (USD/cwt eq.)	52.59	41.34	11.25	0.0003
Milk sold (cwt/cow)	80.36	87.58	−7.22	0.0939

Third, while most grass-fed dairy farms milk twice daily and calve year-round [15], some grass-fed farms are interested in milking less than twice daily or seasonally calving to control labor costs and other on-farm resources. Since altering milking frequency can impact milk production and expenses, these variables were selected along with COP in this comparison. In this dataset, the overall cost of producing milk and cash expenditures per cow did not differ statistically between farms utilizing a typical year-round calving twice daily milking schedule and those using an alternative schedule (Table 5). While farms using alternative schedules were significantly more labor efficient, managing 10 cows per FTE more than the traditional schedule farms, labor earnings per hour were numerically different but statistically similar between the two groups. This is likely due to the alternative schedule farms selling significantly less milk per cow and potential variation in milk price.

Table 5. *t*-test of the mean difference for year-round 2X vs. other schedules.

Variable	Year-Round 2X	Other Schedules	Difference	*p*-Value
Cost of production (USD/cwt eq.)	45.03	47.11	−2.08	0.2709
Land per cow (acre/cow)	5.56	5.47	0.09	0.4500
Labor earnings (USD/hour)	4.96	8.51	−3.55	0.1018
Milk sold (cwt./cow)	88.35	79.57	8.78	0.0515
Cash expense (USD/cow)	2472.00	2442.00	30.00	0.4408
Herd Size (number of cows)	60.95	63.51	−2.56	0.1300
Cows per FTE (cows/worker)	24.01	34.02	−10.01	0.0004

3.3. Regression Results

While the correlation analysis and hypothesis tests reported in the previous subsection provided important information on how the selected variables are correlated in terms of both the direction and degree of correlation, regression analysis provides information on the causality relation as well as the impact of each independent variable on the dependent variable (COP) while holding other variables constant. In our analysis, two regression models are estimated to identify the factors that determine the COP and quantify the marginal impact of each factor while other factors are controlled or held constant. While Model 1 includes four major variables addressing three factor groups as previous discussed, Model 2 includes those variables plus three dummy variables for demographic factors including English vs. Plain status and production scale (herd size). For the four major variables, the land per cow and seed and fertilizer expense per acre represent land utilization efficiency, excluding forage value variations due to purchase differences; the cows per FTE reflects labor efficiency, with its inclusion aimed at isolating its impact on costs while accounting for factors such as experience and community identity in further analyses; and milk sold per cow serves as a direct indicator of output efficiency, allowing for the assessment of cost influences attributable to animal productivity alone, while controlling for additional variables like purchased energy and milking frequency in extended analysis. The estimation results of the two regression models are reported in Table 6.

Table 6. Estimation results of two regression models.

Variable	Model 1	Model 2
Land per cow (acres/cow)	0.9873 * (0.5730)	1.4381 * (0.7820)
Cows per FTE (cows/worker)	−0.3630 *** (0.1110)	−0.1660 (0.1410)
Milk sold (cwt/cow)	−0.0033 *** (0.0006)	−0.0029 *** (0.0006)
Seed and fertilizer expense (USD/acre)	0.0193 (0.0407)	0.0236 (0.0393)
English vs. Plain (English = 1)		1.1870 (3.7170)
Large farm (>70 cows)		−5.8920 (−3.7190)
Small farm (<45 cows)		6.1960 ** (−2.9810)
Constant	78.1200 *** (8.2770)	66.1427 *** (9.3925)
Observations	64	64
R-squared	0.4493	0.5190
Adj. R-squared	0.4126	0.4599
F-statistic	12.2400	8.7900
p-value	0.0000	0.0000

Standard errors in parentheses. *** $p < 0.01$, ** $p < 0.05$, * $p < 0.1$.

The empirical results suggest four major findings: First, the estimated models fit the relatively small dataset well and explain 41.4% of the variation in COP in Model 1 and 45.8% of the variation in COP in Model 2. The F-statistic and its p-value for the two models also suggest that we are 99% confident that the variation in the COP is affected by the independent variables included in the models. Second, regarding the estimated coefficients and their significance levels for the four major independent variables, milk sold per cow showed a negative and significant impact on the COP at the 0.01 significance level in both models; land per cow had a positive impact on the COP at the 0.10 significance level; per acre seed and fertilizer expense showed a positive but insignificant impact; and cows per FTE had a negative impact but was only significant in Model 1. Third, for the two dummy variables for farm size, while small farms (<45 cows) had significantly higher COP at the 0.05 significance level, large farms (>70 cows) had lower COP but it was not statistically significant at the 0.10 significance level. Regarding the dummy variable for English farms, they had higher COP as compared to the COP of Plain farms, but the difference was not statistically significant at the 0.10 significance level. Fourth, the relatively small absolute values of the estimated coefficients of the independent variables suggest that the COP is determined by this set of variables together and the marginal impact of each individual variable is small when other variables are controlled. For example, when the land per cow increases by one acre, the COP is expected to increase by only USD 0.987 according to Model 1 and USD 1.43 according to Model 2, when other variables in each model are held constant. The relatively small marginal impacts of the independent variables suggest that, while each factor influences COP, their individual impacts are relatively small compared to their combined effect when controlling for other factors. In application, this finding suggests that farms need to pay attention to all the factors simultaneously to control their COP.

Like many other empirical studies, this study is limited by the small dataset with only 64 observations (22 from 2019, 21 from 2020, and 21 from 2022) and the empirical results from the data analysis should be interpreted with caution. Additionally, the methods used to calculate COP were chosen to facilitate comparisons across farms with differing debt loads, family living draws, and other external factors that influence COP but are not directly related to farm management. While this allows for investigation into management factors' influences on COP generally, the COP values presented here may not reflect the true experience of some farmers and therefore should be used with caution outside of this context. Another limitation is that some potentially important variables affecting the COP were not included in the regression analysis due to the lack of available data or multicollinearity concerns. Finally, there is the inherent challenge in standardizing farms for comparison. All farms are uniquely different, and while DairyTrans was a necessary tool for creating comparability, the standardization process can never fully account for all farm-level differences. Since there are very limited studies on grass-fed dairy, as it is a relatively new sector in the literature, it is a challenge to compare and validate findings from this study or to discuss to what extent the findings from this can be applied to other regions.

4. Conclusions and Recommendations

As one of the first empirical studies on the COP of grass-fed dairy in the Northeastern United States, this study suggests three major conclusions. First, COP varies widely from USD 25.51 to USD 92.38 with an average of USD 45.91. The cost to produce grass-fed milk appears to be substantially higher than that of conventional and organic production models. Second, COP is highly positively correlated with acres per cow and key expense categories including fuel, supplies, repairs, and supplemental energy. COP is negatively correlated with the quantity of milk sold per cow, cows managed per FTE, and labor earnings per hour. Farms producing their own stored forages had significantly higher COP, acres per cow, and expenses related to forage production including fuel, supplies, and repairs. Farms feeding supplemental energy also had significantly higher COP and did not sell more milk. Farms using alternative milking strategies were more labor efficient but had similar COP due to lower milk sales. These suggest that farm acreage, herd size, cow production efficiency, and labor efficiency highly influence COP. This was further supported by regression analyses that found that over 45% of the variation in COP can be explained by these factors.

While many grass-fed dairy farms are actively searching for information and strategies to enhance their production efficiency and economic viability, this study suggests four major recommendations based on the data analyses as well as field observations of grass-fed farm operations and communication with many farms: (1) optimize the number of replacements being raised to maximize milk sales from milk produced, (2) enhance milk productivity per cow by optimizing cow nutrition, reproductive performance, comfort, and health to aid in maximizing milk sales, (3) invest in labor efficiency enhancements in milking, feeding, and other aspects of farm management to maximize output relative to labor resources, and (4) optimize acreage to balance high-quality forage production with available economic, labor, and other resources. Future COP studies will collect additional management data to better elucidate the impacts of specific farm management factors and practices on COP and farm profitability. In addition, existing grass-fed farmers need access to technical assistance to help calculate their own farms' COP to inform management decisions that enhance farm viability. Furthermore, prospective grass-fed dairy farmers need access to technical assistance to understand whether they are able to transition to this production model successfully and if it will support farm viability.

Author Contributions: Conceptualization, H.D., S.Z., Q.W. and S.F.; Methodology, Q.W., S.Z. and H.U.; Validation, S.Z. H.U. and A.A.; Formal analysis, H.U., Q.W. and S.Z.; Resources, H.D., S.F. and S.Z.; Data curation, S.F. and S.Z.; Writing—original draft, S.Z. Q.W., S.F. and H.U.; Writing—review and editing, S.Z., Q.W., A.A., S.F., H.U. and H.D.; Supervision, S.Z., Q.W. and H.D.; Funding acquisition, H.D. All authors have read and agreed to the published version of the manuscript.

Funding: This work was supported by the U.S. Department of Agriculture (USDA) Organic Agriculture Research and Extension Initiative (OREI) (Grant award number 2023-51300-40985 and 2018-51300-28515).

Institutional Review Board Statement: The study was conducted in accordance with the Declaration of Helsinki, and approved by the University of Vermont Institutional Review Board (protocol code 00000036 and date of approval 12 December 2018).

Informed Consent Statement: Informed consent was obtained from all subjects involved in the study.

Data Availability Statement: The original contributions presented in this study are included in the article. Further inquiries can be directed to the corresponding author.

Acknowledgments: The authors would like to thank the editors and three anonymous reviewers for their valuable comments and suggestions. We also would like to thank other members of our research team for their comments and suggestions.

Conflicts of Interest: Author Sarah Flack has been employed by Sarah Flack Consulting. The remaining authors are employed by the University of Vermont and declare that the research was conducted in the absence of any commercial or financial relationships that could be construed as a potential conflict of interest. The authors declare that this study received funding from the U.S. Department of Agriculture and the government funding organization had no direct involvement with the study.

Appendix A

Table A1. Correlation matrix.

	Cost of Production (USD/cwt eq.)	Land per Cow (acres/cow)	Seed and Fertilizer Expense (USD/acre)	Labor Earning (USD/hour)	Milk per Cow (cwt eq./cow)	Cost of Molasses (USD/cwt eq.)	Cost of Supplies (USD/cwt eq.)	Fuel Cost (USD/cwt eq.)	Repair Cost (USD/cwt eq.)	Total Cash Expense (USD/cow)	Herd Size (cows)	Cows per FTE
Cost of production (USD/cwt eq.)	1											
Land per cow (acres/cow)	0.3751 ** (0.1178)	1										
Seed and fertilizer expense (USD/acre)	−0.1509 (0.1254)	−0.2672 ** (0.1228)	1									
Labor earning (USD/hour)	−0.6473 *** (0.0980)	−0.3010 ** (0.1211)	−0.0400 (0.1268)	1								
Milk per cow (cwt eq./cow)	−0.5545 *** (0.1052)	−0.3335 *** (0.1197)	0.1224 (0.1260)	0.2234 *** (0.1238)	1							
Molasses cost (USD/cwt eq.)	0.6107 *** (0.1008)	0.3482 *** (0.1190)	0.1062 (0.1263)	−0.4501 *** (0.1134)	−0.3480 *** (0.1191)	1						
Supply cost (USD/cwt eq.)	0.3422 *** (0.1193)	0.3253 *** (0.1201)	−0.1165 (0.1261)	−0.2887 ** (0.1216)	−0.3196 ** (0.1203)	0.4333 ** (0.1144)	1					
Fuel cost (USD/cwt eq.)	0.6365 *** (0.0981)	0.4559 *** (0.1130)	−0.1542 (0.1254)	−0.3200 *** (0.1203)	−0.3260 *** (0.1201)	0.4063 ** (0.1160)	0.2204 * (0.1239)	1				
Repair cost (USD/cwt eq.)	0.4642 *** (0.1138)	0.3681 *** (0.1181)	0.0782 (0.1266)	−0.3973 *** (0.1165)	−0.4826 *** (0.1112)	0.5021 *** (0.1098)	0.3408 ** (0.1194)	0.2918 ** (0.1215)	1			
Total cash expense (USD/cow)	0.0178 (0.1269)	−0.0706 (0.1267)	0.2305 * (0.1236)	−0.2702 ** (0.1222)	0.5300 *** (0.1069)	0.1715 (0.1251)	0.0987 (0.1263)	−0.0363 (0.1269)	0. 0675 (0.1267)	1		
Herd size (cows)	−0.2710 ** (0.1222)	0.0300 (0.1269)	0.4104 *** (0.1154)	0.1402 (0.1256)	−0.0095 (0.1269)	−0.0523 (0.1268)	−0.0223 (0.1269)	−0.0808 (0.1265)	0.0943 (0.1263)	0. 0203 (0.1269)	1	
Cows per FTE	−0.2912 ** (0.1220)	−0.0913 (0.1265)	0.2630 ** (0.1225)	0.4140 *** (0.1156)	−0.0841 (0.1266)	−0.0746 (0.1266)	0.0200 (0.1269)	−0.1316 (0.1259)	0.0520 (0.1268)	−0. 0001 (0.1269)	0.6812 *** (0.0925)	1

Standard errors in parentheses. *** $p < 0.01$, ** $p < 0.05$, * $p < 0.1$.

References

1. Walsh, J.; Parsons, R.; Wang, Q.; Conner, D. What makes an organic dairy farm profitable in the United States? Evidence from 10 years of farm level data in Vermont. *Agriculture* **2020**, *10*, 17. [CrossRef]
2. Wang, Q.; Liu, C.-Q.; Zhao, Y.-F.; Kitsos, A.; Cannella, M.; Wang, S.-K.; Han, L. Impacts of the COVID-19 pandemic on the dairy industry: Lessons from China and the United States and policy implications. *J. Integr. Agric.* **2020**, *19*, 2903–2915. [CrossRef]
3. MacDonald, J.M.; Law, J.; Mosheim, R. *Consolidation in U.S. Dairy Farming (ERS Report No. 274)*; U.S. Department of Agriculture, Economic Research Service: Washington, DC, USA, 2020. Available online: https://ers.usda.gov/sites/default/files/_laserfiche/-publications/98901/ERR-274.pdf?v=65010 (accessed on 13 October 2025).
4. U.S. Department of Agriculture (USDA) Agricultural Marketing Service. Dairy Market News at a Glance. *Dairy Market News*, 11–15 November 2024. Available online: https://mymarketnews.ams.usda.gov/filerepo/sites/default/files/2998/2024-11-15/1172059/ams_2998_00251.pdf (accessed on 13 October 2025).
5. U.S. Bureau of Labor Statistics. Consumer Price Index. U.S. Department of Labor; 2024. Available online: https://www.bls.gov/cpi/ (accessed on 13 October 2025).
6. U.S. Department of Agriculture, Agricultural Marketing Service. Advertised Prices for Dairy Products at Major Retail Supermarket Outlets ending during the period of 9/19/2025 to 9/25/2025. *Dairy Market News*, 14 November 2025. Available online: https://www.ams.usda.gov/mnreports/dybretail.pdf (accessed on 13 October 2025).
7. Brito, L.; Bedere, N.; Douhard, F.; Oliveira, H.; Arnal, M.; Peñagaricano, F.; Schinckel, A.; Baes, C.; Miglior, F. Review: Genetic selection of high-yielding dairy cattle toward sustainable farming systems in a rapidly changing world. *Animal* **2021**, *15*, 100292. [CrossRef] [PubMed]
8. Clay, N.; Garnett, T.; Lorimer, J. Dairy intensification: Drivers, impacts and alternatives. *Ambio* **2020**, *49*, 35–48. [CrossRef] [PubMed]
9. Nozari, H.; Szmelter-Jarosz, A.; Ghahremani-Nahr, J. The Ideas of Sustainable and Green Marketing Based on the Internet of Everything—The Case of the Dairy Industry. *Futur. Internet* **2021**, *13*, 266. [CrossRef]
10. Schiano, A.; Harwood, W.; Gerard, P.; Drake, M. Consumer perception of the sustainability of dairy products and plant-based dairy alternatives. *J. Dairy Sci.* **2020**, *103*, 11228–11243. [CrossRef] [PubMed]
11. Dimitri, C.; Nehring, R. Thirty years of organic dairy in the United States: The influences of farms, the market and the organic regulation. *Renew. Agric. Food Syst.* **2022**, *37*, 588–602. [CrossRef]
12. Organic Trade Association. Organic Trade Association Pegs Sales of Organic Products Near $70 Billion [Press Release]. 2024. Available online: https://ota.com/about-ota/press-releases/us-organic-marketplace-posts-record-sales-2023 (accessed on 13 October 2025).
13. Hochul, K.; Ball, R.A. New York State Dairy Statistics 2023 Annual Summary. 2023. Available online: https://agriculture.ny.gov/-system/files/documents/2024/07/2023dairystatisticsannualsummary.pdf#: :text=New%20York%20State%20has%20nearly,state%20-in%20the%20United%20States (accessed on 13 October 2025).
14. MacKenzie, M.K.; Karszes, J. New York Organic Dairy Cost of Production: Benchmarks and Financial Performance 2023. 2024. Available online: https://dyson.cornell.edu/wp-content/uploads/sites/5/2024/07/E.B._2024-04_Organic_Financial_Performance-VD.pdf (accessed on 13 October 2025).
15. Snider, M.A.; Ziegler, S.E.; Darby, H.M.; Soder, K.J.; Brito, A.F.; Beidler, B.; Flack, S.; Greenwood, S.L.; Niles, M.T. An overview of organic, grassfed dairy farm management and factors related to higher milk production. *Renew. Agric. Food Syst.* **2022**, *37*, 624–632. [CrossRef]
16. Organic Plus Trust Inc. OPT Certified Grass-Fed Organic Livestock 2025 Program Manual. 2025. Available online: https://img1.wsimg.com/blobby/go/7dfcba80-cf7b-4ce3-a4de-b2df6219663c/downloads/487d3c3d-3fd1-4131-8ff5-89cc9429442d/Program%20Manual_OPT%20Certified%20Grass-fed%20v5.0_20.pdf?ver=1754330482891 (accessed on 13 October 2025).
17. Alothman, M.; Hogan, S.A.; Hennessy, D.; Dillon, P.; Kilcawley, K.N.; O'donovan, M.; Tobin, J.; Fenelon, M.A.; O'callaghan, T.F. The "Grass-Fed" milk story: Understanding the impact of pasture feeding on the composition and quality of bovine milk. *Foods* **2019**, *8*, 350. [CrossRef] [PubMed]
18. van Vliet, S.; Provenza, F.D.; Kronberg, S.L. Health-promoting phytonutrients are higher in grass-fed meat and milk. *Front. Sustain. Food Syst.* **2021**, *4*, 555426. [CrossRef]
19. Elgersma, A. Grazing increases the unsaturated fatty acid concentration of milk from grass-fed cows: A review of the contributing factors, challenges and future perspectives. *Eur. J. Lipid Sci. Technol.* **2015**, *117*, 1345–1369. [CrossRef]
20. Gammone, M.A.; Riccioni, G.; Parrinello, G.; D'orazio, N. Omega-3 polyunsaturated fatty acids: Benefits and endpoints in sport. *Nutrients* **2019**, *11*, 46. [CrossRef] [PubMed]
21. Martin, B.; Fedele, V.; Ferlay, A.; Grolier, P.; Rock, E.; Gruffat, D.; Chilliard, Y. Effects of grass-based diets on the content of micronutrients and fatty acids in bovine and caprine dairy products 20. In Proceedings of the General Meeting of the European Grassland Federation, Luzern, Switzerland, 21–24 June 2004; Available online: https://hal.inrae.fr/hal-02760385 (accessed on 13 October 2025).

22. Șanta, A.; Mierlita, D.; Dărăban, S.; Socol, C.T.; Vicas, S.I.; Șuteu, M.; Maerescu, C.M.; Stanciu, A.S.; Pop, I.M. The effect of sustainable feeding systems, combining total mixed rations and pasture, on milk fatty acid composition and antioxidant capacity in Jersey Dairy cows. *Animals* **2022**, *12*, 908. [CrossRef] [PubMed]
23. Strusińska, D.; Antoszkiewicz, Z.; Kaliniewicz, J. The concentrations of β-carotene, vitamin A and vitamin E in bovine milk in regard to the feeding season and the share of concentrate in the feed ration. *Rocz. Nauk. Pol. Tow. Zoot* **2010**, *6*, 213–220. Available online: https://publisherspanel.com/api/files/view/2364738.pdf (accessed on 13 October 2025).
24. Benbrook, C.M.; Davis, D.R.; Heins, B.J.; Latif, M.A.; Leifert, C.; Peterman, L.; Butler, G.; Faergeman, O.; Abel-Caines, S.; Baranski, M. Enhancing the fatty acid profile of milk through forage-based rations, with nutrition modeling of diet outcomes. *Food Sci. Nutr.* **2018**, *6*, 681–700. [CrossRef] [PubMed]
25. Butler, G.; Nielsen, J.; Larsen, M.; Rehberger, B.; Stergiadis, S.; Canever, A.; Leifert, C. The effects of dairy management and processing on quality characteristics of milk and dairy products. *NJAS Wagening. J. Life Sci.* **2011**, *58*, 97–102. [CrossRef]
26. Daley, C.A.; Abbott, A.; Doyle, P.S.; Nader, G.A.; Larson, S. A review of fatty acid profiles and antioxidant content in grass-fed and grain-fed beef. *Nutr. J.* **2010**, *9*, 10. [CrossRef] [PubMed]
27. O'callaghan, T.F.; Hennessy, D.; McAuliffe, S.; Kilcawley, K.N.; O'donovan, M.; Dillon, P.; Ross, R.; Stanton, C. Effect of pasture versus indoor feeding systems on raw milk composition and quality over an entire lactation. *J. Dairy Sci.* **2016**, *99*, 9424–9440. [CrossRef] [PubMed]
28. Schwendel, B.; Wester, T.; Morel, P.; Tavendale, M.; Deadman, C.; Shadbolt, N.; Otter, D. Invited review: Organic and conventionally produced milk—An evaluation of factors influencing milk composition. *J. Dairy Sci.* **2015**, *98*, 721–746. [CrossRef] [PubMed]
29. Niles, M.T.; Darby, H.; Flack, S.; Ziegler, S.; Brito, A.; Snider, M. Grass-Fed Dairy Farm Production Practices and Demographics in the US. University of Vermont. 2019. Available online: https://www.uvm.edu/d10-files/documents/2024-10/SurveyArticleFinal.pdf (accessed on 13 October 2025).
30. McMullen, E. (Organic Valley, Lafayette, WI, USA). Personal communication, 12 August 2025.
31. U.S Grass Fed Dairy Market Size by Product Type (Milk, Cheese), by Application (Food and Beverages, Nutraceutical), by Distribution Channel (Indirect, Direct) and Forecast. Verified Market Research. 2025. Available online: https://www.verifiedmarketresearch.com/product/us-grass-fed-dairy-market (accessed on 13 October 2025).
32. Global Grass-Fed Milk Market Size by Product Type (Whole Grass-Fed Milk, Reduced-Fat Grass-Fed Milk), by Packaging Type (Plastic Bottles, Glass Bottles), by Distribution Channel (Supermarkets/Hypermarkets, Convenience Stores), by End-User (Households, Restaurants/Cafes), by Nutrition Profiles (High Nutritional Value (Omega-3 and Vitamins) Standard Nutritional Value), by Geographic Scope and Forecast. Verified Market Reports. 2025. Available online: https://www.verifiedmarketreports.com/product/grass-fed-milk-market/ (accessed on 13 October 2025).
33. UVM Extension NW Crops & Soils. Cost of Production on Organic and Grass-Fed Dairy Farms [Video]. *YouTube*, 4 March 2022. Available online: https://www.youtube.com/watch?v=df3jtrGAlFA (accessed on 13 October 2025).
34. Darby, H.; Flack, S.; Ziegler, S. 2023 Cost of Production on Grass-Fed Dairy Farms in the Northeast. University of Vermont Extension Northwest Crops and Soils Program. 2024. Available online: https://vip2.uvm.edu/d10-files/documents/2025-04/2023-COP-article.pdf (accessed on 13 October 2025).
35. Tranel, F. *Dairy TRANS 20.20*; A Software for Dairy Farm Financial Analysis; Iowa State University Extension: Ames, IA, USA, 2020.

sustainability

MDPI

Article

Sustainability Assessment of Austrian Dairy Farms Using the Tool *NEU.rind*: Identifying Farm-Specific Benchmarks and Recommendations, Farm Typologies and Trade-Offs

Stefan Josef Hörtenhuber [1,*], Caspar Matzhold [2,3], Markus Herndl [4], Franz Steininger [2], Kristina Linke [2], Sebastian Wieser [4] and Christa Egger-Danner [2]

1 Institute of Livestock Sciences, BOKU University, Gregor-Mendel-Straße 33/II, 1180 Vienna, Austria
2 ZuchtData EDV-Dienstleistungen GmbH, Dresdner Straße 89, 1200 Vienna, Austria; matzhold@zuchtdata.at (C.M.); steininger@zuchtdata.at (F.S.); linke@zuchtdata.at (K.L.); egger-danner@zuchtdata.at (C.E.-D.)
3 Complexity Science Hub, Metternichgasse 8, 1030 Vienna, Austria
4 HBLFA Raumberg-Gumpenstein, Raumberg 38, 8952 Irdning-Donnersbachtal, Austria; markus.herndl@raumberg-gumpenstein.at (M.H.); sebastian.wieser@raumberg-gumpenstein.at (S.W.)
* Correspondence: stefan.hoertenhuber@boku.ac.at

Abstract

The sustainable future of dairy farming will depend on how trade-offs between environmental impact, economic viability, and animal welfare are managed. Dairy production contributes significantly not only to human nutrition but also to greenhouse gas (GHG) emissions, ammonia release, and water pollution. Comprehensive sustainability assessments are essential for addressing these impacts, also in light of evolving regulations like the EU Corporate Sustainability Reporting Directive. However, existing research on sustainable dairy farming and intensification often overlooks trade-offs with other ecological aspects like biodiversity, economic viability, or animal welfare. This study evaluated the sustainability performance of Austrian dairy farms using a tool called *NEU.rind*, which combines life cycle assessment (LCA) with other indicators. Applied to 170 dairy farms, the tool identified four sustainability clusters across the dimensions of environmental conditions, efficiency, animal health, and sustainability: (1) Alpine farms (high cow longevity, medium-to-high emissions per kg milk), (2) efficient low-input farms (low emissions, high cow longevity), (3) high-output lowland farms (high productivity, lower animal welfare), and (4) input-intensive lowland farms (high emissions, especially per hectare; inefficient use of resources). The analysis revealed fundamental trade-offs between production intensity, environmental impact, and animal welfare, particularly when comparing product-based (per kg milk) versus hectare-based indicators. Key improvement strategies include increasing the use of regional feed and pasture as well as adapting manure management. For policymakers, these findings underline the importance of site-specific sustainability assessments and the need for region-specific incentive schemes that reward both environmental efficiency and animal health performance. In this context, *NEU.rind* provides farm-specific recommendations with minimal data input, making sustainability assessments practical and feasible.

Keywords: dairy production systems; cattle; farm-level analysis; sustainability dimensions; climate and environmental impacts; mitigation; clustering

1. Introduction

Due to their high-quality protein and other nutrients, cattle, particularly dairy cows, play a significant role in human nutrition: Their milk is an important and valuable food for human nutrition, representing an excellent source of both macronutrients (proteins, carbohydrates, fat) and micronutrients (minerals and vitamins) [1,2]. However, (cow) milk production is also associated with various environmental issues and accounts for 4% of total GHG emissions, measured in terms of the Global Warming Potential (GWP_{100}), mainly due to methane (CH_4) and nitrous oxide (N_2O) emissions [3]. Additionally, dairy farming affects air quality, mainly through ammonia (NH_3) emissions, and water quality through nutrient leaching, run-off, and erosion, including nitrates (NO_3) and phosphates [4,5].

Mitigation strategies for reducing these impacts include improving dairy herd fertility, modifying diets, and optimising the use of fertilisers, such as through optimised slurry applications [4,6,7]. Such measures target key emission sources, e.g., enteric CH_4 from rumen fermentation, CH_4, N_2O and NH_3 from manure and its use as fertiliser, and ammonia volatilisation. Combining approaches that link feeding, housing, and manure handling practices is particularly effective and may offer synergies between emission reduction and resource efficiency [7]. Some studies used model farms to gain a better understanding of the systems [7]. However, little is known about how the most important sustainability trade-offs and synergies related to specific practices perform in real-world settings in European dairy production systems. With regard to Austrian dairy farming, the eco-efficiency of organic versus conventional farms as well as the environmental impacts of low-input farms have been analysed [8,9].

LCA has been extensively used to assess the environmental impacts of agricultural and livestock production, including dairy farms. Indicators that are most often evaluated include GHG emissions, fossil energy use, land occupation, and emissions contributing to acidification and eutrophication. Studies conducted in Europe or the USA have often applied LCA to determine the environmental footprints of dairy production. However, although LCA provides robust results for some impacts, a common limitation of such detailed assessments is their narrow focus on a few environmental impacts, such as especially the GWP, overlooking other crucial sustainability risks. These include biodiversity loss, soil health, animal welfare, and socio-economic resilience, which are difficult to capture within conventional LCA system boundaries and functional units [10–14].

An accurate and comprehensive sustainability assessment is essential for improving farm management and complying with new regulatory frameworks, such as the Corporate Sustainability Reporting Directive (CSRD; https://eur-lex.europa.eu/legal-content/EN/ALL/?uri=CELEX:32022L2464, accessed on 5 December 2025) and Product Environmental Footprints [15]. Research on 'sustainable development' or 'sustainable intensification' in dairy farming often focuses on reducing GHG but has not sufficiently explored the potential trade-offs and synergies with other sustainability aspects. There are some insights into the trade-offs with animal health [16], economic performance of farms [17], biodiversity conservation [18], or the feed-food-conflict [19,20]. However, clear pathways that guide dairy farms towards win-win situations within planetary boundaries [21], using LCAs and assessment methods for a comprehensive sustainability view, are still lacking.

Some tools for assessing impacts of dairy farming apply LCA-based approaches but are limited in their scope of impact indicators, often focusing on GWP. Examples include the Danish ARLA tool [22] and the German LfL Klimarechner [23], the latter combining farm economics (gross margin) with carbon footprinting. Other tools such as the ANCA tool [24] and the Irish sustainability measurement platform AgNav [25] cover a broader range of ecological aspects, including nutrient flows, ammonia and GHG emissions, but are still limited to the environmental dimension. Several other tools have been developed

https://doi.org/10.3390/su18010303

to evaluate dairy farms more comprehensively. For instance, Paçarada et al. [26] discuss tools such as SAFA and SMART, which offer multidimensional sustainability assessments that extend beyond the farm gate. These tools offer a comprehensive approach but lack full quantification, limiting their ability to identify specific measures and improvement potentials. While qualitative analyses are valuable in addressing certain details, quantitative evaluations are essential to address trade-offs between indicators and to classify the effects of measures for optimisation and conflict resolution. Some tools like 'MOTIFS' [27] and others described in [28] apply a mixed-method approach that combines quantitative LCA indicators with semi-quantitative indicators to integrate sustainability aspects across multiple dimensions—ecological, economic, social, animal health and welfare—within one single analysis. Similar approaches have been applied in other livestock sectors, for instance, pig farming [29], incorporating additional sustainability aspects related to biodiversity, economic resilience, workplace safety, and ethical concerns, including animal health and welfare, and the feed-food conflict. An effective tool should provide a comprehensive but sufficiently quantitative set of sustainability indicators while minimising data collection demands to ensure broad acceptance [30]. Additionally, it should enable the simulation of farm scenarios and generate optimisation recommendations [31]. Integration into established data management systems enhances user-friendliness while supporting third-party verification for potential certification purposes.

The aims of this study are (i) to evaluate the sustainability performance of Austrian dairy farms using the Digital Farm Assistant *NEU.rind*. This web-based tool integrates a comprehensive set of indicators including LCA approaches. Its theoretical framework is based on the concept of multidimensional sustainability assessment, linking environmental, economic, and animal-welfare dimensions via measurable indicators and data-driven analysis. Applied in a case study with future-oriented dairy farms in Austria, this approach (ii) enables the identification of trade-offs and synergies between sustainability goals, which are essential for guiding both farm management and policy design. The tool and this study address important real-world challenges in sustainable dairy farming, offering practical and policy-relevant insights. By identifying patterns in these data using statistical and machine-learning approaches, the study seeks to offer insights to propose more sustainable pathways for the dairy sector.

2. Materials and Methods

Applying the *NEU.rind* tool, important sustainability aspects of dairy production systems were evaluated using a set of aggregated indicators at farm level. The set of indicators (Table 1) was selected by project stakeholders during the method development phase from a list of 15 candidate indicators compiled by the research team. This selection of indicators also took into account the availability of existing data and the effort required from farmers to collect additional data. Selected indicators covered environmental impacts (e.g., GWP, fossil energy demand, biodiversity aspects), animal health issues, and economic outcomes (costs, revenues, and gross margins). An overview of the main indicators, ranked by their importance as assessed by a group of potential users and stakeholders (farmers, dairy consultants and managers, agricultural advisors), is provided in Table 1.

The *NEU.rind* tool uses LCA methodologies to evaluate impacts of farm-level dairy units across the functional units 'kg of energy corrected milk' (considering the co-products beef and sold cattle) and 'hectare of farmland related to milk production' (land for feed production or land related to the dairy unit via manure application). For some additional figures, the tool also uses key performance indicators (KPIs), partially per 'cow and year' or at farm-level. Further details on the methods are provided in [32].

https://doi.org/10.3390/su18010303

Table 1. Main indicators and functional units used in the *NEU.rind* tool to evaluate the sustainability of milk production on dairy farms.

	Indicators	Functional Units	
		'kg ECM' [1]	'ha Farmland' [2], 'Farm' and 'Cow'
1a	Global Warming Potential (GWP_{100})	kg CO_2-eq	kg CO_2-eq
1b	Methane emissions	kg CH_4	kg CH_4
1c	Di-nitrous oxides emissions	kg N_2O	kg N_2O
1d	Fossil carbon dioxide emissions	kg CO_2	kg CO_2
2	Food/protein supply	Human-edible feed conversion efficiency	kg protein (net/gross)
3	Biodiversity	Potential species losses (feed-dependent)	% HNVF [3] Type 1; endangered livestock breeds (y/n)
4	Fossil energy demand	MJ	GJ
5	Ammonia emission and Acidification (SO_2-eq)	g NH_3, g SO_2-eq	kg NH_3
6	Animal Health Scores	Scores of cows and calves	
7	Profit margin	€	€

[1] ECM = Energy-Corrected Milk. [2] Related to milk production. [3] High Nature Value Farmland.

2.1. LCA Core Module

Analysis was based on the LCA Core Module of the *NEU.rind* tool, programmed using APEX (Oracle Cooperation, version 24.2.2). The module focused on calculating various environmental impacts, including GWP_{100}, both over a 100-year time frame and with characterisation factors from [33], which are currently used in national inventories. One aim in developing the method was to align it as closely as possible with the national (Austrian) inventory method [34] and other relevant guidelines for milk production, e.g., [35,36]. Fossil energy use was calculated using the Cumulative Energy Demand method (v1.1). GWP_{100} calculation included fossil CO_2, methane (CH_4), and di-nitrous oxide (N_2O), while biogenic carbon dioxide (CO_2) was considered 'climate neutral' and thus excluded. LCA followed Tier 2 calculation procedures described in IPCC 2019 [37] and EMEP/EEA [38] guidelines, the latter for NH_3, nitrogen oxides (NO_X), and nitrate (NO_3) emissions. Further input-related emission factors were mainly derived from Ecoinvent version 3 [39]. Potential species losses associated with land occupation are calculated as described in [40].

System boundaries covered on-farm resource inputs (e.g., fertilisers, feed, energy) up to the farm gate ('cradle to farm gate'), ending at milk collection (for processing in dairies). They included emissions related to:

- Enteric fermentation
- Feed production (on-farm and external sources, i.e., purchased feed)
- Manure handling and application (including internal nutrient flows), fertiliser production, and application
- Energy and material input used for dairy farming
- Milk and growth performance of cows, biological data, animal health
- Infrastructure (milk production-related machinery and buildings on farms)

Not included in the assessment were transportation and processing of milk beyond the farm gate, retail and consumption phases, or on-farm environmental effects that were not related to dairy production.

https://doi.org/10.3390/su18010303

The following Energy-Corrected Milk (ECM) calculation formula [41] was used (Equation (1)):

$$\text{kg ECM} = (\text{kg milk yield}) \times [0.38 \times (\text{fat \%}) + 0.21 \times (\text{protein \%}) + 1.05]/3.28 \quad (1)$$

The allocation of impacts to milk and the by-product beef followed a procedure described in an IDF guideline [35], referencing [36,42]. The biophysical allocation applied an avoidance approach, comparing (i) the feed net energy required for milk production to (ii) the feed net energy required for body growth, thereby distinguishing impacts attributed to milk from those assigned to beef by-products, cull cows, and calves.

2.2. *Supplementary Key Performance Indicators*

Additional (non-LCA) key performance indicators (see Table 1), together with their respective evaluation methods, were used to complement the sustainability assessment:

- Proportion of High Nature Value Farmland (HNVF) Type 1 by the method according to [43]
- Keeping of endangered livestock species (assessed categories: yes/no; proportion)
- Animal Health Scores for cows and calves using the Q-Check Animal Welfare Assessment, developed by [44]. The Q-Check Animal Welfare Assessment includes indicators for longevity, udder health, metabolic stability, as well as raising losses related to calves culling rates in cows.
- Profit margins of farms according to the Federal Institute of Agricultural Economics, Rural and Mountain Research [45], accounting for direct and indirect inputs, their costs and revenues, also related to the functional unit '1 cow per year' in addition to product, land and farm level.

2.3. *Data Demand and Collection*

We used activity data from 170 farms, the majority being sourced from existing databases, including the RDV Cattle Data Network (www.rdv-gmbh.net), IACS (Integrated Administration and Control System; in German 'INVEKOS') [46], and other economic data [45]. These datasets provided default values for most parameters. Where farm-specific data were unavailable, default values were applied, for example breed-based cattle body masses, but could be overwritten whenever farm-specific data became available.

Certain parameters had not previously been consistently recorded across all farms (e.g., housing system features like lying areas, walking paths, or slatted floors) and were thus not available for all farms in the RDV dataset. Additionally, data on volumes of slurry and feed stores, or details on machinery, especially their masses (subject to amortisation over multiple years), were not available for any of the farms and thus generally had to be entered manually. In the case of machinery, estimations based on default values were employed. Consequently, a complete farm dataset was drawn from a combination of automated queries and manual entries, which are grouped in Table 2.

Table 2. Data sources and (exemplary) parameters used for the *NEU.rind* sustainability assessments.

Data Group	Data Source	Parameters on…
Animal Data	Cattle Data Network (RDV)	Animal arrivals and departures, milk yields and milk ingredients, reproduction characteristics, body masses and slaughter performances, health records

https://doi.org/10.3390/su18010303

Table 2. *Cont.*

Data Group	Data Source	Parameters on...
Housing & Manure Management	Cattle Data Network (RDV) and manual entries	Barn systems, type of manure storage, manure removal and treatments, manure application
Feeding	Cattle Data Network (RDV) and manual entries	Animal diets, including concentrate amounts and roughage proportions, periods with specific diets
Land Management	Integrated Administration and Control System (IACS)	Grassland and crop type areas with their intensity of use, other biodiversity-related farmland
Economic Data	Federal Institute of Agricultural Economics, Rural and Mountain Research	Default milk and slaughter cattle prices, costs for replacement animals, costs related to feed or energy carriers, and other farm inputs

2.4. Description of Study Farms

For an initial overview, farms were categorised into four groups based on a binary classification: farming type (conventional or organic) and geographical location ('favoured' or 'alpine' areas). The distinction between favoured and alpine farms was made using the severity score frequently applied in Austria (see IACS data), with farms having up to 90 points (out of a possible 570 points) being classified as favoured and farms above that as alpine. The distribution of the farms was as follows:

- Conventional dairy farms in favoured areas: 70 farms
- Conventional alpine dairy farms: 45 farms
- Organic dairy farms in favoured areas: 29 farms
- Organic alpine dairy farms: 26 farms

Regarding the median herd size, conventional farms in favoured areas with a median of 48 cows differed significantly from the other groups with herd sizes (medians) ranging between 26 and 35 cows. The lifetime milk production of cull cows was relatively high across all farm types, averaging 35,000 kg per cow. The annual milk yield per cow, however, varied between the groups, with conventional farms in favoured areas producing the most (Table 3). The productive lifetime of cows also varied among the different farm types. Most of the cows, across all farm types, were housed in slurry-based systems. Grazing practices varied significantly between conventional and organic farms. Conventional dairy cows were rarely allowed to graze, while organic cows spent, on average, more than one sixth of the total time per year on pasture, illustrating a clear difference in management practices concerning animal welfare and land use. In terms of feed imports, conventional farms relied more on external feed sources, importing 31% of the total nitrogen in feed, primarily in the form of (protein) concentrates. In contrast, organic farms imported less nitrogen, about 15% of the total nitrogen in feed, which underscores their greater reliance on home-grown feed sources. The data are summarised in Table 3 below.

This descriptive approach provides a clear overview of the sample based on common groupings and facilitates comparison with existing literature and political frameworks. However, it often fails to capture the complexity of farm typologies.

https://doi.org/10.3390/su18010303

Table 3. Characteristics of farm groups for key sustainability parameters (means ± 1 standard deviation).

Parameter/Characteristic	Conventional Dairy Farms in Favoured Areas	Conventional Alpine Dairy Farms	Organic Dairy Farms in Favoured Areas	Organic Alpine Dairy Farms
Farms (n)	71	45	29	25
Herd size—cows (n)	46 ± 20	35 ± 19	33 ± 20	26 ± 13
Average lifetime performance (kg ECM [1] per cow)	34,749 ± 11,489	34,479 ± 14,326	35,130 ± 14,966	36,695 ± 13,934
Average herd yield (kg ECM per cow and year)	9239 ± 1497	8722 ± 1865	7300 ± 1625	7229 ± 1728
Average productive lifetime [2] (years)	3.65 ± 1.24	3.52 ± 1.56	5.07 ± 2.26	4.41 ± 1.94
Proportion of annual time budget on pasture (%)	2.8 ± 6.5	7.5 ± 13.0	23.4 ± 21.8	23.0 ± 15.7
Imported feed-nitrogen (%)	31.0 ±10.3	31.8 ±13.2	15.4 ± 10.6	17.1 ±14.2

[1] Energy-Corrected Milk. [2] For cull cows.

2.5. Data-Driven Clustering Approach

To provide an alternative perspective that goes beyond binary classification, we applied a machine learning-based clustering approach to identify farm types based on multiple features. Binary classifications (e.g., organic vs. conventional) are limited in their ability to capture the inherent heterogeneity of farming systems [47]. Hypothesis-guided clustering approaches have the potential to better capture meaningful patterns than rigid dichotomies [48], as they allow for the identification of naturally occurring gradients and trade-offs rather than forcing farms into predefined categories.

The clustering approach applied here integrates domain expertise (knowledge-driven feature selection) with unsupervised machine learning to assess environmental conditions, efficiency, animal health, and sustainability. Combining expertise-guided variable selection with data-driven pattern recognition, our approach has redefined Austrian farm typology by making latent conflicts and trade-offs explicit, providing a perspective that extends beyond binary classification.

2.6. Statistical Analysis Methods

Correlation and regression analyses were conducted using SPSS version 29.0.2.0 (SPSS Inc., Chicago, IL, USA). The statistical analyses investigated relationships between sustainability indicators (GWP, fossil energy demand, and economic performance) and explored the impact of specific farm management practices on these indicators.

To identify distinct sustainability profiles among the surveyed dairy farms, we applied an integrated clustering approach that extends previous methodologies [49] by combining knowledge-driven feature selection with unsupervised pattern discovery. Key features were systematically selected to characterise farms across the dimensions of environmental conditions, efficiency, animal health, and sustainability. Specifically, severity score (site conditions), kg ECM per cow and year (production efficiency), productive lifetime as number of lactations (animal health), and kg nitrogen balance per hectare and year (environmental sustainability) established the feature space for cluster analysis. Several multivariate approaches were explored, including PCA, similarity-based metrics (Jaccard for binary and cosine for numeric features), and the integrated UMAP–HDBSCAN clustering pipeline. In the initial large dataset (>60 features), UMAP–HDBSCAN outperformed the other approaches, capturing the nonlinear and heterogeneous structure of farm profiles most effectively. Since it delivered the most stable and interpretable results, UMAP–HDBSCAN was selected as the analytical framework and consequently applied throughout the research process. In subsequent iterations, the feature set was gradually

https://doi.org/10.3390/su18010303

reduced through expert-guided and statistical selection to maintain comparability and interpretability across analyses.

Dimensionality reduction was performed using Uniform Manifold Approximation and Projection (UMAP), followed by clustering via Hierarchical Density-Based Spatial Clustering of Applications with Noise (HDBSCAN) for threshold-free farm grouping. The methodology applied here was enhanced by penalty-based parameter optimisation, testing various constrained settings (final: 3–8 clusters, ≥10 farms/cluster, ≤5% noise), and validated across 2000 seeds for reproducibility. These variations were explored to assess the stability and interpretability of the resulting farm typologies in relation to sustainability outcomes. Cluster characterisation was conducted using z-score analysis ($z = (x - \mu)/\sigma$) to identify key variables distinguishing each cluster. In addition, means and standard deviations were evaluated, along with correlations between individual features and cluster assignments. The identified structures were validated through expert evaluation to ensure relevance and interpretability. All cluster analyses were conducted in R (v4.4.2) using the umap (v0.2.10.0), dbscan (v1.1-11), and cluster (v2.1.4) packages. The R code for a clustering is provided in Appendix A.

3. Results & Discussion

This section presents and discusses the main findings of the sustainability assessment of Austrian dairy farms using the *NEU.rind* tool. The results derive from a cluster analysis aimed at identifying distinct farm types and their associated sustainability profiles. Moreover, key trade-offs observed between important indicators are shown, and strategies are derived to support holistic and synergistic pathways for sustainable development. The section concludes with methodological reflections on the suitability and limitations of the applied assessment approach.

3.1. Dairy Farm Clusters and Their Sustainability Profiles

An integrated analysis of the dairy farms identified four distinct farm clusters, each characterised by specific sustainability profiles. These clusters emerged from a combination of unsupervised cluster analysis (UMAP-HDP) based on key sustainability indicators and subsequent expert interpretation. This methodological approach surpasses binary categorisations such as 'organic vs. conventional', translating complex data into meaningful farm typologies relevant for sustainable dairy farming (see Figures 1 and 2).

The four clusters yielded the following profiles, see Table 4:

Table 4. Key sustainability parameters by farm cluster (means ± standard deviation).

Characteristic	Alpine Farms	Efficient Low-input Farms	Output-Oriented Farms	Input-Intensive Farms
Number of farms (n)	41	60	33	36
Altitude (m a.s.l.)	846 ± 253	622 ± 154	554 ± 203	564 ± 118
Gross margin (€)	3755 ± 1185	3923 ± 1157	4009 ± 771	4277 ± 987
Average lactations [1] (n)	3.19 ± 0.72	3.50 ± 0.86	2.53 ± 0.30	3.05 ± 0.36
Average productive lifetime [2] (years)	3.94 ± 1.95	4.45 ± 2.08	3.12 ± 1.18	4.01 ± 0.91
kg ECM/cow & year	8900 ± 1350	8655 ± 1310	10,027 ± 1117	9819 ± 924
kg CO_2/kg ECM – incl. infrastructure	1.17 ± 0.20	1.10 ± 0.17	1.09 ± 0.13	1.16 ± 0.50
kg CO_2/hectare – incl. infrastructure	11,601 ± 7910	13,605 ± 5409	16,183 ± 4276	18,555 ± 4516

https://doi.org/10.3390/su18010303

Table 4. Cont.

Characteristic	Alpine Farms	Efficient Low-input Farms	Output-Oriented Farms	Input-Intensive Farms
kg CO_2/kg ECM – excl. infrastructure	1.07 ± 0.15	1.03 ± 0.14	1.04 ± 0.11	1.11 ± 0.47

[1] Average lactations of live cows at farms. [2] For cull cows.

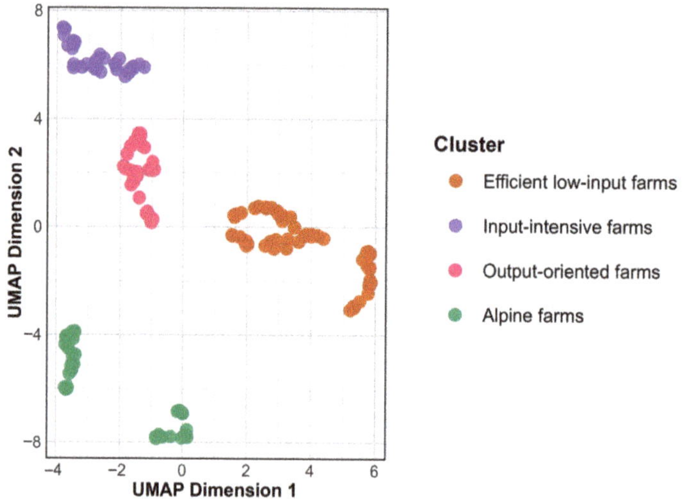

Figure 1. The second UMAP-HDP clustering distinguished between four clusters. Cluster 1 contains 60 of 170 farms (35.3%), Cluster 2 contains 36 farms (21.2%), Cluster 3 contains 33 farms (19.4%), and Cluster 4 contains 41 farms (24.1%).

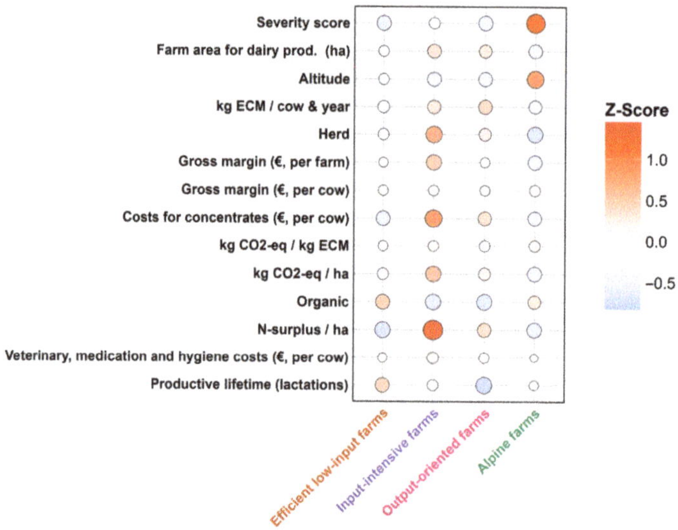

Figure 2. Cluster characteristics based on standardised z-scores comparing each cluster mean to the overall sample mean. The z-score analysis complements the UMAP projection by quantifying how

https://doi.org/10.3390/su18010303

clusters differ along the main latent gradients (production intensity and environmental–management conditions). The x-axis represents farm clusters, and the y-axis shows standardised deviations from the global mean. Point size indicates the magnitude of deviation, and colour intensity reflects direction (red = above average, blue = below average).

Cluster 1: Alpine farms. Farms in high altitudes with high cow longevity, average to low milk output per cow and year, characterised by low emissions per hectare, but increased CO_2-eq emissions per kg of milk. However, their contribution margin per cow and year is low. This system reflects adapted small-scale milk production in the Alpine region.

Cluster 2: Efficient low-input farms. This cluster comprises farms mostly located in medium altitude regions. They are marked by rather low annual milk yields per cow and year, moderate economic returns, and quite low CO_2-eq emissions, both per kg of milk and per hectare. Their cows show high longevity. These—often organic (57%)—farms combine low resource use with high efficiency and represent a sustainable approach under moderately favourable site conditions.

Cluster 3: Output-oriented lowland farms. Located in lowland areas, these farms produce the highest average milk volumes per cow and year and achieve good gross margins. While their emissions per kg of milk are low, they show rather high hectare-based emissions and the lowest cow longevity. This system maximises productivity and results in trade-offs regarding longevity and some ecological parameters.

Cluster 4: Input-intensive lowland farms. Also located in lowland regions, these farms demonstrate good production levels but show comparably high emissions per kg of milk and per hectare due to higher levels of external inputs. Their moderate gross margins and average cow longevity reflect partial inefficiencies in input resource use and environmental impact management.

The initial binary classification of farm groups into 'organic' and 'conventional' was intended primarily to describe sample characteristics and provide straightforward insights. The cluster analysis offers an alternative perspective, showing where this binary view aligns with the clustering results and where it fails to capture relevant aspects of farm heterogeneity.

Cluster 1 (Alpine extensive farms) aligns with the geographic dimension of binary classification, containing farms located in alpine regions that differ substantially from others across multiple variables including severity score, farm size (ha), number of dairy cows, altitude (m a.s.l.), and economic outcomes. This suggests that geographic-environmental constraints represent a dominant structural driver shaping farm characteristics beyond management choices alone. In contrast, Cluster 2 (efficient low-input farms) reveals a dimension missed by binary classification: environmental performance differences that transcend both conventional and organic systems, irrespective of location. In this case, management strategy emerges as the defining characteristic, rather than certification status or geography, representing a distinct form of extensification compared to that observed in alpine farms.

As an unsupervised method, the HDBSCAN clustering algorithm identifies patterns based on density and cluster stability [50,51]. In our case, the geographic location—particularly alpine conditions—emerges as the primary driver for Cluster 1, while other clusters differ primarily in production intensity and management approaches. This demonstrates that site conditions and management choices operate as separate but interacting axes of farm differentiation, which binary classifications cannot fully capture.

3.2. Identified Sustainability Trade-Offs

This study emphasises critical trade-offs between production intensity, environmental performance, and animal welfare. A prominent finding is a general contrast be-

https://doi.org/10.3390/su18010303

tween product-based (per kg milk) and area-based (per hectare) environmental indicators (Table 5). High-intensity systems with a high stocking density and high feed imports (see clusters 3 and 4; Table 4, Figure 2) commonly perform better per kg of milk due to dilution of emissions over higher milk (and beef) outputs, but exert higher impacts per hectare. Conversely, extensive systems typically show favourable hectare-based indicator results but higher emissions per kg of milk, consistent with findings from other authors, e.g., [52–54]. For instance, 'intensive' farms in Clusters 3 and 4 provide more protein per hectare of farmland, but they use more potentially human-edible concentrated feed. In this respect, the net amount of protein provided does not differ significantly between the 4 farm clusters, and the conversion efficiency of producing food-grade protein in milk and beef from the protein in feed is higher on the more extensive farms.

Table 5. Significant trade-offs between paired environmental indicator results related to the different product- and area-related functional units.

Indicator Pair	Product-Related Unit	Area-Related Unit	*p*-Value
Human-edible feed conversion efficiency vs. protein yield	heFCE	kg protein from milk/ha	<0.001
Acidification potential	kg SO_2-eq/kg ECM	kg SO_2-eq/ha	<0.001
Global warming potential	kg CO_2-eq/kg ECM	kg CO_2-eq/ha	<0.01
Gross margin (€)	€/kg ECM	€/kg ECM	<0.001
Fossil energy demand	MJ/kg ECM	MJ/kg ECM	<0.05
Potential species loss vs. HNVF proportion	Pot. species losses/kg ECM	% HNVF/ total farmland (ha)	<0.05

GHG emissions per kg ECM did not differ significantly between clusters, with conventional and organic farms showing similar average performances (means: conventional milk 1.11 kg CO_2-eq/kg ECM, organic milk 1.15 kg CO_2-eq/kg ECM). One reason for this small difference was that conventional production with higher milk output per cow and year was better able to dilute the environmental impact of infrastructure (machines and buildings); means of conventional and organic infrastructure-related GWP impacts were 0.094 kg and 0.124 kg CO_2-eq per kg ECM, respectively. The fact that the environmental impact results presented here generally include construction of infrastructure must be taken into account when comparing them with the results of other studies. Unlike GHG emissions, NH_3 emissions and fossil energy use per kg of ECM were generally lower in intensive systems, reflecting their higher productivity.

Studies on northern Italian dairy farms, partly located in the alpine region [54,55], used similar (solely environmental) indicators, functional units and cluster analyses and reported comparable patterns: for example, differences in GHG intensity per kg ECM between farm clusters were small and statistically insignificant, whereas traditional (extensive) farms showed substantially lower impacts per hectare, especially in [54], or a much better feed-food conversion [55].

More intensive farms, which purchased more external concentrated feed, had higher gross margins per ha and lower gross margins per kg ECM ($p < 0.05$), but insignificant differences per cow and year. Additionally, the gross margin is influenced by both the revenue per kg ECM (milk price; $p < 0.001$) and costs, such as those for (often purchased) concentrated feed ($p < 0.01$). Similarly, a strong influence of concentrate feed costs on farm profitability was found in Southern Germany, leading to the conclusion that lowering concentrate feed levels can create a win-win situation for both environmental performance

https://doi.org/10.3390/su18010303

and economic outcomes [56]. Larger proportions of permanent grassland, such as in the farms in Clusters 1 and 2, supported higher biodiversity scores in terms of proportions of HNVF land area ($p < 0.001$), but showed a lower feed conversion efficiency, resulting in higher land occupation. Consequently, an increase in the proportion of permanent grassland tended to result in higher potential species losses per kg ECM ($p = 0.094$; see also Table 4, Figure 2).

Conflicting trade-offs occur when sustainability is specifically increased in only one dimension. For instance, increasing lactation yields and income, and reducing product-related GHG emissions may reduce impacts per unit of product in general, but can also diminish cow longevity and health, and increase, according to our data, fossil resource use and dependency ($p < 0.05$). We found a tendency towards a negative relationship between location disadvantages of a farm and udder health ($p = 0.06$), showing that lowland farms tended to have more problems with udder health. This could be due to a combination of management and genetic effects, resulting from high milk yields and larger farms. However, no other correlations with animal health could be found in the data. Poor animal health can reduce both economic profitability and ecological sustainability, whereas good health and high cow longevity positively affects both. For instance, higher longevity improved GWP-related scores, with averages of product- and hectare-related scores being shown, and increased gross margins per cow and year ($p < 0.05$) as well as per kg ECM ($p < 0.01$). This was also observed in other studies [57,58].

3.3. *Strategic Pathways for Sustainable Development*

The following section presents measures for improving the sustainability performance of dairy farms derived from the cluster analysis and other analyses, such as correlations.

3.3.1. General Improvement Options

Analysis revealed that a higher reliance on regionally produced forage and a reduction of external concentrate feed were consistently associated with lower GHG scores ($p < 0.001$) and reduced production costs per cow and year ($p < 0.01$), suggesting that self-sufficiency in feed contributes positively to both environmental and economic performances. Providing (more) pasture emerged as an effective optimisation measure for animal welfare and environmental sustainability. Especially when (medium to) intensively managed grazing systems allowed for reductions in concentrate use, and thus in reduced environmental burdens (e.g., $p < 0.001$ for both kg CO_2-eq/ha and the averaged product- and hectare-related scores, or $p < 0.05$ for kg CO_2-eq per kg ECM) and costs per cow and year ($p < 0.01$). In more extensive systems, pasture—particularly alpine pasture—also provided biodiversity benefits, enhancing habitat quality and the landscape conservation value. Analysis revealed that the top 10% of farms with the healthiest cattle (based on the Q-Check health score) use 20% less concentrate feed per kilogram milk (185 g/kg ECM) compared to the average *NEU.rind* farm (230 g/kg ECM). However, this relationship is not straightforward or one-dimensional. Although there is limited direct influence, which is often overruled by management practices, improved animal health and longer productive lifespans, which are influenced by genetics and various management factors, can contribute to reducing resource use and environmental impacts [59]. For instance, increasing the number of lactations from, for example, 3 to 5 lactations—without any other changes—resulted in a 2% reduction in GHG emissions on average across all *NEU.rind* farms, when accounting for co-products such as beef from culled cows and calves and not including a better health status.

Improved manure handling practices showed a clear potential to reduce NH_3 emissions and support soil fertility. Farms with covered slurry stores ($p < 0.05$) and improved spreading techniques (such as trailing hoses, trailing shoes, and slurry injection; $p < 0.001$) demonstrated significantly lower acidifying emissions per kg of ECM. However, these reduction effects were not significant for kg NH_3 per ha, which is mainly influenced by livestock densities ($p < 0.01$).

3.3.2. Cluster- and Farm-Specific Recommendations for Improvement

The most extensive farms, particularly those in Cluster 1 (alpine farms), may face negative nutrient balances. These can occur on (Austrian) organic farms, especially concerning phosphate [60], and may require adjustments in feeding strategies such as increased feed imports or targeted fertilisation of land [61]. In such cases, the use of nitrogen- and phosphorus-rich by-products from food or bioenergy industries could improve animal performance and, via manure, enhance grassland and arable land productivity and thus economic returns. Given the low or even negative nitrogen surplus in this cluster (see Figure 2), efforts should focus on fine-tuning manure distribution and timing to further improve nutrient-use efficiency and pasture quality. However, care must be taken to ensure that this intensification of nutrient cycles does not come at the expense of biodiversity [62]. Specific measures such as diversified (zoned) grassland use, staggered mowing, or retention of tall grass refuges can be implemented to actively promote biodiversity [63,64]. Additionally, if higher nitrogen throughput occurs at farm level, more attention should be paid to reducing NH_3 emissions, for instance, by adapting the application of slurry with trailing hoses, trailing shoes or injection into soil [65]; this is relevant for all farms (clusters), especially those in the lowlands.

In both Clusters 1 and 2, particularly under extreme site conditions (e.g., higher altitude), and where dual-purpose genetics prevail, an integration of beef production and milk production may enable good utilisation of local resources and might thus increase economic efficiency [66]. On these farms, biodiversity and ecosystem services are comparably high, area-based and product-based environmental impacts are rather low, and animal health and longevity performances are good, while economic results are rather low. Development of and farm participation in ecosystem services and biodiversity payment schemes (e.g., for alpine pasture conservation) would offer more appropriate compensation for the ecosystem services provided by such alpine farms [67,68], thereby enhancing profitability. Although cow longevity in Clusters 1 and 2 is already good, there is still some possibility for improvement with potential synergistic economic and environmental effects (see Sections 3.2 and 3.3.1).

Contrarily, Cluster 3 (output-oriented lowland farms) and Cluster 4 (input-intensive lowland farms) benefit from favourable site conditions. However, they should pay special attention to animal health, cow longevity, and nitrogen surpluses, especially in the case of Cluster 4 (Figure 2). These farms often maintain high stocking densities enabled by significant external feed inputs, which bring substantial nutrient loads onto the farms. As a result, hectare-based environmental impacts can be high [10,69]. Due to high concentrate feed use per kg ECM, which can reduce enteric methane emissions, farms may achieve below-average values for CO_2-eq per kg ECM. However, hectare-based GHG emissions are clearly above the average (Figure 2).

Our findings underline the complexity of improving sustainability in dairy farming, which has been detected in other studies as well, e.g., in [24,56]. Multiple and sometimes conflicting factors influence sustainability indicators, and their interpretation depends heavily on the chosen reference unit (e.g., per kg milk, per hectare, per animal). Optimal solutions are therefore highly context-dependent and farm-specific, varying with production systems and site conditions. For every sustainability aspect in *NEU.rind* that is assessed per hectare and per kg milk, approximately a quarter of all farms show results below

https://doi.org/10.3390/su18010303

medians in both functional reference units. These farms are not restricted to less favourable locations. For such farms and the respective indicators, the *NEU.rind* tool provides tailored recommendations based on individual farm profiles, enabling targeted improvements in sustainability outcomes (see Section 3.4).

3.4. Methodological Reflections

In contrast to other tools, the digital farm assistant *NEU.rind* provides farm-specific recommendations for a comprehensive set of indicators, supported by numerous pre-recorded values to simplify data entry and ensure user-friendliness for routine use. By integrating powerful databases (e.g., Cattle Data Network RDV, IACS), it minimises additional manual data collection. Some tools require substantial time and effort for data collection because the analyses are highly detailed (e.g., SALCAsustain, [26]). They do not use extensive and highly effective databases such as IACS-data and especially the Cattle Data Network (RDV) containing, inter alia, milk recording data, but they often include detailed calculations in other areas such as crop production [26]. Several tools have adopted holistic frameworks to assess dairy systems, highlighting the importance of considering factors beyond CO_2-eq (GWP) when addressing sustainability challenges (e.g., SAFA, SMART, SALCA-sustain, FARMLIFE or MOTIFS, see [26]). However, they either do not evaluate quantitatively, are not easy to use or complex regarding data collection, do not present easily understandable analysis results, do not provide benchmarking options with other farms, do not give clear recommendations, or do not comprehensively analyse sustainability with regard to the dimension of animal health and welfare [26]. In contrast, *NEU.rind* provides an on-farm assessment tool seeking compromises between these methodological trade-offs. Furthermore, its comprehensive dataset enables detailed sustainability assessments across different farm typologies. While binary classification of typologies can provide straightforward insights, clustering approaches based on such a holistic dataset offer an alternative, more exploratory perspective for identifying patterns and relationships within the data. Nevertheless, the *NEU.rind* tool also has limitations and has potential for improvement, for instance with respect to an integrated quantitative assessment across different scores. DEXi-Dairy, for example, includes a decision rule-based, hierarchical multi-attribute model for deriving holistic improvements, however, it has difficulties assessing actual farms and shows better performance in the evaluation of model production systems with statistical data [70]. Future developments of the *NEU.rind* tool should aim to tailor recommendations more specifically to farm site conditions, taking into account individual development potentials of each dairy farm and asking for farmers' preferences. This would enable the identification of the measures with the highest improvement potential across different sustainability aspects and dimensions. Asking for preferences could enhance the implementation of high-priority measures in everyday farming practices.

4. Conclusions

The application of the *NEU.rind* tool on 170 farms identified four distinct farm clusters, each representing distinct locations and management strategies with trade-offs between productivity, environmental impacts, and economic performance. Alpine and efficient low-input farms (Clusters 1 and 2), for example, emphasise health and longevity, but they face economic limitations. In contrast, output-oriented and input-intensive lowland farms (Clusters 3 and 4) prioritise productivity and economic returns, achieving lower impacts per kg milk but at the expense of health and longevity. Moreover, fossil resource use and environmental loads per ha are high in these clusters. Higher dependence on regional forage, reduced concentrate feed, and increased use of pasture may improve welfare, reduce the use of resources and enhance environmental outcomes. Despite differences

https://doi.org/10.3390/su18010303

in production systems, the net protein output per ha of farmland was similar across all clusters, with more efficient food-grade protein conversion in extensive systems.

Strategies for sustainable development must be tailored to farm type: alpine and efficient farms (Clusters 1 & 2) provide high levels of ecosystem services, including biodiversity conservation, but lower economic performance. Output-oriented and input-intensive farms (Clusters 3 & 4) need to address nutrient surpluses or cow longevity and health issues.

The comprehensive *NEU.rind* dataset supports advanced analytical approaches, including farm typology classification and multidimensional clustering, which enable both benchmarking and deeper exploration of sustainability patterns across farm types. This allows stakeholders to gain additional insights into complex trade-offs and relationships that would otherwise remain hidden. While *NEU.rind* already provides valuable support for sustainability management, ongoing development should address current limitations, such as integrating a more quantitative assessment across sustainability aspects and further tailoring recommendations to specific farm conditions and farmer preferences. This will enhance the tool's contribution to sector-wide and policy-relevant sustainability goals, supporting adaptive management and continuous improvement at both the farm and national levels.

Overall, improving dairy farm sustainability requires context-specific measures that enhance efficiency, economic performance, animal health and welfare, while balancing productivity and environmental impact. *NEU.rind* has proven to be a valuable tool for deriving tailored sustainability measures. Additionally, *NEU.rind* meets external reporting requirements, such as sustainability indicators demanded by dairy processors. Repeated application of *NEU.rind* enables consistent monitoring over time and supports alignment with frameworks such as the Science Based Targets initiative (SBTi). Its standardised multidimensional evaluation and benchmarking, based on existing farm records, make *NEU.rind* a practical and robust tool for guiding Austria's dairy sector toward more sustainable practices.

Author Contributions: Conceptualisation, S.J.H. and all authors; methodology, S.J.H., M.H., F.S., S.W. and C.M.; software, F.S., C.M. and S.J.H.; validation, S.J.H., S.W., M.H. and F.S.; formal analysis, S.J.H. and C.M.; investigation, S.J.H. and M.H.; resources, F.S. and C.E.-D.; data curation, F.S., S.J.H., S.W. and M.H.; writing—original draft preparation, S.J.H. and C.M.; writing—review and editing, C.E.-D., K.L., M.H., S.J.H., C.M. and S.W.; visualisation, C.M.; supervision, C.E.-D.; project administration, K.L. and C.E.-D.; project idea for *NEU.rind*, C.E.-D., S.J.H. and F.S.; funding acquisition, C.E.-D., S.J.H., M.H. and K.L. All authors have read and agreed to the published version of the manuscript.

Funding: The project *NEU.rind* was funded by the Federal Ministry of Agriculture and Forestry, Climate and Environmental Protection, Regions and Water Management of the Republic of Austria and the European Union within the framework of the European Innovation Partnership for Agricultural Productivity and Sustainability (EIP-AGRI). The grant numbers are LE 14-20; 16.1.1-S2-46/21 and 16.2.1-S2-46/21.

Institutional Review Board Statement: Not applicable.

Informed Consent Statement: Not applicable.

Data Availability Statement: The farm-level datasets generated and analysed in this study contain confidential farm business information and personal data. In accordance with data protection regulations and confidentiality agreements with participating farmers, these data are not publicly available and cannot be shared.

Acknowledgments: The authors gratefully acknowledge the contributions of our project partners, including the Environment Agency Austria (Umweltbundesamt), the Federal Institute of Agricultural Economics, Rural and Mountain Research (Bundesanstalt für Agrarwirtschaft und Bergbauernfragen), the Federal Ministry of Agriculture and Forestry, Climate and Environmental Protection, Regions

https://doi.org/10.3390/su18010303

and Water Management (BMLUK), the Austrian Milk Association (MVÖ), the participating Austrian dairies, Rinderzucht AUSTRIA, LKV Austria, the Chamber of Agriculture and the six farmers in the project operational group. Special thanks are extended to the employees of LKV Austria for motivating farmers and assisting in data collection. We are especially grateful to the nearly 200 farmers who provided farm-level data essential for developing the methodology and conducting the case-study analyses. Furthermore, we would like to thank four reviewers who helped to improve the original submission of the manuscript, and we gratefully acknowledge Mag. Martina Kichler for English language editing.

Conflicts of Interest: Authors Caspar Matzhold, Franz Steininger, Kristina Linke and Christa Egger-Danner were employed by the ZuchtData EDV-Dienstleistungen GmbH. The remaining authors declare that the research was conducted in the absence of any commercial or financial relationships that could be construed as a potential conflict of interest.

Appendix A. R Code for a Clustering

```
# ---- Package versions ----
# R version 4.4.2
# data.table: 1.17.8
# umap:       0.2.10.0
# dbscan:     1.2.2
# cluster:    2.1.8.1


# Robust UMAP-HDBSCAN Clustering Workflow
# Step 1: Grid search for optimal parameters
# Step 2: Test robustness across multiple seeds

library(umap)
library(dbscan)
library(cluster)
library(data.table)


# Step 1: Grid Search for Parameter Optimisation

set.seed(142)
#dt_cluster <- read.csv()


# Step 1.1: Data Preparation

# Impute missing values with column mean and convert to numeric
dt_cluster[, (key_features) := lapply(.SD, function(col)
ifelse(is.na(col), mean(col, na.rm = TRUE), col)), .SDcols =
key_features]
dt_cluster[, (key_features) := lapply(.SD, as.numeric), .SDcols =
key_features]

# Scale the data
data_matrix <- scale(dt_cluster)

# Step 1.2: Define Parameter Grids
n_neighbors <- c(3, 5, 10, 15, 20, 30, 50)
min_dist <- c(0.001, 0.01, 0.1, 0.3, 0.5, 0.8)
```

https://doi.org/10.3390/su18010303

```
metrics <- c("euclidean", "cosine")
minPts_values <- c(3, 5, 10, 15, 20, 30, 50)

# Constraints for valid clustering
min_cluster_size <- 10
min_stable_clusters <- 3
max_stable_clusters <- 8
penalty_factor <- 0.1
max_unclustered_points <- 0.05 * nrow(data_matrix)

# Initialize storage for grid search results
integrated_grid_results <- list()

# Step 1.3: Perform Grid Search
perform_grid_search <- function(n_neighbors, min_dist, metric, minPts)
{
  tryCatch({
    # Apply UMAP
    umap_result  <-  umap(data_matrix,  n_neighbors  =  n_neighbors,
min_dist = min_dist, metric = metric, verbose = FALSE)
    if (is.null(umap_result$layout) || anyNA(umap_result$layout)) {
      stop("UMAP embedding failed")
    }
    umap_embedding <- as.data.frame(umap_result$layout)
    colnames(umap_embedding) <- c("UMAP1", "UMAP2")

    # Apply HDBSCAN
    clusters <- hdbscan(umap_embedding, minPts = minPts)
    if (is.null(clusters$cluster) || length(clusters$cluster) == 0) {
      stop("HDBSCAN clustering failed")
    }

    # Evaluate clustering
    all_clusters <- unique(clusters$cluster)
    actual_clusters <- all_clusters[all_clusters != 0]
    num_clusters <- length(actual_clusters)
    num_noise <- sum(clusters$cluster == 0)
    cluster_sizes <- table(clusters$cluster)
    small_clusters      <-      sum(cluster_sizes[cluster_sizes      <
min_cluster_size])

    # Compute silhouette score
    avg_silhouette <- if (num_clusters > 1) {
      sil <- silhouette(clusters$cluster, dist(umap_embedding))
      mean(sil[, 3], na.rm = TRUE)
    } else {
      NA
    }

    # Compute penalties
    cluster_penalty    <-    penalty_factor    *    (small_clusters    +
```


https://doi.org/10.3390/su18010303

```
abs(num_clusters - min_stable_clusters))
    effective_score     <-     ifelse(is.na(avg_silhouette),     -Inf,
avg_silhouette - cluster_penalty)

    # Validation
    is_valid_clustering <- (
      num_clusters >= min_stable_clusters &&
      num_clusters <= max_stable_clusters &&
      small_clusters == 0 &&
      num_noise <= max_unclustered_points
    )

    result <- list(
      n_neighbors = n_neighbors,
      min_dist = min_dist,
      metric = metric,
      minPts = minPts,
      num_clusters = num_clusters,
      num_noise = num_noise,
      silhouette = avg_silhouette,
      effective_score = if(is_valid_clustering) effective_score else
-Inf
    )
    return(result)

  }, error = function(e) {
    return(NULL)
  })
}

# Perform grid search
for (n in n_neighbors) {
  for (d in min_dist) {
    for (m in metrics) {
      for (minPts in minPts_values) {
        result <- perform_grid_search(n, d, m, minPts)
        if (!is.null(result)) {
          integrated_grid_results <- append(integrated_grid_results,
list(result))
        }
      }
    }
  }
}

# Convert results to DataFrame
results_df        <-        rbindlist(lapply(integrated_grid_results,
as.data.frame))
results_df <- results_df[
  num_clusters >= min_stable_clusters &
    num_clusters <= max_stable_clusters &
```

https://doi.org/10.3390/su18010303

```
    effective_score > -Inf
]

results_df <- results_df[order(-effective_score), ]

# Extract best parameters
best_params <- results_df[1, ]
best_n_neighbors <- best_params$n_neighbors
best_min_dist <- best_params$min_dist
best_metric <- best_params$metric
best_minPts <- best_params$minPts

# Step 2: Test Robustness Across Multiple Seeds

seeds <- 1:2000

# Initialize storage for seed results
seed_results <- data.frame(
  seed = integer(),
  num_clusters = integer(),
  num_noise = integer()
)

run_clustering_for_seed  <-  function(seed,  n_neighbors,  min_dist,
metric, minPts) {
  set.seed(seed)
  tryCatch({
    # Apply UMAP
    umap_result  <-  umap(data_matrix,  n_neighbors  =  n_neighbors,
min_dist = min_dist, metric = metric, verbose = FALSE)
    if (is.null(umap_result$layout) || anyNA(umap_result$layout)) {
      stop("UMAP embedding failed")
    }
    umap_embedding <- as.data.frame(umap_result$layout)
    colnames(umap_embedding) <- c("UMAP1", "UMAP2")

    # Apply HDBSCAN
    clusters <- hdbscan(umap_embedding, minPts = minPts)
    if (is.null(clusters$cluster) || length(clusters$cluster) == 0) {
      stop("HDBSCAN clustering failed")
    }

    # Evaluate clustering
    num_clusters   <-   length(unique(clusters$cluster))   -   (0   %in%
clusters$cluster)
    num_noise <- sum(clusters$cluster == 0)

    return(list(
      seed = seed,
      num_clusters = num_clusters,
      num_noise = num_noise
```


https://doi.org/10.3390/su18010303

```
      ))
    }, error = function(e) {
      return(NULL)
    })
  }

  # Run clustering across seeds
  for (seed in seeds) {
    result    <-    run_clustering_for_seed(seed,    best_n_neighbors,
  best_min_dist, best_metric, best_minPts)
    if (!is.null(result)) {
      seed_results <- rbind(seed_results, as.data.frame(result))
    }
  }

  # Convert to data.table
  setDT(seed_results)

  # Get best seed
  original_best_clusters <- best_params$num_clusters
  seed_results_consistent     <-     seed_results[num_clusters     ==
  original_best_clusters]

  if (nrow(seed_results_consistent) > 0) {
    best_seed <- seed_results_consistent[which.min(num_noise), seed]
  } else {
    seed_results_valid <- seed_results[
      num_clusters >= min_stable_clusters &
        num_clusters <= max_stable_clusters
    ]

    if (nrow(seed_results_valid) == 0) {
      cluster_distance     <-     abs(seed_results$num_clusters     -
  original_best_clusters)
      closest_seeds <- seed_results[order(cluster_distance, num_noise)]
      best_seed <- closest_seeds[1, seed]
    } else {
      best_seed <- seed_results_valid[which.min(num_noise), seed]
    }
  }


  # Step 3: Apply best parameters + best seed


  # Set the chosen seed for reproducibility
  set.seed(best_seed)

  # Apply UMAP with best parameters
  best_umap_result <- umap(data_matrix, n_neighbors = best_n_neighbors,
  min_dist = best_min_dist, metric = best_metric, verbose = FALSE)
```

https://doi.org/10.3390/su18010303

```
umap_embedding <- as.data.frame(best_umap_result$layout)
colnames(umap_embedding) <- c("UMAP1", "UMAP2")

# Apply HDBSCAN with best parameters
best_hdbscan <- hdbscan(umap_embedding, minPts = best_minPts)

# Add cluster labels to UMAP embedding
umap_embedding$Cluster <- as.factor(best_hdbscan$cluster)

# Assign cluster labels to original dataset
dt[, cluster := umap_embedding$Cluster]

# Filter clusters
final_clusters <- dt[cluster != "0", ]
noise_points <- dt[cluster == "0", ]

# Summary
cluster_summary <- final_clusters[, .N, by = cluster]

# Print results
cat("Best parameters:\n")
cat("n_neighbors:", best_n_neighbors, "\n")
cat("min_dist:", best_min_dist, "\n")
cat("metric:", best_metric, "\n")
cat("minPts:", best_minPts, "\n")
cat("best_seed:", best_seed, "\n")
```

References

1. Cimmino, F.; Catapano, A.; Petrella, L.; Villano, I.; Tudisco, R.; Cavaliere, G. Role of Milk Micronutrients in Human Health. *Front. Biosci. (Landmark Ed.)* **2023**, *28*, 41. [CrossRef] [PubMed]
2. Weaver, C.M. Role of Dairy Beverages in the Diet. *Physiol. Behav.* **2010**, *100*, 63–66. [CrossRef] [PubMed]
3. Singaravadivelan, A.; Sachin, P.B.; Harikumar, S.; Vijayakumar, P.; Vindhya, M.V.; Farhana, F.M.B.; Rameesa, K.K.; Mathew, J. Life Cycle Assessment of Greenhouse Gas Emission from the Dairy Production System—Review. *Trop. Anim. Health Prod.* **2023**, *55*, 320. [CrossRef] [PubMed]
4. Misselbrook, T.; del Prado, A.; Chadwick, D. Opportunities for Reducing Environmental Emissions from Foragebased Dairy Farms. *Agric. Food Sci.* **2013**, *22*, 93–107. [CrossRef]
5. Douglas, C.; Pandey, P. Nitrate and Bacterial Loads in Dairy Cattle Drinking Water and Potential Treatment Options for Pollutants—A Review. *Appl. Sci.* **2025**, *15*, 3017. [CrossRef]
6. Gislon, G.; Ferrero, F.; Bava, L.; Borreani, G.; Prà, A.D.; Pacchioli, M.T.; Sandrucci, A.; Zucali, M.; Tabacco, E. Forage Systems and Sustainability of Milk Production: Feed Efficiency, Environmental Impacts and Soil Carbon Stocks. *J. Clean. Prod.* **2020**, *260*, 121012. [CrossRef]
7. Díaz De Otálora, X.; Del Prado, A.; Dragoni, F.; Balaine, L.; Pardo, G.; Winiwarter, W.; Sandrucci, A.; Ragaglini, G.; Kabelitz, T.; Kieronczyk, M.; et al. Modelling the Effect of Context-Specific Greenhouse Gas and Nitrogen Emission Mitigation Options in Key European Dairy Farming Systems. *Agron. Sustain. Dev.* **2024**, *44*, 4. [CrossRef]
8. Grassauer, F.; Herndl, M.; Nemecek, T.; Guggenberger, T.; Fritz, C.; Steinwidder, A.; Zollitsch, W. Eco-Efficiency of Farms Considering Multiple Functions of Agriculture: Concept and Results from Austrian Farms. *J. Clean. Prod.* **2021**, *297*, 126662. [CrossRef]
9. Grassauer, F.; Herndl, M.; Iten, L.; Gaillard, G. Environmental Assessment of Austrian Organic Dairy Farms With Closed Regional Production Cycles in a Less Favorable Production Area. *Front. Sustain. Food Syst.* **2022**, *6*, 817671. [CrossRef]
10. Eisert, J.; Sahraei, A.; Knob, D.A.; Lambertz, C.; Zollitsch, W.; Hörtenhuber, S.; Kral, I.; Breuer, L.; Gattinger, A. Transforming the Feeding Regime towards Low-Input Increases the Environmental Impact of Organic Milk Production on a Case Study Farm in Central Germany. *Int. J. Life Cycle Assess.* **2025**, *30*, 79–92. [CrossRef]

https://doi.org/10.3390/su18010303

11. Zanon, T.; Hörtenhuber, S.; Fichter, G.; Peratoner, G.; Zollitsch, W.; Gatterer, M.; Gauly, M. Effect of Management System and Dietary Seasonal Variability on Environmental Efficiency and Human Net Food Supply of Mountain Dairy Farming Systems. *J. Dairy Sci.* **2025**, *108*, 597–610. [CrossRef]
12. Drews, J.; Czycholl, I.; Krieter, J. A Life Cycle Assessment Study of Dairy Farms in Northern Germany—The Development of Environmental Impacts throughout a Decade. *Züchtungskunde* **2020**, *92*, 236–256.
13. Rotz, C.A.; Stout, R.C.; Holly, M.A.; Kleinman, P.J.A. Regional Environmental Assessment of Dairy Farms. *J. Dairy Sci.* **2020**, *103*, 3275–3288. [CrossRef] [PubMed]
14. Van Der Werf, H.M.G.; Knudsen, M.T.; Cederberg, C. Towards Better Representation of Organic Agriculture in Life Cycle Assessment. *Nat. Sustain.* **2020**, *3*, 419–425. [CrossRef]
15. European Commission. Joint Research Centre. *Understanding Product Environmental Footprint and Organisation Environmental Footprint Methods*; Publications Office: Luxembourg, 2022. [CrossRef]
16. Von Soosten, D.; Meyer, U.; Flachowsky, G.; Dänicke, S. Dairy Cow Health and Greenhouse Gas Emission Intensity. *Dairy* **2020**, *1*, 20–29. [CrossRef]
17. O'Brien, D.; Hennessy, T.; Moran, B.; Shalloo, L. Relating the Carbon Footprint of Milk from Irish Dairy Farms to Economic Performance. *J. Dairy Sci.* **2015**, *98*, 7394–7407. [CrossRef]
18. Balmford, A.; Amano, T.; Bartlett, H.; Chadwick, D.; Collins, A.; Edwards, D.; Field, R.; Garnsworthy, P.; Green, R.; Smith, P.; et al. Author Correction: The Environmental Costs and Benefits of High-Yield Farming. *Nat. Sustain.* **2019**, *2*, 339–341. [CrossRef]
19. Ertl, P.; Klocker, H.; Hörtenhuber, S.; Knaus, W.; Zollitsch, W. The Net Contribution of Dairy Production to Human Food Supply: The Case of Austrian Dairy Farms. *Agric. Syst.* **2015**, *137*, 119–125. [CrossRef]
20. Ertl, P.; Knaus, W.; Zollitsch, W. An Approach to Including Protein Quality When Assessing the Net Contribution of Livestock to Human Food Supply. *Animal* **2016**, *10*, 1883–1889. [CrossRef]
21. Steffen, W.; Richardson, K.; Rockstrom, J.; Cornell, S.E.; Fetzer, I.; Bennett, E.M.; Biggs, R.; Carpenter, S.R.; de Vries, W.; de Wit, C.A.; et al. Planetary Boundaries: Guiding Human Development on a Changing Planet. *Science* **2015**, *347*, 1259855. [CrossRef]
22. Schmidt, J.; Dalgaard, R. *Arla Foods FarmTool V2021—Updates and Adding New Technologies*; LCA Consultants: Aarhus, Denmark, 2021; p. 58. Available online: https://lca-net.com/files/FarmTool-v2021_20211221.pdf (accessed on 5 December 2025).
23. Bayrische Landesanstalt für Landwirtschaft (LfL) Deckungsbeiträge Und Kalkulationsdaten—Milchkuhhaltung. Available online: https://www.stmelf.bayern.de/idb/milchkuhhaltung.html (accessed on 5 December 2025).
24. Vries, d.M.; van Dijk, W.; de Boer, J.A.; de Haan, M.H.A.; Oenema, J.; Verloop, J.; Lagerwerf, L.A. *Calculation Rules of the Annual Nutrient Cycling Assessment (ANCA) 2019; Background Information about Farm-Specific Environmental Performance Parameters. Report 1279*; Wageningen Livestock Research: Wageningen, The Netherlands, 2020; p. 112. Available online: https://edepot.wur.nl/533905 (accessed on 5 December 2025).
25. Herron, J.; O'Brien, D.; Jordan, S.; Shalloo, L. AgNav: The New Digital Sustainability Platform for Agriculture in Ireland. 2025. Available online: https://teagasc.ie/wp-content/uploads/2025/05/AgNav-The-new-digital-sustainability-platform-for-agriculture-in-Ireland.pdf (accessed on 5 December 2025).
26. Paçarada, R.; Hörtenhuber, S.; Hemme, T.; Wurzinger, M.; Zollitsch, W. Sustainability Assessment Tools for Dairy Supply Chains: A Typology. *Sustainability* **2024**, *16*, 4999. [CrossRef]
27. Meul, M.; Passel, S.; Nevens, F.; Dessein, J.; Rogge, E.; Mulier, A.; Hauwermeiren, A. MOTIFS: A Monitoring Tool for Integrated Farm Sustainability. *Agron. Sustain. Dev.* **2008**, *28*, 321–332. [CrossRef]
28. de Otalora, X.; del Prado, A.; Dragoni, F.; Estelles, F.; Amon, B. Evaluating Three-Pillar Sustainability Modelling Approaches for Dairy Cattle Production Systems. *Sustainability* **2021**, *13*, 6332. [CrossRef]
29. Ruckli, A.K.; Hörtenhuber, S.J.; Ferrari, P.; Guy, J.; Helmerichs, J.; Hoste, R.; Hubbard, C.; Kasperczyk, N.; Leeb, C.; Malak-Rawlikowska, A.; et al. Integrative Sustainability Analysis of European Pig Farms: Development of a Multi-Criteria Assessment Tool. *Sustainability* **2022**, *14*, 5988. [CrossRef]
30. Triste, L.; Marchand, F.; Debruyne, L.; Meul, M.; Lauwers, L. Reflection on the Development Process of a Sustainability Assessment Tool: Learning from a Flemish Case. *Ecol. Soc.* **2014**, *19*, art47. [CrossRef]
31. Siebrecht, N. Sustainable Agriculture and Its Implementation Gap—Overcoming Obstacles to Implementation. *Sustainability* **2020**, *12*, 3853. [CrossRef]
32. Hörtenhuber, S.J.; Steininger, F.; Herndl, M.; Wieser, S.; Linke, K.; Egger-Danner, C. *NEU.Rind-Tool—Method Description*; Rinderzucht Austria: Vienna, Austria, 2022; Available online: https://www.rinderzucht.at/projekt/neu-rind.html (accessed on 5 December 2025).
33. IPCC. *Climate Change 2013: The Physical Science Basis. Contribution of Working Group I to the Fifth Assessment Report of the Intergovernmental Panel on Climate Change*; Stocker, T.F., Qin, D., Plattner, G.-K., Tignor, M., Allen, S.K., Boschung, J., Nauels, A., Xia, Y., Bex, V., Midgley, P.M., Eds.; Cambridge University Press: Cambridge, UK; New York, NY, USA, 2013; ISBN 978-1-107-05799-1.

https://doi.org/10.3390/su18010303

34. Umweltbundesamt. *Austria's National Inventory Report 2023*; Umweltbundesamt (Austrian Environment Agency): Vienna, Austria, 2023; p. 833. Available online: https://www.umweltbundesamt.at/fileadmin/site/publikationen/rep0852.pdf (accessed on 5 December 2025).
35. International Dairy Federadtion (IDF). *The IDF Global Carbon Footprint Standard for the Dairy Sector. Bulletin of the IDF 520(2022*; The International Dairy Federation: Brussels, Belgium, 2022. Available online: https://shop.fil-idf.org/products/the-idf-global-carbon-footprint-standard-for-the-dairy-sector (accessed on 5 December 2025).
36. Ineichen, S.; Schenker, U.; Nemecek, T.; Reidy, B. Allocation of Environmental Burdens in Dairy Systems: Expanding a Biophysical Approach for Application to Larger Meat-to-Milk Ratios. *Livest. Sci.* **2022**, *261*, 104955. [CrossRef]
37. IPCC. *2019 Refinement to the 2006 IPCC Guidelines for National Greenhouse Gas Inventories*; IPCC: Kyoto, Japan, 2019. Available online: https://www.ipcc.ch/report/2019-refinement-to-the-2006-ipcc-guidelines-for-national-greenhouse-gas-inventories/ (accessed on 5 December 2025).
38. Amon, B.; Hutchings, N.; Dämmgen, U.; Sommer, S.; Webb, J.; Mellios, G.; Juhrich, K. EMEP/EEA Air Pollutant Emission Inventory Guidebook 2023. 2023; p. 64. Available online: https://www.eea.europa.eu/en/analysis/publications/emep-eea-guidebook-2023 (accessed on 5 December 2025).
39. Wernet, G.; Bauer, C.; Steubing, B.; Reinhard, J.; Moreno-Ruiz, E.; Weidema, B. The Ecoinvent Database Version 3 (Part I): Overview and Methodology. *Int. J. Life Cycle Assess.* **2016**, *21*, 1218–1230. [CrossRef]
40. Chaudhary, A.; Brooks, T.M. Land Use Intensity-Specific Global Characterization Factors to Assess Product Biodiversity Footprints. *Environ. Sci. Technol.* **2018**, *52*, 5094–5104. [CrossRef]
41. Sjaunja, L.O.; Baevre, L.; Junkkarinen, L.; Pedersen, J.; Setälä, J. A Nordic Proposal for an Energy-Corrected Milk Standard. In *Proceedings of the 11th Session of the International Committee for Recording and Use of Animal Productivity (ICAR)*; International Dairy Federation (IDF): Paris, France, 1990.
42. Nemecek, T.; Thoma, G. Allocation between Milk and Meat in Dairy LCA: Critical Discussion of the International Dairy Federation's Standard Methodology. In Proceedings of the 12th International Conference on Life Cycle Assessment of Food 2020, Berlin, Germany, 13–16 October 2020; DIL: Quakenbrück, Germany, 2020; pp. 86–89.
43. Bartel, A.; Süßenbacher, E.; Sedy, K. *Weiterentwicklung des Agrarumweltindikators "High Nature Value Farmland" für Österreich*; Umweltbundesamt (Austrian Environment Agency): Wien, Austria, 2011; p. 90. Available online: https://www.umweltbundesamt.at/fileadmin/site/publikationen/rep0348.pdf (accessed on 5 December 2025).
44. Thünen Institute. Q Check: Tierwohl Mit System—Von Der Betrieblichen Eigenkontrolle Zum Nationalen Monitoring. 2021. Available online: https://www.thuenen.de/media/publikationen/project_brief/Project_brief_2021_35.pdf (accessed on 5 December 2025).
45. Federal Institute of Agricultural Economics, Rural and Mountain Research IDB—Interaktive Deckungsbeiträge und Kalkulationsdaten 2025. Available online: https://idb.agrarforschung.at/verfahren/konventionell/milchkuhhaltung (accessed on 5 December 2025).
46. Bundesministerium für Land- und Forstwirtschaft, Regionen und Wasserwirtschaft (BML) Integrated Administration and Control System (IACS) Data (Integriertes Verwaltungs- und Kontrollsystem, INVEKOS). 2024. Available online: https://www.bmluk.gv.at/themen/landwirtschaft/gemeinsame-agrarpolitik-foerderungen/nationaler-strategieplan/direktzahlungen-ab-2023/invekosinvekosgis.html (accessed on 5 December 2025).
47. Slagboom, M.; Kargo, M.; Edwards, D.; Sørensen, A.C.; Thomasen, J.R.; Hjortø, L. Organic Dairy Farmers Put More Emphasis on Production Traits than Conventional Farmers. *J. Dairy Sci.* **2016**, *99*, 9845–9856. [CrossRef]
48. Alvarez, S.; Timler, C.J.; Michalscheck, M.; Paas, W.; Descheemaeker, K.; Tittonell, P.; Andersson, J.A.; Groot, J.C.J. Capturing Farm Diversity with Hypothesis-Based Typologies: An Innovative Methodological Framework for Farming System Typology Development. *PLoS ONE* **2018**, *13*, e0194757. [CrossRef]
49. Matzhold, C.; Schodl, K.; Klimek, P.; Steininger, F.; Egger-Danner, C. A Key-Feature-Based Clustering Approach to Assess the Impact of Technology Integration on Cow Health in Austrian Dairy Farms. *Front. Anim. Sci.* **2024**, *5*, 1421299. [CrossRef]
50. Campello, R.J.G.B.; Moulavi, D.; Sander, J. Density-Based Clustering Based on Hierarchical Density Estimates. In *Advances in Knowledge Discovery and Data Mining*; Lecture Notes in Computer Science; Pei, J., Tseng, V.S., Cao, L., Motoda, H., Xu, G., Eds.; Springer: Berlin/Heidelberg, Germany, 2013; Volume 7819, pp. 160–172, ISBN 978-3-642-37455-5.
51. McInnes, L.; Healy, J. Accelerated Hierarchical Density Clustering. *arXiv* **2017**. [CrossRef]
52. Baldini, M.; Da Borso, F.; Rossi, A.; Taverna, M.; Bovolenta, S.; Piasentier, E.; Corazzin, M. Environmental Sustainability Assessment of Dairy Farms Rearing the Italian Simmental Dual-Purpose Breed. *Animals* **2020**, *10*, 296. [CrossRef] [PubMed]
53. Mazzetto, A.M.; Falconer, S.; Edwards, P.J.; Glassey, C.B.; Neal, M.B.; Ledgard, S.F. Effects of Management and Technology Scenarios on the Carbon Footprint of Milk from Pasture-Based Dairy Farm Systems. *J. Dairy Sci.* **2025**, *108*, 12407–12420. [CrossRef]
54. Bava, L.; Sandrucci, A.; Zucali, M.; Guerci, M.; Tamburini, A. How Can Farming Intensification Affect the Environmental Impact of Milk Production? *J. Dairy Sci.* **2014**, *97*, 4579–4593. [CrossRef]

https://doi.org/10.3390/su18010303

55. Berton, M.; Bittante, G.; Zendri, F.; Ramanzin, M.; Schiavon, S.; Sturaro, E. Environmental Impact and Efficiency of Use of Resources of Different Mountain Dairy Farming Systems. *Agric. Syst.* **2020**, *181*, 102806. [CrossRef]
56. Mennig, P.; Szigeti, Z. Synergies and Trade-Offs between Environmental Impacts and Farm Profitability: The Case of Pasture-Based Dairy Production Systems. *J. Agric. Food Res.* **2025**, *21*, 101798. [CrossRef]
57. Han, R.; Kok, A.; Mourits, M.; Hogeveen, H. Effects of Extending Dairy Cow Longevity by Adjusted Reproduction Management Decisions on Partial Net Return and Greenhouse Gas Emissions: A Dynamic Stochastic Herd Simulation Study. *J. Dairy Sci.* **2024**, *107*, 6902–6912. [CrossRef]
58. Grandl, F.; Furger, M.; Kreuzer, M.; Zehetmeier, M. Impact of Longevity on Greenhouse Gas Emissions and Profitability of Individual Dairy Cows Analysed with Different System Boundaries. *Animal* **2019**, *13*, 198–208. [CrossRef]
59. Hörtenhuber, S. Wir wirken Langlebigkeit und Tiergesundheit auf die Klima- und Umweltbilanz? In *Proceedings of the Rinderzucht Austria Seminar. Nutzungsdauer—Ein traditionelles Konzept mit Zukunft?* Rinderzucht Austria: Vienna, Austria, 2025; pp. 61–68. Available online: https://www.rinderzucht.at/downloads/seminarunterlagen.html?file=files/rinderzucht-austria/01-rinderzucht-austria/downloads/rza-seminar/2025-03-06-tagungsband-rinderzucht-austria-seminar-2025-web.pdf (accessed on 5 December 2025).
60. Cooper, J.; Reed, E.Y.; Hörtenhuber, S.; Lindenthal, T.; Løes, A.-K.; Mäder, P.; Magid, J.; Oberson, A.; Kolbe, H.; Möller, K. Phosphorus Availability on Many Organically Managed Farms in Europe. *Nutr. Cycl. Agroecosystems* **2018**, *110*, 227–239. [CrossRef]
61. Haas, G.; Deittert, C.; Köpke, U. Farm-Gate Nutrient Balance Assessment of Organic Dairy Farms at Different Intensity Levels in Germany. *Renew. Agric. Food Syst.* **2007**, *22*, 223–232. [CrossRef]
62. Hu, Z.; Delgado-Baquerizo, M.; Fanin, N.; Chen, X.; Zhou, Y.; Du, G.; Hu, F.; Jiang, L.; Hu, S.; Liu, M. Nutrient-Induced Acidification Modulates Soil Biodiversity-Function Relationships. *Nat. Commun.* **2024**, *15*, 2858. [CrossRef] [PubMed]
63. Parmentier, L. "Three-Strip Management": Introducing a Novel Mowing Method in Perennial Flower Strips and Grass Margins to Increase Habitat Complexity and Attractiveness for Pollinators. *J. Pollinat. Ecol.* **2023**, *34*, 267–283. [CrossRef]
64. Buri, P.; Humbert, J.-Y.; Arlettaz, R. Promoting Pollinating Insects in Intensive Agricultural Matrices: Field-Scale Experimental Manipulation of Hay-Meadow Mowing Regimes and Its Effects on Bees. *PLoS ONE* **2014**, *9*, e85635. [CrossRef]
65. Aguirre-Villegas, H.A.; Besson, C.; Larson, R.A. Modeling Ammonia Emissions from Manure in Conventional, Organic, and Grazing Dairy Systems and Practices to Mitigate Emissions. *J. Dairy Sci.* **2024**, *107*, 359–382. [CrossRef]
66. Bang, R.; Samsonstuen, S.; Hansen, B.G.; Guajardo, M.; Møller, H.; Sommerseth, J.K.; Goez, J.C.; Flaten, O. Win-Win or Lose-Win? Economic-Climatic Synergies and Trade-Offs in Dual-Purpose Cattle Systems. *Agric. Syst.* **2025**, *222*, 104189. [CrossRef]
67. Hagemann, N.; Gerling, C.; Hölting, L.; Kernecker, M.; Markova-Nenova, N.N.; Wätzold, F.; Wendler, J.; Cord, A.F. Improving Result-Based Schemes for Nature Conservation in Agricultural Landscapes—Challenges and Best Practices from Selected European Countries. *Reg. Environ. Change* **2025**, *25*, 12. [CrossRef]
68. Zabel, A. Biodiversity-Based Payments on Swiss Alpine Pastures. *Land Use Policy* **2019**, *81*, 153–159. [CrossRef]
69. Guerci, M.; Knudsen, M.T.; Bava, L.; Zucali, M.; Schönbach, P.; Kristensen, T. Parameters Affecting the Environmental Impact of a Range of Dairy Farming Systems in Denmark, Germany and Italy. *J. Clean. Prod.* **2013**, *54*, 133–141. [CrossRef]
70. Wilfart, A.; Baillet, V.; Balaine, L.; De Otálora, X.D.; Dragoni, F.; Krol, D.J.; Frątczak-Müller, J.; Rychła, A.; Rodriguez, D.G.P.; Breen, J.; et al. DEXi-Dairy: An Ex Post Multicriteria Tool to Assess the Sustainability of Dairy Production Systems in Various European Regions. *Agron. Sustain. Dev.* **2023**, *43*, 82. [CrossRef]

https://doi.org/10.3390/su18010303

sustainability

MDPI

Article

Quantification of CH_4 and N_2O Fluxes from Piggery Wastewater Treatment System for Emission Factor Development

Anthony Kintu Kibwika , Il-Hwan Seo and In-Sun Kang *

Department of Rural Construction Engineering, Jeonbuk National University, Baekje-daero 567, Jeonju-si 54896, Republic of Korea; anthony@jbnu.ac.kr (A.K.K.); ihseo@jbnu.ac.kr (I.-H.S.)
* Correspondence: insunk@jbnu.ac.kr

Abstract

Piggery farming is the largest source of livestock manure in South Korea, yet greenhouse gas (GHG) data from piggery wastewater treatment systems remain limited. This study quantified methane (CH_4) and nitrous oxide (N_2O) fluxes from a full-scale treatment facility to develop stage-, seasonal-, and diurnal-specific emission factors. Continuous laser-based monitoring using a PVC air-pool chamber was applied across raw wastewater storage, an anoxic nitrogen-conversion reactor, and strongly aerated nitrification units. Mean CH_4 fluxes ranged from 1.1 to 15.6 mg s^{-1} m^{-2} peaking in summer, while N_2O fluxes ranged from 0.01 to 17,971 mg s^{-1} m^{-2}, with maxima in fall. Emissions were dominated by two functional zones: aerated basins where vigorous mixing enhanced CH_4 stripping, and an upstream anoxic reactor where oxygen instability and nitrite accumulation produced extreme N_2O peaks. Derived emission factors were 0.11 kg CH_4 $head^{-1}$ yr^{-1} and 45.2 kg N_2O $head^{-1}$ yr^{-1}, equivalent to 3.1 and 12,300 kg CO_2-eq $head^{-1}$ yr^{-1}. CH_4 variability was controlled mainly by treatment stage and temperature, whereas N_2O was governed by internal redox conditions. These results refine emission factors for inventories and underscore the need for improved aeration stability and denitrification control to reduce GHG emissions from piggery wastewater systems.

Keywords: piggery wastewater; methane; nitrous oxide; emission factors; temperature; aeration; denitrification; South Korea

1. Introduction

Greenhouse gas (GHG) emissions, particularly methane (CH_4) and nitrous oxide (N_2O), are major contributors to global climate change. Agriculture accounts for approximately 9 to 18% of global anthropogenic GHG emissions, with livestock manure management recognized as a significant source of both CH_4 and N_2O [1]. International frameworks such as the Paris Agreement emphasize the need for accurate GHG quantification to guide mitigation strategies [2], yet emissions from livestock wastewater remain difficult to characterize due to methodological inconsistencies, high spatial and temporal variability, and limited emission factor (EF) datasets [3,4].

South Korea has pledged carbon neutrality by 2050 [5], with livestock-sector mitigation identified as a key priority. As of 2023, the national livestock population reached roughly 190 million head, including 11.09 million pigs [6]. Piggery operations generate the largest share of manure in Korea, contributing about 39% ($\approx$19.7 million tons yr^{-1}) of the national total [7]. Despite this, comprehensive CH_4 and N_2O emission factors for piggery wastewater treatment processes remain scarce, as most previous studies have focused on barn-level or slurry-pit emissions. This gap limits the refinement of national inventories and assessment of mitigation opportunities [8,9].

Quantifying GHG emissions from pig manure systems remains challenging because emissions vary widely with chamber configuration, sampling frequency, environmental conditions, and treatment-stage operation [8,10–12]. Flux-gradient and micrometeorological approaches can be difficult to apply in complex wastewater environments such as aeration basins and storage tanks [13,14]. Emissions also respond strongly to operational and seasonal conditions: CH_4 generally increases during warm periods, while N_2O often peaks in cooler, oxygen-limited conditions [13,14]. Existing empirical models only partially capture these dynamics, contributing to uncertainty in current inventories [15,16].

Advances in laser-based gas analyzers and automated monitoring systems now allow high-resolution CH_4 and N_2O measurements with improved stability and sensitivity [17–20]. However, their application in piggery wastewater treatment systems remains limited, and localized emission factors reflecting diurnal and seasonal variability are lacking [21].

This study quantifies CH_4 and N_2O emissions from a commercial piggery wastewater treatment facility and evaluates how fluxes vary across treatment stages, diurnal periods, and seasons. By deriving stage- and time-resolved emission factors, this work fills a critical data gap in Korea's GHG inventory framework and provides empirical guidance for designing low-emission wastewater treatment strategies under warm, humid agricultural conditions.

2. Materials and Methods

2.1. Waste Water Treatment Plant and Weather Conditions

The study was conducted at a commercial pig farm located in Baeksan-myeon, Jojong-ri, Kimje City, South Korea (35°52′1.31″ N, 126°52′41.31″ E). Figure 1 illustrates the four-season classification system applied in this study for seasonal flux analysis. Table 1 presents the average monthly weather characteristics of the study area during the monitoring period. The experimental site geographical location and layout of the swine wastewater treatment system used for GHG monitoring are represented in Figure 2. The pig farm houses approximately 9000 pigs, generating an average 16.47 t day^{-1} of solid waste and 25.65 t day^{-1} of effluent. The treatment plant is a self-purification wastewater treatment system designed based on aerobic and anaerobic processes to degrade organic content at multiple treatment stages (Figure 2B).

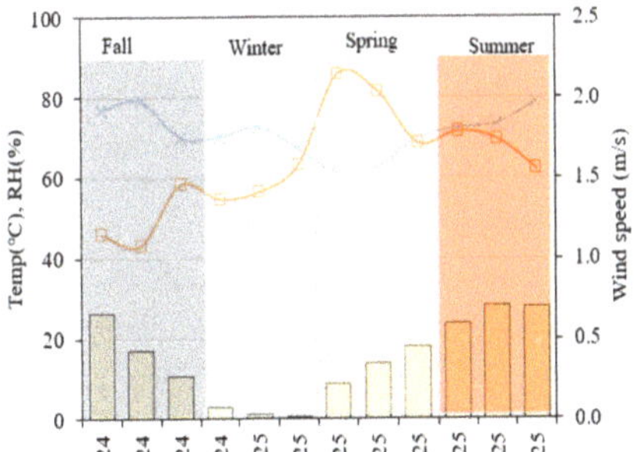

Figure 1. Seasonal variations in ambient environmental conditions (temperature, relative humidity, and wind speed) obtained from the nearest Korea Meteorological Administration (KMA) station during the measurement period. Wind speed corresponds to the standard 10 m measurement height.

https://doi.org/10.3390/su18010321

Table 1. Average annual weather conditions of the study area during the experiment period.

Index	Fall	Winter	Spring	Summer
Wind speed (m/s)	1.2	1.5	2.0	1.7
Temperature (°C)	18.0	1.5	13.4	26.5
Relative humidity (%)	75.3	69.9	64.5	74.9

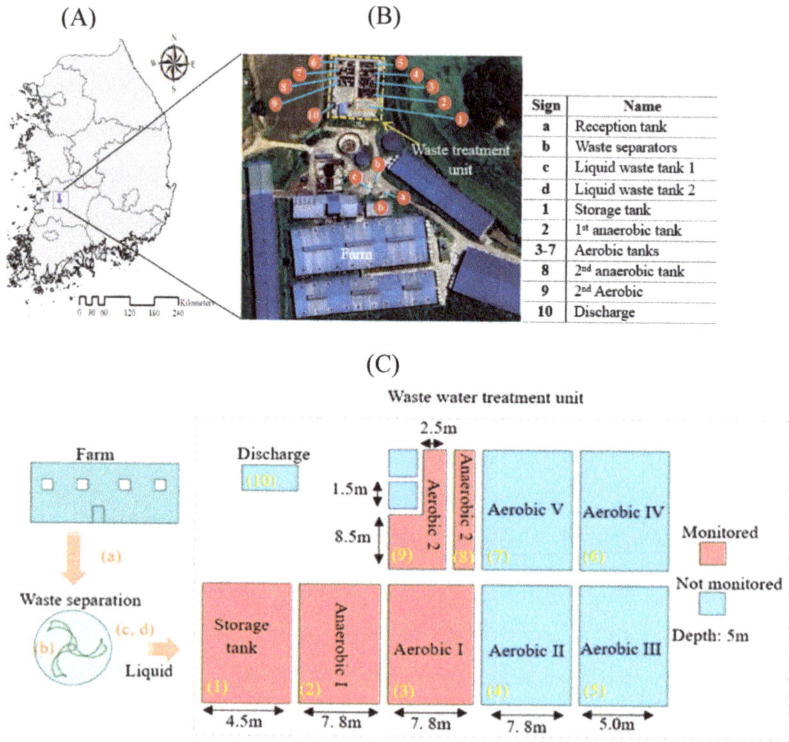

Figure 2. Experimental site location and layout of the wastewater treatment system used for GHG monitoring. (**A**) Geographical location of the study site in Korea. (**B**) Aerial view of the farm and waste treatment facility showing sampling points (1 to10) and key units (a to d). (**C**) Illustration of the treatment sequence, tank dimensions, and sampling stages.

From Figure 2, the treatment process begins at Tank (a), where raw waste is received before undergoing a separation process at (b) to divide solid and liquid waste fractions. The liquid waste is then stored in liquid waste tanks (c) and (d) before being pumped into the treatment unit. The wastewater first accumulates in an underground storage tank (Stage 1) and then moves to the anaerobic treatment stage (Stage 2), where microorganisms break down organic matter under oxygen-deficient conditions, releasing GHG as a byproduct. Following this, the waste undergoes aerobic treatment (Stages 3 to 7), where organic matter is further degraded in aeration tanks (Aerobic 1–5) under oxygen-rich conditions. Secondary anaerobic treatment occurs in Anaerobic Tank 2 (Stage 8), followed by secondary aerobic treatment in Aerobic Tank 2 (Stage 9) before final discharge.

Figure 2C provides a detailed layout of the wastewater handling process, technical specifications of the treatment unit, and GHG monitoring points considered in this study. The CH_4 and N_2O emissions were monitored at three main stages: pre-treatment (raw

https://doi.org/10.3390/su18010321

waste storage), primary treatment (anaerobic and aerobic stages), and secondary treatment (anaerobic and aerobic stages).

2.2. Equipment and GHGs Collection

2.2.1. Equipment

A custom floating PVC air-pool chamber was installed at Stages 2, 3, 8, and 9 to enable continuous gas capture while preventing corrosion from wastewater emissions (Figure 3). The chamber measured 1.22 m × 2.44 m × 0.60 m (collection area 2.98 m^2) and was connected to flexible air ducts, an air blower, and 0.25-inch Teflon tubing for transport to the analyzer (Figure 4a). At Stage 1, where chamber deployment was not possible due to the underground location and restricted access, a stainless-steel cover with two inlet ports and one outlet port was used, and gas was extracted using 117 $m^3\ s^{-1}$ blowers (Figure 4b).

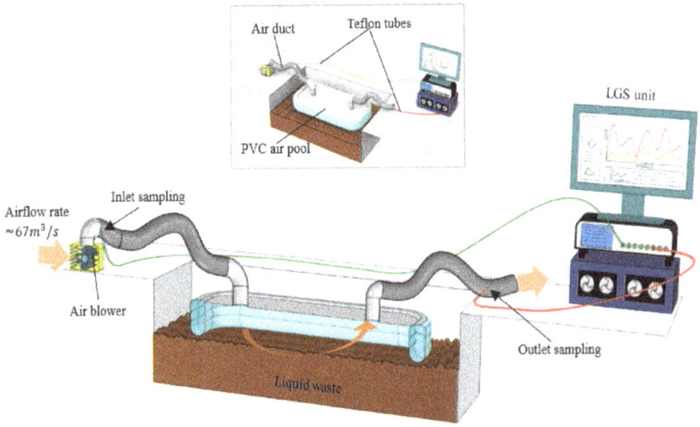

Figure 3. Schematic illustration of gas collection and monitoring setup, showing the isometric and cross-sectional views of the PVC air pool system. The setup includes an air blower airflow, Teflon sampling tubes for inlet and outlet gas collection, and a laser gas spectrometer (LGS) unit for real-time CH_4 and N_2O measurement.

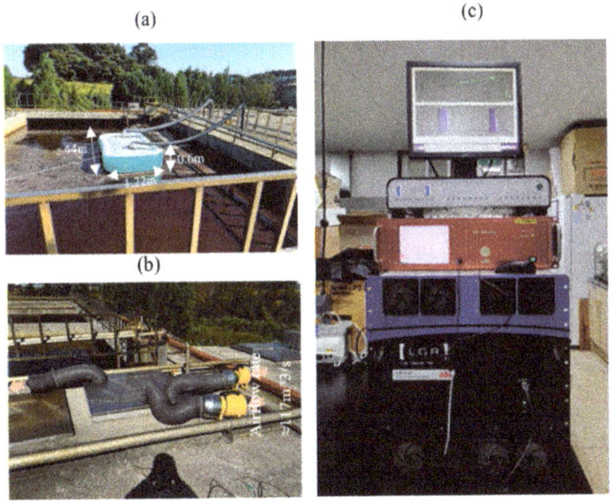

Figure 4. Field installation and analytical system used for GHG flux monitoring. (**a**) Floating PVC air chamber deployed at monitoring stages 2, 3, 8, and 9; (**b**) gas collection and transfer setup at the raw-water storage tank (stage 1) and (**c**) Laser Gas Analyzer (LGR) unit.

https://doi.org/10.3390/su18010321

Gas concentrations were measured using a laser-based analyzer (ABB Measurement & Analytics, Quebec City, QC, Canada; LGR-ICOS™ GLA351-N2OM1; Figure 4c). The analyzer alternated between CH_4 and N_2O using its internal optical-switching mechanism. All sampled gas was conveyed through Teflon tubing to minimize adsorption or chemical loss. Sensor detection ranges and measurement methods are summarized in Table 2.

Table 2. Measurement range, interval, and method of the gas sensors.

Sensor Type	Measurement Range	Measurement Cycle	Measurement Method
Methane (CH_4)	0–1000 ppm	1 s	Laser analysis
Nitrous oxide (N_2O)	0–40 ppm	1 s	Laser analysis

Before each monthly monitoring event, the analyzer was calibrated using certified low- and high-concentration CH_4 and N_2O standards. Calibration checks included zero stability, span accuracy, and spectral-fit performance according to manufacturer guidelines.

2.2.2. Gas Collection

Monitoring was performed for ten 24-h campaigns from 6 September 2024 to 30 August 2025, covering autumn, winter, spring, and summer. During each campaign, continuous sampling was conducted at all five treatment stages. The PVC chamber inlets and outlets were monitored simultaneously, with the inlet measurements serving as the blank reference. The detailed monitoring days and seasonal coverage are represented in Table 3.

Table 3. Monitoring period and seasonal coverage.

Item	Description
Monitoring period	6 September 2024–30 August 2025
Seasons covered	Autumn (Sep–Nov), Winter (Dec–Feb), Spring (Mar–May), Summer (Jun–Aug)
Monitoring days	6 Sep 2024; 24 Sep 2024; 24 Oct 2024; 9 Nov 2024; 17 Dec 2024; 13 Feb 2025; 13 Mar 2025; 29 May 2025; 30 Jun 2025; 30 Aug 2025

CH_4 and N_2O were recorded at 1-s intervals producing up to 86,400 measurements per stage per day. Approximately 90% of all measurements were retained after quality control and used in flux calculation. Measurements were grouped into morning (06:00 to 12:00), afternoon (12:00 to 18:00), evening (18:00 to 00:00), and night (00:00 to 06:00) for seasonal flux aggregation. Negative values were treated as zeros while occasional gaps due to operational interruptions were treated as missing data and excluded from averaging.

Environmental variables, including temperature and humidity, were obtained from the Korea Meteorological Administration (KMA) for each monitoring day and were used to assess the influence of weather conditions on GHG flux patterns.

2.3. Data Analysis and Emission Flux Calculation

2.3.1. CH_4 and N_2O Analysis

The collected GHG emission data were analyzed to assess seasonal and treatment stage variations in CH_4 and N_2O fluxes. Seasonal classification was based on local meteorological records, and data were grouped into spring, summer, fall and winter to study environmental effects on emission trends. To characterize diurnal behavior, all hourly fluxes were assigned to four 6 h periods including morning (06:00 to 12:00), afternoon (12:00 to 18:00), evening (18:00 to 00:00), and night (00:00 to 06:00). For treatment-stage characterization, emission data were

https://doi.org/10.3390/su18010321

categorized according to pre-treatment (raw waste storage), primary treatment (anaerobic tank I and aerobic I), and secondary treatment (anaerobic II and aerobic II) stages.

Descriptive statistics (mean, standard deviation, coefficient of variation) were used to characterize greenhouse gas. The differences among groups were evaluated using one-way ANOVA, followed by Tukey's HSD multiple comparison test ($p < 0.05$) where appropriate. To quantify the influence of environmental factors, linear regression was applied with temperature and relative humidity as covariates based on the model as shown in Equation (1). All statistical analyses and graphical visualizations were performed using OriginPro 2025b SR1 (OriginLab Corporation, Northampton, MA, USA).

$$F_i = \beta_0 + \beta_1 T_i + \beta_2 RH_i + \varepsilon_i \quad (1)$$

2.3.2. CH_4 and N_2O Flux Calculation

Flux rates were calculated using standard gas flux equations, integrating measured gas concentrations, air flow rates, and environmental parameters (Equation (2)).

$$F_{GHG} = \frac{(C_o - C_i) \times p_{air} \times \dot{m}}{T \times R \times A \times 1000} \quad (2)$$

where F_{GHG} is the greenhouse house gas flux (mg s m^{-2}), C_o and C_i are the outlet and inlet greenhouse gas concentrations $(CH_4 : \text{ppm}, N_2O : \text{ppb})$, p_{air} is the atmospheric air pressure (pa), $\dot{m}$ is the air flow rate through the chamber system ($m^3\ s^{-1}$), T is the ambient air temperature (K), R is the molar gas constant ($J\ mol^{-1}\ K^{-1}$) and A is the effective area of the chamber (m^2).

2.3.3. Emission Factors Calculation

Measured fluxes for each treatment stage were aggregated across the four daily measurement periods and seasons. For each stage s and season k the daily representative flux was obtained as the Mean of the four period-level fluxes [9,22]. The daily emissions were calculated as:

$$E_{s,k,day} = F_{s,k,day} \times A_s \times 86{,}400 \times 10^{-6},$$

where $F_{s,k,day}$ is the daily representative flux, A_s is the stage emitting surface area (m^2), 86,400 converts seconds to days and 10^{-6} converts mg to kg. Seasonal total emissions were obtained by multiplying $E_{s,k,day}$ by the number of days in each season ($E_{s,k,season}$). The annual emissions per stage were then computed as follows.

$$E_{s,year} = \sum_k E_{s,k,season}.$$

Overall treatment plant annual emissions ($E_{total,year}$) were the sum of all stage totals. Per animal emission factors were calculated by dividing $E_{total,year}$ by the average number of pigs on the farm. CO_2-equivalent emission factors were derived using IPCC AR6 GWP_{100} values (CH_4 = 27.2, N_2O = 273) [9].

3. Results and Discussion

3.1. Diurnal GHG Flux Characteristics

Diurnal CH_4 and N_2O fluxes showed clear 6-h interval patterns across the monitoring period (Figure 5). CH_4 fluxes were highest during the evening and night (7.2 and 7.0 mg s^{-1} m^{-2} respectively), with secondary peak in the morning hours (5.8 mg s^{-1} m^{-2}). The lowest mean Flux occurred during the afternoon (3.2 mg s^{-1} m^{-2}). CH_4 variability

https://doi.org/10.3390/su18010321

remained comparatively narrow, with coefficient of variation (CV) ranging from 1.7 at night to 3.4 in the morning.

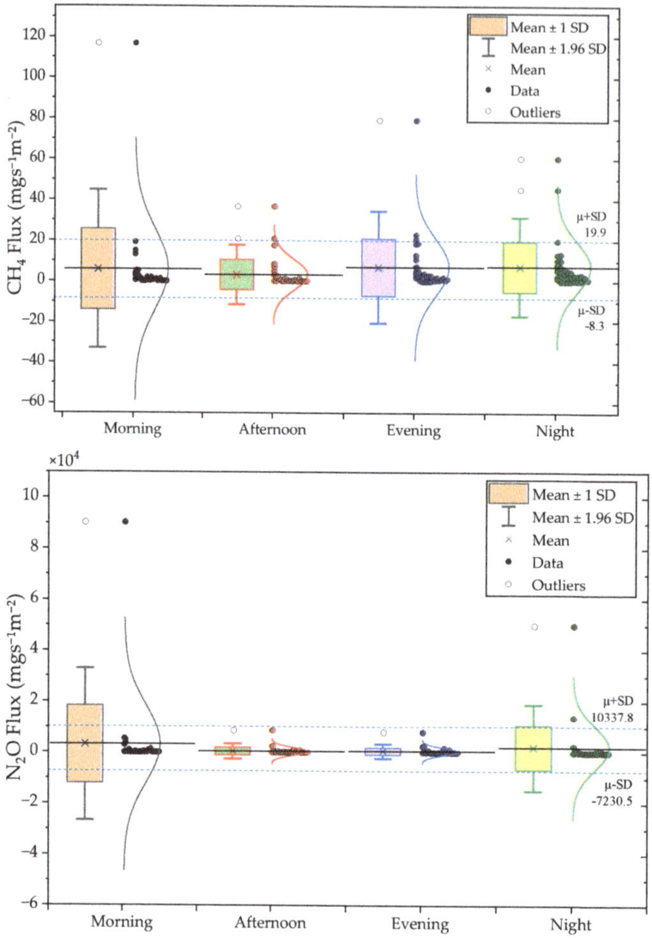

Figure 5. (**Top**) CH_4 and (**Bottom**) N_2O diurnal flux distributions for Morning, Afternoon, Evening, and Night periods.

N_2O fluxes exhibited far greater diurnal variability. The highest mean Flux occurred during morning hours (3174.6 mg s^{-1} m^{-2}), followed by a secondary night peak (2056.5 mg s^{-1} m^{-2}). Afternoon emissions were substantially lower, being 7.2- and 4.6-fold lower than morning and evening Fluxes, respectively. Table 4 summarizes the of diurnal CH_4 and N_2O flux statistics. The high CV values (2.6–4.8) underline the instability of N_2O production and release, which is strongly influenced by rapid operational and biochemical transitions such as changes in oxygen availability and nitrification–denitrification cycling. These findings are consistent with previous studies showing that N_2O emissions from livestock wastewater systems often occur in short-lived bursts linked to intermittent aeration and redox shifts [23,24].

https://doi.org/10.3390/su18010321

Table 4. Summary of diurnal CH_4 and N_2O flux statistics from pig wastewater treatment system. Values represent the Mean, standard deviation (SD), standard error (SE), coefficient of variation (CV), and maximum observed flux (n =35 per period).

Period	GHG	Mean	SD	SE of Mean	CV	Max
Morning	CH_4	5.8	19.8	3.3	3.4	116.8
	N_2O	3174.6	15,199.2	2569.1	4.8	90,162.9
Afternoon	CH_4	3.2	7.5	1.3	2.3	36.5
	N_2O	439.9	1501.4	253.8	3.4	8671.8
Evening	CH_4	7.0	14.1	2.4	2.0	79.1
	N_2O	543.6	1434.7	242.5	2.6	7900.4
Night	CH_4	7.2	12.4	2.1	1.7	60.5
	N_2O	2056.5	8650.5	1462.2	4.2	49,913.1

3.2. Treatment Stage Specific GHG Flux Characteristics

Treatment-stage flux distributions are shown in Figure 6 and summarized in Table 5. CH_4 emissions varied markedly across the system, with the highest mean flux observed in Aeration Tank II (27.8 mg m^{-2} s^{-1}). Short-term CH_4 peaks (up to 119.8 mg m^{-2} s^{-1}) coincided with strong turbulence events, which likely released CH_4 accumulated within sludge flocs. Such aeration-driven stripping has been reported in both municipal and livestock wastewater systems [22,25–27]). Lower CH_4 fluxes in Anaerobic Tank II and the raw-water tank (2.8–3.7 mg m^{-2} s^{-1}) align with reduced substrate availability, higher redox potential, and possible ammonia inhibition, conditions known to suppress methanogenesis [28–31].

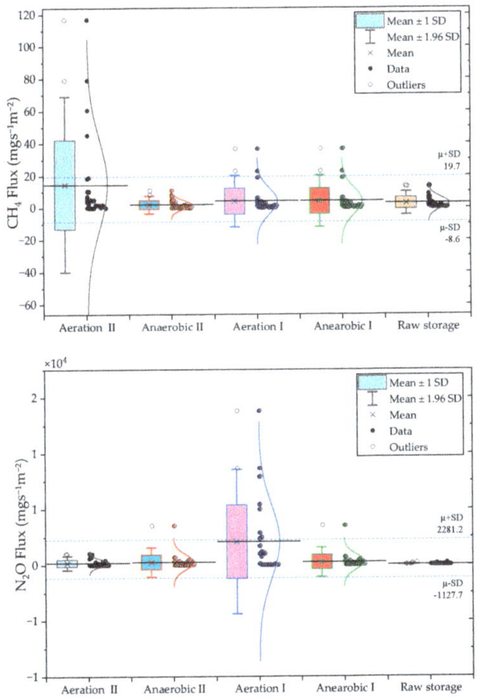

Figure 6. Treatment stage variability in CH_4 and N_2O fluxes measured across the swine wastewater process.

https://doi.org/10.3390/su18010321

Table 5. Summary of treatment stage specific CH_4 and N_2O flux statistics from pig wastewater treatment system. Values represent the Mean, standard deviation (SD), standard error (SE), coefficient of variation (CV), and maximum observed flux (n = 35 per stage).

Period	GHG	Mean	SD	SE of Mean	CV	Max
Aeration tank I	CH_4	4.3	8.3	1.6	1.9	36.5
	N_2O	2108.4	3297.3	623.1	1.6	13,803.1
Aeration tank II	CH_4	14.5	27.8	5.3	1.9	116.8
	N_2O	202.9	323.2	61.1	1.6	1039.3
Anaerobic tank II	CH_4	2.0	2.8	0.5	1.4	10.8
	N_2O	279.5	666.0	125.9	2.4	3522.4
Anaerobic Tank I	CH_4	5.5	7.2	1.4	1.3	20.8
	N_2O	5163.9	19,129.6	3615.2	3.7	90,162.9
Raw water storage tank	CH_4	2.7	3.7	0.7	1.4	13.3
	N_2O	13.5	39.9	7.5	2.9	181.8

N_2O emissions were highly stage dependent. Anaerobic Tank I produced the highest mean and peak fluxes (5163.9 mg m^{-2} s^{-1} and 90,162.9 mg m^{-2} s^{-1}, respectively), reflecting unstable nitrification–denitrification processes driven by oxygen fluctuations, nitrite accumulation, and low C/N ratios. Similar extreme N_2O peaks have been observed in pig-slurry and intermittently aerated reactors, where up to 18% of influent nitrogen may be emitted as N_2O [32,33]. The high CV in Anaerobic Tank I (3.7) reinforces the episodic nature of N_2O production, consistent with wastewater studies showing large temporal variability in N_2O yields [22,34].

By contrast, Aeration Tank II and the raw-water tank exhibited minimal N_2O emissions (<300 mg s^{-1} m^{-2}), likely due to more stable redox environments where nitrite does not accumulate and denitrification proceeds fully to N_2 [24,35]. These contrasting emission behaviors indicate that CH_4 and N_2O mitigation strategies must target different stages and processes. CH_4 mitigation should focus on mixing and turbulence control, whereas N_2O reduction requires improving biochemical stability through controlled aeration, adequate carbon supply, and prevention of nitrite accumulation.

3.3. Seasonal GHG Flux Characteristics

Seasonal CH_4 and N_2O flux patterns are shown in Figure 7 and Table 6. CH_4 fluxes exhibited moderate but consistent seasonal variability, with the lowest values in summer (3.2 mg s^{-1} m^{-2}) and the highest in late fall and winter (7.0 to 7.2 mg s^{-1} m^{-2}). These patterns reflect temperature-driven changes in methanogenic and methanotrophic activity, as well as differences in gas–liquid transfer efficiency under cooler conditions [36]. Persistent CH_4 generation in winter likely arises from thermally buffered sludge layers [37].

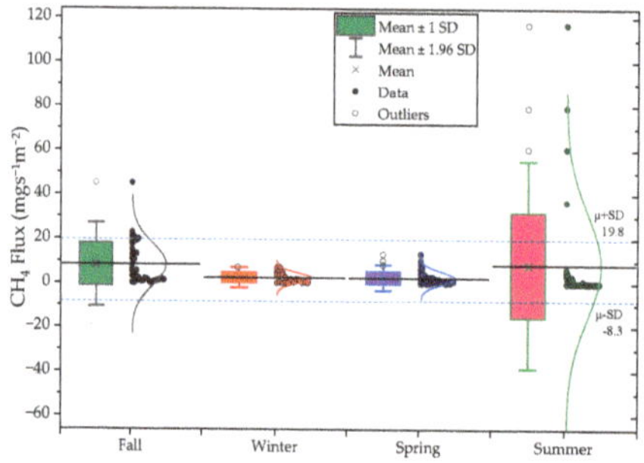

Figure 7. *Cont.*

https://doi.org/10.3390/su18010321

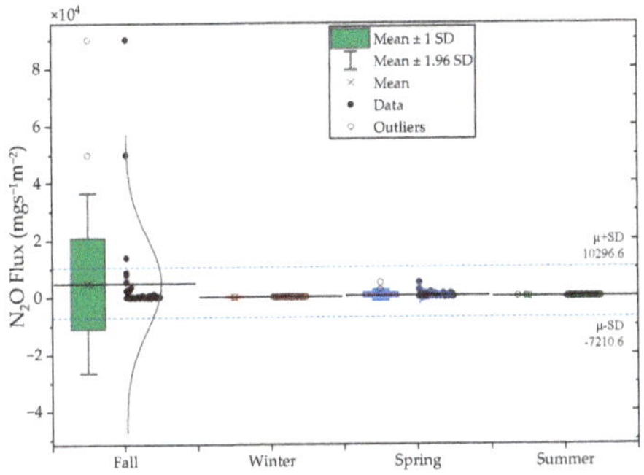

Figure 7. Seasonal variations in CH_4 (**top**) and N_2O (**bottom**) fluxes across the monitoring period.

Seasonal N_2O fluxes were again dominated by Anaerobic Tank I (mean = 5163.9 mg s^{-1} m^{-2}, max = 90,162.9 mg s^{-1} m^{-2}), consistent with nitrogen transformation instability rather than climatic influence. Aeration Tank II and the raw-water tank continued to exhibit low fluxes (<300 mg s^{-1} m^{-2}). These results reaffirm that N_2O emissions in this system are governed primarily by operational and biochemical factors rather than seasonal temperature or humidity conditions.

Table 6. Summary of seasonal CH_4 and N_2O flux statistics from the swine wastewater treatment system. Values represent the Mean, standard deviation (SD), standard error (SE), coefficient of variation (CV), and maximum observed flux for each 6-h monitoring interval (n = 35 per period).

Seasons	GHG	Mean	SD	SE	CV	Max
Spring	CH_4	5.8	19.8	3.3	3.4	116.8
	N_2O	3174.6	15,199.2	2569.1	4.8	90,162.9
Summer	CH_4	3.2	7.5	1.3	2.3	36.5
	N_2O	439.9	1501.4	253.8	3.4	8671.8
Fall	CH_4	7.0	14.1	2.4	2.0	79.1
	N_2O	543.6	1434.7	242.5	2.6	7900.4
Winter	CH_4	7.2	12.4	2.1	1.7	60.5
	N_2O	2056.5	8650.5	1462.2	4.2	49,913.1

3.4. Effect of Environmental Factors on GHG Emission Fluxes

The seasonal relationships between GHG fluxes and environmental variables are shown in Figure 8. CH_4 fluxes exhibited a very strong positive correlation with temperature (r = 0.97; R^2 = 0.95), confirming that methanogenesis intensifies under warmer conditions. Relative humidity (RH) showed a moderate negative correlation with CH_4 flux (r = −0.63; R^2 = 0.39), indicating that drier, more ventilated conditions enhance gas transfer and promote CH_4 release. These patterns are consistent with previous studies showing strong microclimatic control over CH_4 emissions from stored and aerated pig manure [36–38].

https://doi.org/10.3390/su18010321

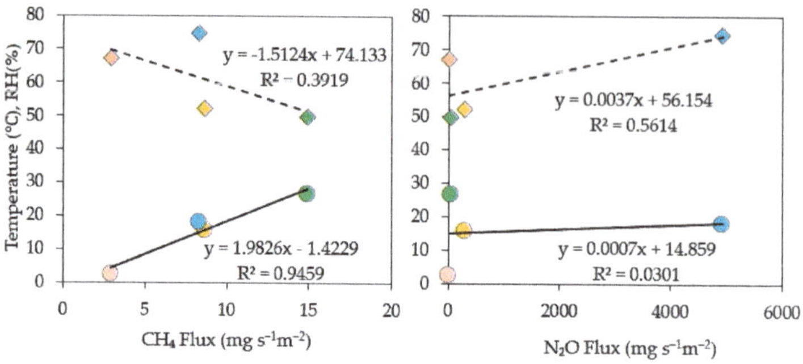

Figure 8. Seasonal scale linear regressions between GHG flux and temperature (○), and relative humidity (◇). Each color represents a seasonal mean: Spring, Summer, Fall, and Winter. Each point represents a seasonal mean (n = 4: Spring, Summer, Fall, Winter). Regression equations and R^2 values are shown for each relationship.

For N_2O, the seasonal relationship with temperature was weak (r = 0.17; R^2 = 0.03), suggesting that temperature plays only a minor role at the seasonal scale. In contrast, RH displayed a strong positive correlation with N_2O flux (r = 0.75; R^2 = 0.56), implying greater N_2O accumulation under humid, oxygen-limited conditions that favor incomplete denitrification. This finding aligns with studies showing that N_2O production in wastewater systems is driven primarily by internal biochemical dynamics rather than external temperature changes [34,35,39].

3.5. Greenhouse Gas Emission Factors

Emission factor calculation yielded annual emissions of 1.01×10^3 kg CH_4 yr^{-1} and 4.07×10^5 kg N_2O yr^{-1} for the wastewater treatment system. When normalized by the average herd population of 9000 pigs, the corresponding emission factors were 0.11 kg CH_4/head/year and 45.2 kg N_2O $head^{-1}$ yr^{-1}. Expressed in CO_2-equivalent units using IPCC AR6 100-year global warming potentials (CH_4 = 27.2; N_2O = 273), the resulting emission factors were 3.1 kg CO_2-eq $head^{-1}$ yr^{-1} for CH_4 and 1.23×10^4 kg CO_2-eq $head^{-1}$ yr^{-1} for N_2O.

Stage-level emissions showed different spatial patterns for CH_4 and N_2O within the treatment system (Table 7). The largest CH_4 emissions originated from Aeration Tank I, Aeration Tank II, and Anaerobic Tank I. In contrast, N_2O emissions were highly concentrated, as Anaerobic Tank I and Aeration Tank I collectively accounted for more than 90 percent of the total annual N_2O released. This concentration of N_2O emissions indicates that a small portion of the treatment process is responsible for most of the system's greenhouse-gas output and reflects unstable nitrogen transformation conditions in those units.

Table 7. Annual CH_4 and N_2O emissions by treatment stage, percentage contributions, and system-level emission factors.

Treatment Stage	CH_4 Emission (kg yr^{-1})	CH_4 Share (%)	N_2O Emission (kg yr^{-1})	N_2O Share (%)
Aeration Tank I	252.8	25.0	115,434.3	28.3
Aeration Tank II	315.3	31.2	4258.6	1.0
Anaerobic Tank II	42.0	4.1	4906.8	1.2
Anaerobic Tank I	316.1	31.3	281,934.7	69.4
Raw Water Storage Tank	87.7	8.7	426.7	0.1
Total Annual Emissions	1014.0	100	406,961.0	100
Per-Head EF (9000 pigs)	0.11 kg CH_4 $head^{-1}$ yr^{-1}	–	45.2 kg N_2O $head^{-1}$ yr^{-1}	–
CO_2-eq EF (IPCC AR6 GWP)	3.1 kg CO_2-eq $head^{-1}$ yr^{-1}	–	1.23×10^4 kg CO_2-eq $head^{-1}$ yr^{-1}	–

https://doi.org/10.3390/su18010321

The CH_4 emission factor obtained in this study (0.11 kg CH_4 head^{-1} yr^{-1}) lies at the lower edge of values reported for swine wastewater treatment systems, which typically range from 0.1–1.0 kg CH_4 head^{-1} yr^{-1} in full-scale biological treatment processes [22]. This consistency reflects the predominantly aerobic conditions in the treatment train, which suppress methanogenesis. In contrast, the N_2O emission factor measured here (45.2 kg N_2O head^{-1} yr^{-1}) is several orders of magnitude higher than values reported for optimized nitrogen removal systems, where per-head N_2O emissions generally fall between 0.001 and 0.01 kg N_2O head^{-1} yr^{-1}, and only 0.5–3% of influent nitrogen is typically emitted as N_2O-N. These elevated emissions align with the exceptionally high N_2O fluxes observed in Anaerobic Tank I and Aeration Tank I. Overall, the system's greenhouse-gas footprint is dominated by these localized N_2O hotspots, indicating that focused improvements in aeration control, carbon availability, and nitrogen transformation processes in these units could substantially reduce total emissions.

4. Conclusions

This study quantified CH_4 and N_2O emissions from a full-scale piggery wastewater treatment system and developed treatment stage and time-resolved emission factors. CH_4 emissions were strongly influenced by temperature and gas transfer dynamics, with higher fluxes during warm periods and in aerated units, particularly during morning and night when mixing intensity and thermal gradients favored release. In contrast, N_2O emissions were dominated by biochemical instability in Anaerobic Tank I, where oxygen fluctuations, nitrite accumulation, and incomplete denitrification produced extreme and irregular peaks.

Annual emission factors were 0.11 kg CH_4 head^{-1} yr^{-1} and 45.2 kg N_2O head^{-1} yr^{-1}, equivalent to 3.1 and 1.23 × 10^{-4} kg CO_2-eq head^{-1} yr^{-1}, respectively. More than 90% of N_2O emissions originated from a single treatment stage, underscoring the disproportionate influence of nitrification–denitrification instability on system-wide greenhouse-gas release.

These findings indicate that mitigation must be focused on the distinct mechanisms driving each gas. CH_4 reductions should focus on minimizing the physical release of accumulated gas, for example, by moderating aeration-induced turbulence, improving mixing uniformity to limit anaerobic microsites within aerobic reactors, and applying surface covers or gas-collection hoods in areas prone to vigorous stripping. In contrast, N_2O mitigation requires stabilization of the nitrogen transformation pathway. This involves maintaining consistent dissolved oxygen conditions to prevent abrupt redox oscillations, ensuring adequate carbon availability for complete denitrification, and employing real-time monitoring of oxygen and intermediate nitrogen species to avoid nitrite accumulation. Improving the operational stability of Anaerobic Tank I, including reducing unintended oxygen intrusion and homogenizing mixing, is likely to yield substantial reductions given its outsized contribution to total N_2O emissions.

The emission factors developed in this work enhance the accuracy of livestock greenhouse-gas inventories and offer process-specific insights that can inform low-emission wastewater treatment design. Future research will incorporate wastewater chemical characteristics and microbial community analysis to strengthen mechanistic understanding, and on developing dynamic emission factor models that integrate operational data and climatic variability to support climate-smart livestock facility management.

Author Contributions: Conceptualization, A.K.K., I.-H.S. and I.-S.K.; methodology, A.K.K. and I.-S.K.; software, A.K.K.; validation, A.K.K., I.-H.S. and I.-S.K.; formal analysis, A.K.K.; investigation, A.K.K.; resources, I.-H.S. and I.-S.K.; data curation, I.-S.K.; writing—original draft preparation, A.K.K.; writing—review and editing, A.K.K.; visualization I.-S.K.; supervision I.-H.S. and I.-S.K.; project administration, I.-S.K.; funding acquisition, I.-H.S. and I.-S.K. All authors have read and agreed to the published version of the manuscript.

https://doi.org/10.3390/su18010321

Funding: This research was supported by the Rural Development Administration's 'Climate Change Mitigation and Low-Carbon Agricultural Technology Development' research project (project number: RS-2023-00221189).

Institutional Review Board Statement: Not applicable.

Informed Consent Statement: Not applicable.

Data Availability Statement: Research data available upon request.

Conflicts of Interest: The authors declare no conflicts of interest.

Abbreviations

The following abbreviations are used in this manuscript:

GHG	Greenhouse Gas
CH_4	Methane
N_2O	Nitrous Oxide
EF	Emission Factor
RH	Relative Humidity
FTIR	Fourier Transform Infrared Spectroscopy
LGR	Los Gatos Research (laser-based gas analyzer)
AGSC	Automated Gas Sampling Chamber
ANOVA	Analysis of Variance
ANCOVA	Analysis of Covariance
SD	Standard Deviation
SE	Standard Error
CV	Coefficient of Variation
R^2	Coefficient of Determination
p	Probability Value (Significance Level)
η^2	Partial Eta Squared (Effect Size)
PVC	Polyvinyl Chloride
COD	Chemical oxygen demand

References

1. Gerber, P.; Steinfeld, H.; Henderson, B.; Mottet, A.; Opio, C.; Dijkman, J.; Tempio, G. *Reducing Enteric Methane and Livelihoods Win-Win Opportunities for Farmers*; FAO: Rome, Italy, 2013; Volume 14, ISBN 9789251079201.
2. Delbeke, J.; Runge-Metzger, A.; Slingenberg, Y.; Werksman, J. The Paris Agreement. In *Towards a Climate-Neutral Europe Curbing the Trend*; Routledge: Oxfordshire, UK, 2015; pp. 24–45. [CrossRef]
3. Rivera Moncada, A.; Dupas, M.C.; Tempio, G.; Lanzoni, L.; Li, Y.; Rakotovao, N.; Wisser, D.; Gilbert, M. Sensitivity Analysis of Parameters, Emission Factors, and Coefficients for Estimating Animal Emissions of Ruminant Species in the Global Livestock Environmental Assessment Model (GLEAM). *Int. J. Life Cycle Assess.* **2025**. [CrossRef]
4. Graham, M.W.; Butterbach-Bahl, K.; du Toit, C.J.L.; Korir, D.; Leitner, S.; Merbold, L.; Mwape, A.; Ndung'u, P.W.; Pelster, D.E.; Rufino, M.C.; et al. Research Progress on Greenhouse Gas Emissions From Livestock in Sub-Saharan Africa Falls Short of National Inventory Ambitions. *Front. Soil Sci.* **2022**, *2*, 927452. [CrossRef]
5. Government of Republic of Korea. *The Government of the Republic of Korea 2050 Carbon Neutral Strategy of the Republic of Korea: Towards a Sustainable and Green Society*; Government of Republic of Korea: Seoul, Republic of Korea, 2020; pp. 1–131.
6. Statistics Korea. *Livestock Statistics in the Fourth Quarter of 2013*; Statistics Korea: Daejeon, Republic of Korea, 2013; Volume 2024, pp. 11–12. Available online: https://unfccc.int/sites/default/files/resource/LTS1_RKorea.pdf (accessed on 8 December 2025).
7. Ministry of Agriculture, Food and Rural Affairs (MAFRA). *2023 Livestock Manure Generation and Management Statistics Report 2024*; MAFRA: Mafra, Portugal, 2024.
8. Symeon, G.K.; Akamati, K.; Dotas, V.; Karatosidi, D.; Bizelis, I.; Laliotis, G.P. Manure Management as a Potential Mitigation Tool to Eliminate Greenhouse Gas Emissions in Livestock Systems. *Sustainability* **2025**, *17*, 586. [CrossRef]
9. Eska, N.; Lee, J.S.; Park, K.H. Greenhouse Gas Emissions from Livestock: Sources, Estimation, and Mitigation. *J. Anim. Sci. Technol.* **2024**, *66*, 1083–1098. [CrossRef] [PubMed]

https://doi.org/10.3390/su18010321

10. Al Zahra, W.; Ikhsan Shiddieqy, M.; Anisa, R.; Yani, A.; Priyo Purwanto, B. The Dynamics of Nitrous Oxide and Methane Emissions from Various Types of Dairy Manure at Smallholder Dairy Farms as Affected by Storage Periods. *Waste Manag.* **2024**, *183*, 10–20. [CrossRef]
11. Philippe, F.X.; Nicks, B. Review on Greenhouse Gas Emissions from Pig Houses: Production of Carbon Dioxide, Methane and Nitrous Oxide by Animals and Manure. *Agric. Ecosyst. Environ.* **2015**, *199*, 10–25. [CrossRef]
12. Montes, F.; Meinen, R.; Dell, C.; Rotz, A.; Hristov, A.N.; Oh, J.; Waghorn, G.; Gerber, P.J.; Henderson, B.; Makkar, H.P.S.; et al. SPECIAL TOPICS-Mitigation of Methane and Nitrous Oxide Emissions from Animal Operations: II. A Review of Manure Management Mitigation Options. *J. Anim. Sci.* **2013**, *91*, 5070–5094. [CrossRef] [PubMed]
13. Ro, K.S.; Johnson, M.H.; Stone, K.C.; Hunt, P.G.; Flesch, T.; Todd, R.W. Measuring Gas Emissions from Animal Waste Lagoons with an Inverse-Dispersion Technique. *Atmos. Environ.* **2013**, *66*, 101–106. [CrossRef]
14. Gu, D.; Liu, Y.; Zhao, W.; Qiu, S.; Cui, N.; Hu, X.; Zhao, P. Status of Research on Greenhouse Gas Emissions from Wastewater Collection Systems. *Water* **2023**, *15*, 2512. [CrossRef]
15. Jensen, S.A.; Webb, J.R.; Simpson, G.L.; Baulch, H.M.; Leavitt, P.R.; Finlay, K. Seasonal Variability of CO_2, CH_4, and N_2O Content and Fluxes in Small Agricultural Reservoirs of the Northern Great Plains. *Front. Environ. Sci.* **2022**, *10*, 895531. [CrossRef]
16. Vechi, N.T.; Falk, J.M.; Fredenslund, A.M.; Edjabou, M.E.; Scheutz, C. Methane Emission Rates Averaged over a Year from Ten Farm-Scale Manure Storage Tanks. *Sci. Total Environ.* **2023**, *904*, 166610. [CrossRef]
17. Wilkinson, J.; Bors, C.; Burgis, F.; Lorke, A.; Bodmer, P. Measuring CO_2 and CH_4 with a Portable Gas Analyzer: Closed-Loop Operation, Optimization and Assessment. *PLoS ONE* **2018**, *13*, E0193973, Correction in *PLoS ONE* **2019**, *14*, e0206080. https://doi.org/10.1371/journal.pone.0206080. [CrossRef] [PubMed]
18. Stiefvater, G.; Hespos, Y.; Wiedenmann, D.; Lambrecht, A.; Brunner, R.; Wöllenstein, J. A Portable Laser Spectroscopic System for Measuring Nitrous Oxide Emissions on Fertilized Cropland. *Sensors* **2023**, *23*, 6686. [CrossRef] [PubMed]
19. Briz, S.; Barrancos, J.; Nolasco, D.; Melián, G.; Padrón, E.; Pérez, N. New Method for Estimating Greenhouse Gas Emissions from Livestock Buildings Using Open-Path FTIR Spectroscopy. *Remote Sens. Clouds Atmos. XIV* **2009**, *7475*, 747510. [CrossRef]
20. Cattaneo, M.; Tayà, C.; Burgos, L.; Morey, L.; Noguerol, J.; Provolo, G.; Cerrillo, M.; Bonmatí, A. Assessing Ammonia and Greenhouse Gas Emissions from Livestock Manure Storage: Comparison of Measurements with Dynamic and Static Chambers. *Sustainability* **2023**, *15*, 15987. [CrossRef]
21. Hassouna, M.; van der Weerden, T.J.; Beltran, I.; Amon, B.; Alfaro, M.A.; Anestis, V.; Cinar, G.; Dragoni, F.; Hutchings, N.J.; Leytem, A.; et al. DATAMAN: A Global Database of Methane, Nitrous Oxide, and Ammonia Emission Factors for Livestock Housing and Outdoor Storage of Manure. *J. Environ. Qual.* **2023**, *52*, 207–223. [CrossRef]
22. Osada, T.; Shiraishi, M.; Hasegawa, T.; Kawahara, H. Methane, Nitrous Oxide and Ammonia Generation in Full-Scale Swine Wastewater Purification Facilities. *Front. Environ. Sci. Eng.* **2017**, *11*, 10. [CrossRef]
23. Su, J.J.; Chen, Y.J. Monitoring of Greenhouse Gas Emissions from Farm-Scale Anaerobic Piggery Waste-Water Digesters. *J. Agric. Sci.* **2018**, *156*, 739–747. [CrossRef]
24. Yamashita, T.; Shiraishi, M.; Yokoyama, H.; Ogino, A.; Yamamoto-Ikemoto, R.; Osada, T. Evaluation of the Nitrous Oxide Emission Reduction Potential of an Aerobic Bioreactor Packed with Carbon Fibres for Swine Wastewater Treatment. *Energies* **2019**, *12*, 1013. [CrossRef]
25. Lens, P.N.; De Poorter, M.P.; Cronenberg, C.C.; Verstraete, W.H. Sulfate Reducing and Methane Producing Bacteria in Aerobic Wastewater Treatment Systems. *Water Res.* **1995**, *29*, 871–880. [CrossRef]
26. Litti, Y.V.; Nekrasova, V.K.; Kulikov, N.I.; Siman'kova, M.V.; Nozhevnikova, A.N. Detection of Anaerobic Processes and Microorganisms in Immobilized Activated Sludge of a Wastewater Treatment Plant with Intense Aeration. *Microbiology* **2013**, *82*, 690–697. [CrossRef]
27. Zhang, D.; Li, H.; Li, X.; Ao, D.; Wang, N. Source of Methanogens and Characteristics of Methane Emission in Two Wastewater Treatment Plants in Xi'an. *Water* **2024**, *16*, 2101. [CrossRef]
28. Wang, J.; Zhang, J.; Xie, H.; Qi, P.; Ren, Y.; Hu, Z. Methane Emissions from a Full-Scale A/A/O Wastewater Treatment Plant. *Bioresour. Technol.* **2011**, *102*, 5479–5485. [CrossRef]
29. Belmonte, M.; Hsieh, C.F.; Figueroa, C.; Campos, J.L.; Vidal, G. Effect of Free Ammonia Nitrogen on the Methanogenic Activity of Swine Wastewater. *Electron. J. Biotechnol.* **2011**, *14*. [CrossRef]
30. Pu, Y.; Tang, J.; Zeng, T.; Hu, Y.; Yang, J.; Wang, X.; Huang, J.; Abomohra, A. Pollutant Removal and Energy Recovery from Swine Wastewater Using Anaerobic Membrane Bioreactor: A Comparative Study with Up-Flow Anaerobic Sludge Blanket. *Water* **2022**, *14*, 2438. [CrossRef]
31. Riaño, B.; García-González, M.C. Greenhouse Gas Emissions of an On-Farm Swine Manure Treatment Plant—Comparison with Conventional Storage in Anaerobic Tanks. *J. Clean. Prod.* **2015**, *103*, 542–548. [CrossRef]
32. Béline, F.; Martinez, J. Nitrogen Transformations during Biological Aerobic Treatment of Pig Slurry: Effect of Intermittent Aeration on Nitrous Oxide Emissions. *Bioresour. Technol.* **2002**, *83*, 225–228. [CrossRef]

https://doi.org/10.3390/su18010321

33. Kishida, N.; Kim, J.H.; Kimochi, Y.; Nishimura, O.; Sasaki, H.; Sudo, R. Effect of C/N Ratio on Nitrous Oxide Emission from Swine Wastewater Treatment Process. *Water Sci. Technol.* **2004**, *49*, 359–365. [CrossRef]
34. Law, Y.; Ye, L.; Pan, Y.; Yuan, Z. Nitrous Oxide Emissions from Wastewater Treatment Processes. *Philos. Trans. R. Soc. B Biol. Sci.* **2012**, *367*, 1265–1277. [CrossRef] [PubMed]
35. Xie, Y.; Jiang, C.; Kuai, B.; Xu, S.; Zhuang, X. N2O Emission Reduction in the Biological Nitrogen Removal Process for Wastewater with Low C/N Ratios: Mechanisms and Strategies. *Front. Bioeng. Biotechnol.* **2023**, *11*, 1247711. [CrossRef] [PubMed]
36. Hilgert, J.E.; Amon, B.; Amon, T.; Belik, V.; Dragoni, F.; Ammon, C.; Cárdenas, A.; Petersen, S.O.; Herrmann, C. Methane Emissions from Livestock Slurry: Effects of Storage Temperature and Changes in Chemical Composition. *Sustainability* **2022**, *14*, 9934. [CrossRef]
37. Sommer, S.G.; Petersen, S.O.; Møller, H.B. Algorithms for Calculating Methane and Nitrous Oxide Emissions from Manure Management. *Nutr. Cycl. Agroecosyst.* **2004**, *69*, 143–154. [CrossRef]
38. Park, K.H.; Thompson, A.G.; Marinier, M.; Clark, K.; Wagner-Riddle, C. Greenhouse Gas Emissions from Stored Liquid Swine Manure in a Cold Climate. *Atmos. Environ.* **2006**, *40*, 618–627. [CrossRef]
39. Massara, T.M.; Malamis, S.; Guisasola, A.; Baeza, J.A.; Noutsopoulos, C.; Katsou, E. A Review on Nitrous Oxide (N2O) Emissions during Biological Nutrient Removal from Municipal Wastewater and Sludge Reject Water. *Sci. Total Environ.* **2017**, *596–597*, 106–123. [CrossRef] [PubMed]

https://doi.org/10.3390/su18010321

sustainability

MDPI

Article

Leaf–Air Temperature Difference as a Nondestructive Indicator of Waterlogging Tolerance in Cassava Genotypes

Lado Aquilino [1,2], Ten Naito [3], Alex Tamu [1,4], Peter Ssenyonga [1,5], Rael Chepkoech [1], Ibrahim Soe [1,4] and Jun-Ichi Sakagami [1,3,*]

1 The United Graduate School of Agricultural Sciences, Kagoshima University, Kagoshima City 890-0065, Japan; aquilinowani@gmail.com (L.A.); tamualex1@gmail.com (A.T.); sspeter2@gmail.com (P.S.); chepkoechrael06@gmail.com (R.C.); soeibrahim@gmail.com (I.S.)

2 Directorate of Research and Training, Ministry of Agriculture and Food Security, Ministries Complex, Parliament Road, Juba City P.O. Box 33, South Sudan

3 Faculty of Agriculture, Kagoshima University, Kagoshima City 890-0065, Japan; 7110ten@gmail.com

4 Sierra Leone Agricultural Research Institute, Freetown PMB 1313, Sierra Leone

5 Rwebitaba Zonal Agricultural Research and Development Institute, National Agricultural Research Organization, Entebbe, Kampala P.O. Box 295, Uganda

* Correspondence: sakagami@agri.kagoshima-u.ac.jp; Tel.: +81-099(285)8543

Abstract

Cassava plants' response to waterlogging must be monitored in an accurate and timely manner to mitigate the adverse effects of waterlogging stress. Under waterlogging conditions, root hypoxia reduces water uptake and stomatal closure limits transpiration, which often results in increased leaf temperature due to reduced evaporative cooling. However, how this relationship changes in cassava leaves under waterlogged conditions remains poorly understood. This study hypothesized that more negative ΔT values reflect enhanced transpirational cooling, which is a key determinant of superior physiological performance under waterlogging stress among cassava genotypes. Two cassava cultivars were subjected to twelve days of waterlogging. Results revealed a significant decrease in photosynthetic rate ($p < 0.001$), stomatal conductance ($p < 0.001$), and transpiration rate ($p < 0.001$), as well as an increase in leaf temperature ($p < 0.001$) and ΔT ($p < 0.001$), reflecting impaired stomatal regulation and reduced evaporative cooling. Strong negative correlations between ΔT and photosynthetic parameters (Pn ($p < 0.001$, r = −0.91), gs ($p < 0.001$, r = −0.91), and E ($p < 0.001$, r = −0.87)) were observed, presenting ΔT as a reliable, nondestructive indicator of cassava's physiological responses under hypoxic conditions. Findings indicate that maintaining cooler canopies may contribute to waterlogging-tolerant cassava genotypes, and that ΔT can act as a screening parameter for waterlogging-tolerant genotypes. However, further studies with contrasting genotypes and additional parameters are recommended for validation.

Keywords: leaf temperature; waterlogging stress; cassava; stomatal conductance; photosynthesis; genotypic tolerance

1. Introduction

Understanding cassava's physiological responses to waterlogging is essential for sustaining and improving food production in regions facing rapid population growth [1] and increasing climatic stress. In sub-Saharan Africa, the population is projected to increase by >120%. Within this region, cassava (*Manihot esculenta*) is the second-most important source of calories, providing ~30% of the daily dietary energy per person. Because of its starchy roots, cassava is widely used for starch extraction, animal feed, and feedstock production in China and other Southeast Asian countries [2]. Cassava is typically grown as an upland crop with a long maturation period of 8–12 months, which often overlaps with the rainy season. However, in West and Central Africa, it is preferred for inland valleys during the dry season, where it benefits from residual soil moisture [3]. Inland valleys, "as described by Andriesse et al. (1994), comprise the upper reaches of river systems, including valley bottoms, their hydromorphic fringes, and adjacent upland slopes [4]". Because these valleys offer better water availability throughout the growing season compared to the surrounding uplands, they hold strong potential for intensified and sustainable land use [5].

Climate change is exerting a serious toll on human life, largely due to human activities [6]. To date, global mean surface temperature (GMST) has risen by ~1 °C and, if current warming trends persist, is projected to increase by 1.5 °C by 2050 and by 2 °C by the end of the 21st century [7]. These changes are expected to intensify the frequency of extreme climatic events (floods, drought, hailstorms, salinity, and heat stress) [6] and that will pose serious constraints to food production.

Among these stresses, waterlogging stress poses a particular threat. Once soils become saturated with water, root and microbial activity rapidly deplete oxygen in the rhizosphere [8]. Under anaerobic conditions, plants undergo various survival responses, including changes in physiology, anatomy, and morphology to adapt to the anoxic environment [9,10]. Waterlogging alters plant–water relations and carbon fixation at the physiological level. Leaf water content, which depends on root uptake capacity and leaf transpiration, is affected by such stress. "The work of Setter, T. et al. (2009) [11] indicated that waterlogging leads to a reduction in plant water content mainly due to the inhibition of root water uptake resulting from oxygen deprivation". A reduction in leaf water content frequently influences physiological and metabolic processes within the leaf [12]. Leaf water deficits primarily influence the stomatal aperture, leading to reduced transpiration and gas exchange. Consequently, reduction in photosynthesis and transpiration rates associated with decreased leaf water content have been documented across several crop species, such as sorghum [13], maize [14], olives [15], and hibiscus [16].

Our previous study showed an 82.6% reduction in net photosynthetic rate (*Pn*) under waterlogging. This decline was attributed to a 96.7% decrease in stomatal conductance (*gs*) and a 21% reduction in transpiration rate (*E*). In the same study on three-month-old cassava plants, *Pn*, *gs*, and *E* deceased progressively with increasing waterlogging duration. In contrast, the *SPAD* value showed no significant change compared with the control across all observation periods, while *Fv*/*Fm* declined significantly three days after treatment but later recovered. This is likely due to increased energy dissipation rather than chlorophyll degradation, as *SPAD* values remained stable [17]. Overall, cassava exhibited a functional stay-green phenotype under waterlogging, which was characterized by stable *SPAD* values, maintained photosynthetic nitrogen use efficiency, and coordinated stomatal and non-stomatal regulation of photosynthesis [17].

Therefore, precise and timely monitoring of plant responses to waterlogging is essential to alleviate its detrimental impacts, particularly on cassava production [13]. "Blum A. J. et al. (1992) assessed canopy temperature in wheat and identified a strong correlation between canopy temperature and leaf water content [18]". Leaf temperature is strongly associated with plant water status, as it is regulated by water loss through transpiration [19]. Under water deficit conditions, alterations in leaf water content and transpiration are often indirectly manifested through variations in leaf temperature [20]. Although a correlation exits between leaf water content and temperature, leaf temperature is additionally influenced by both transpiration rate and ambient air temperature [21]. Hence, the temperature

https://doi.org/10.3390/su18010405

differential between the leaf temperature and the surrounding air (ΔT) may serve as a more precise indicator of leaf water status than leaf temperature alone [22]. Elevated transpiration rates coupled with reduced stomatal conductance contribute to lower leaf temperatures [19], thereby resulting in a greater ΔT [23]. Thus, ΔT has been extensively employed as a tool for monitoring leaf gas exchange and water status. Wang, Zhang [13] further demonstrated that waterlogging reduces leaf relative water content (RWC), free water content, and leaf gas exchange parameters, while also inducing anatomical modifications, all of which are associated changes in leaf temperatures and ΔT. These results indicate that ΔT may serve as a reliable indicator of plant water status and photosynthetic adjustments under waterlogging. "According to Pineda M.M. et al. (2020) [24], the increases in leaf temperatures may be due to stomatal closure". Stomatal closure is a non-specific plant defense response to both abiotic and biotic stresses. Elevated temperature may also arise from reduced evaporative cooling or vegetation loss [25]. In such cases, temperature increases are closely linked to limited transpiration caused by stomatal closure. Consequently, abiotic stressors such as drought, soil salinity, or extreme ambient temperatures are known to increase canopy temperature [24]. This is particularly important under natural conditions, where these stressors are difficult to quantify or control.

Moreover, leaf temperature measurement has been suggested as a cost-effective, indirect criterion for selecting plant genotypes with resistance to drought [19] and heat stress. However, its application to cassava, particularly under waterlogging conditions, remains unknown. Leaf temperature is indirectly associated with stomatal conductance and carbon exchange [26], and photosynthetic activity is influenced by elevated leaf temperature, either alone or in conjunction with drought, as a result of reduced stomatal conductance [27]. Furthermore, under suboptimal soil water conditions, higher ΔT and crop yield have been linked to increased stomatal conductance and enhanced water-use efficiency [28].

In extensively studied major crops and cereals, the relationship between leaf temperature changes and photosynthetic performance has been studied. However, this relationship changes remain poorly understood in cassava, particularly regarding anaerobic responses under waterlogging conditions. The objectives of this study were to (1) evaluate changes in leaf gas exchange, photosynthetic performance, and growth parameters in cassava during the early growth stage under waterlogging; (2) determine the relationship between leaf temperature, ΔT, and gas exchange parameters in cassava; and (3) elucidate the effects of waterlogging stress on leaf temperature and test the hypothesis that more negative ΔT values reflect enhanced transpirational cooling, which is a key determinant of superior physiological performance under waterlogging stress among cassava genotypes. Plants exhibiting cooler canopies are able to more effectively regulate stomatal conductance, resulting in lower leaf temperatures to ambient conditions [29]. Timely and precise monitoring of cassava leaf temperature, ΔT, and leaf gas exchange parameters can facilitate the development of strategies to mitigate waterlogging stress. This study aims to establish a novel approach for identifying waterlogging-tolerant genotypes to support future breeding programs. Our findings align with those of similar studies found online, with different crops (Sorghum) [13] and sources of stress (drought) [19] acting to reinforce ΔT utility as a nondestructive indicator of stress (waterlogging) response. The proposal of this index (ΔT) will enable a wide range of future research.

2. Materials and Methods

2.1. Study Site and Plant Materials

The study was performed in a greenhouse at Kagoshima University, Faculty of Agriculture, Japan. The average daily temperature and humidity values during the experimental period were 28.04 °C and 81.56%, respectively. Two cassava cultivars (Tokunoshima-white

https://doi.org/10.3390/su18010405

and Tokunoshima-yellow) were used; the two cassava cultivars were sourced from cassava grown at Kizuna farm in Tokunoshima, Kagoshima prefecture [30].

2.2. *Experimental Design*

Cassava stem cuttings (20 cm in length) were obtained from freshly harvested plants and immersed in a Hyponica nutrient solution (Kyowa Co., Ltd., Osaka, Japan) for 24 h to enhance shoot and root development. The treated cuttings were subsequently planted in small polythene bag containing sawdust as the propagation medium. After a 60-day growth period, six seedlings each per polyethene bag and per cultivar were selected based on their SPAD values and transplanted into small buckets containing a soil mixture of loam and sand in a 4:1 ratio (v/v). The transplanted seedlings were acclimatized for two weeks before the initiation of treatments.

A total of twelve buckets containing uniform cassava seedlings were selected for the experiment. Six buckets per cultivar were maintained under well-watered (WW) conditions by daily irrigation to field capacity with free drainage (control treatment), whereas the remaining six buckets were subjected to waterlogging (WL) by sealing the drainage holes and maintaining a water level of 3–5 cm above the soil surface. The treatments were applied for a period of 12 days. The experiment was arranged in a completely randomized design (CRD) with three replicates per treatment.

2.3. *Data Collection*

2.3.1. Microclimate Data

The microclimate data, including average daily air temperature and relative humidity, as well as soil temperature and moisture content, were monitored using a sensor (5TE, METER Group Inc., Pullman, WA, USA) equipped with a data logger (RTR-503, T&D Corporation, Matsumoto, Nagano, Japan). The average daily air temperature and relative humidity were used to calculate air vapor pressure deficit (VPD_{air}), using the formula: $VPD = es\ (1 - RH/100)$. VPD_{air} varied between 1.42 and 3.19 kPa (average 2.39 kPa). The average daily light in the greenhouse was measured at 8:00 am, 12:00 noon, and 4:00 pm using a light sensor ("Quantum sensor meter, apogee instrument" Inc., QMSS; Sun-13231, Logan, UT, USA).

2.3.2. Leaf Gas Exchange Parameters

A portable leaf gas exchange system ("LI-6400XT" Inc., Lincoln, NE, USA) was used for measuring Leaf gas exchange parameters, such as *Pn*, *E*, *gs*, and intercellular CO_2 concentration (Ci), and leaf temperatures. Measurements were taken from the central leaflet of the youngest fully expanded leaf on one plant/replicate (n = 1) at 75 days after planting (DAP). Central leaflet of the youngest fully expanded leaf is physiologically an active leaf that minimizes variation due to leaf age. Prior to data collection, leaves were acclimated to a saturating light intensity of 1500 μmol m^{-2} s^{-1} and a CO_2 concentration of 400 μmol mol^{-1} inside the cuvette until steady states for *Pn* and *gs* were obtained. The block temperature was maintained to 28 °C, with a vapor pressure deficit (VPD) corresponding to an airflow rate of 500 "μmol s^{-1}." Values of *gs* and *Ci* were obtained from the data points collected at 400 "μ mol^{-1}," (CO_2). Photosynthetic water-use efficiency (PWUE) was calculated by dividing *Pn* by *E* at this same CO_2 concentration, "while ΔT was calculated according to Jones, H.G. (1999) [31], using the following formula: ΔT = Leaf temperature − Air temperature."

Leaf gas exchange parameters and leaf temperature were measured before treatment (0 DAT), then measured periodically at four-day intervals (4, 8, and 12 DAT) on the same plant leaves from three biological replicates per treatment on a clear day between 8:00 and 11:00 AM.

https://doi.org/10.3390/su18010405

2.3.3. Chlorophyll Fluorescence (Fv/Fm) Measurement

AquaPen ("AP 100-P", Photon Systems Instruments, Prumyslova, Drasov, Czech Republic) was used to measure *Fv/Fm* under dark adaptive conditions on fully developed leaves. *Fv/Fm* ratio, which is the maximum yield of PSII photochemistry, was obtained by applying actinic light through the instrument's quantum yield protocol. Chlorophyll fluorescence measurements were carried out concurrently with leaf gas exchange assessments. Three replicates per treatment were used for *Fv/Fm* data collection. *Fv/Fm* were collected from three biological replicates per treatment at 0, 4, 8, and 12 DAT on the same plant leaves.

2.3.4. Soil–Plant Analysis Development (SPAD)

A *SPAD* meter ("SPAD502"; Konica Minolta, Tokyo, Japan) was utilized to record the *SPAD* values from three plants per treatment on the same day as the leaf gas exchange assessment. *SPAD* values were recorded at three consecutive positions along the leaf of the youngest fully expanded leaf of each plant. The same leaves used for chlorophyll fluorescence measurements were used for *SPAD* value measurement. *SPAD* values were measured at 0, 4, 8, and 12 days after treatment (DAT) on the same plant leaves from three biological replicates per treatment.

2.3.5. Morphological Data

Morphological data, including plant height, number of leaves per plant, and number of branches per plant, were counted. Plant height was measured from the base of the stem to the highest point of a shoot using a tape measure. The number of leaves per plant and branches was counted manually per plant for each treatment. These morphological data were collected at 0, 4, 8, and 12 DAT on the same plant from three biological replicates per treatment.

2.4. Statistical Analyses

All the data were subjected to Q-Q plot, Kolmogorov–Smirnvo, and Shapiro–Wilk tests of normality using IBM SPSS Statistics (Version 30.0.0.0 [172]). The results showed normal distribution with few outliers in some data, and as a result, an analysis of variance (ANOVA) was chosen. An ANOVA was conducted with IBM SPSS Statistics (Version 30.0.0.0 [172]) using a univariate analysis of variance to determine single and interaction effects of the treatment day interval (treatment duration) (D0–12) and treatment (WW and WL). The mean results of replicates per treatment were compared using "Turkey's HSD test" at $p < 0.05$. Pearson's correlation matrix analysis was conducted to test correlations between parameters using the multi-environment trial analysis package ("Metan") V1,19.0 in R version 4.4.0. All results are presented as means of three replicates ± standard deviation.

3. Results

3.1. Growth Environments, Average Daily Temperature, Daily Percentage of Relative Humidity, Soil Temperature, Daily Soil Moisture Content, and Average Daily Light Intensity

The temperature and relative humidity ranged from 20 °C to 31.5 °C and 56.0–97.0%, with an average of 28.0 °C and 81.6%, respectively, during the experimental period (Figure 1A). The average soil temperature for Tokunoshima-yellow cultivar buckets under WW and WL conditions was 20.7–30.0 °C and 20.5–29.8 °C, with averages of 26.8 °C and 26.8 °C, respectively. Meanwhile, the average soil temperature for Tokunoshima-white cultivar buckets under WW and WL conditions was 20.8–30.4 °C and 21–30.6 °C, with averages of 27.1 °C and 27.4 °C, respectively. The percentage difference between the WW

https://doi.org/10.3390/su18010405

and WL treatment buckets was 1.3% for the Tokunoshima-white cultivar and −0.01% for the Tokunoshima-yellow cultivar (Figure 1B).

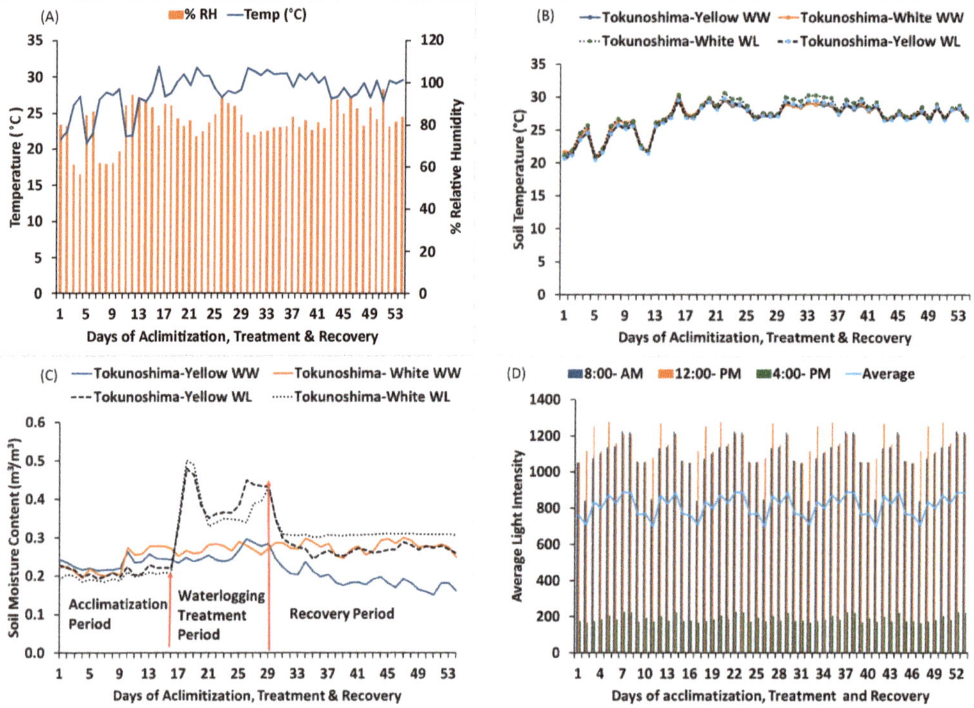

Figure 1. Average temperature and average relative humidity: (**A**) soil temperature, (**B**) soil moisture content, (**C**) and average light intensity (**D**) during experimental duration.

The average soil moisture content for the Tokunoshima-yellow cultivar under WW and WL conditions was 0.15–0.30 and 0.19–0.30 m^3/m^3, with averages of 0.22 and 0.26 m^3/m^3, respectively. The average soil moisture content for the Tokunoshima-white cultivar under WW and WL conditions was 0.15–0.30 and 0.19–0.30 m^3/m^3, with averages of 0.22 and 0.29 m^3/m^3, respectively. The percentage differences between the WW and WL buckets were 24.14% for the Tokunoshima-yellow cultivar and 10.34% for the Tokunoshima-white cultivar (Figure 1C). The average light intensity ranged from 840.5 to 1227.5 at 8:00 am, 1053.5 to 1274.5 at 12:00 noon, and 700.2 to 889.8 as the daily average (Figure 1D).

3.2. *Effect of Waterlogging Treatment on Leaf Temperature and ΔT*

The effects of waterlogging on leaf temperature and ΔT are presented in Figure 2A,B. The waterlogging treatment and waterlogging duration significantly affected leaf temperature and ΔT ($p < 0.001$); the interaction of treatment and waterlogging duration also significantly affected leaf temperature and ΔT ($p < 0.01$) (Table 1). Meanwhile, no significant difference was observed between the two cassava cultivars (Table 1). The leaf temperature of the two cassava cultivars increased immediately under waterlogging treatment at 4, 8, and 12 DAT compared to the WW condition (Figure 2A). The same trend was observed for ΔT in the two cassava cultivars (Figure 2B). The average leaf temperature increased by 27.2%, 14.7%, and 24.0% in the Tokunoshima-white cultivar and by 27.5%, 10.2%, and 15.3% in the Tokunoshima-yellow cultivar under WL conditions at 4, 8, and

https://doi.org/10.3390/su18010405

12 DAT, respectively (Figure 2A). The average ΔT value increased by 3.2%, 1.7%, and 2.1% in the Tokunoshima-yellow cultivar and by 4.0%, 2.2%, and 3.6% in the Tokunoshima-white cultivar at 4, 8, and 12 DAT compared to WW conditions (Figure 2B). Additionally, the effects of leaf temperature and ΔT on cassava cultivars increased under WL conditions with the duration of treatment (Figure 2A,B).

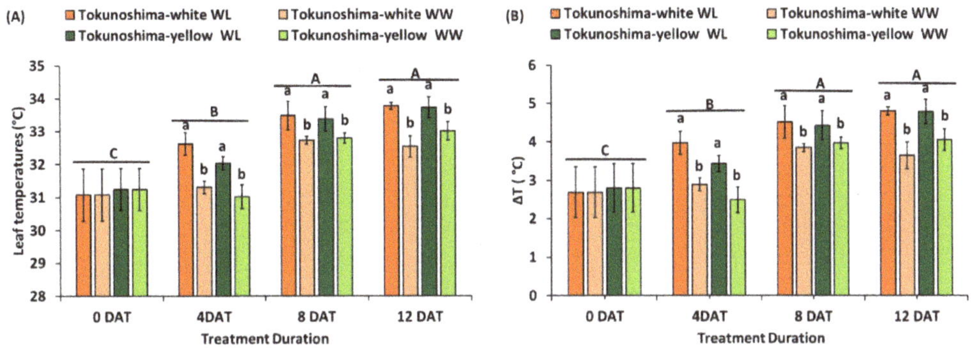

Figure 2. Effects of waterlogging durations on the leaf temperature (**A**) and difference in leaf and air temperatures (**B**) of two cassava cultivars. Uppercase letters above bars indicate significant differences between waterlogging durations, while those in lower case indicate differences among the waterlogging treatment. Bars are mean ± SD.

Table 1. ANOVA results of the effect of waterlogging duration on morphological traits (plant height, number of branches, and number of leaves), photosynthetic traits (SPAD and Fv/Fm), and leaf gas exchange parameters (Pn, gs, E, Ci, and PWUE) of cassava cultivars at the early growth stage.

Parameters	Cultivars (C)	Treatment (Trt)	Treatment Duration (TrtD)	C × Trt	TrtD × Trt
DF	**1**	**1**	**3**	**1**	**3**
Plant height	0.66 ^{ns}	47.68 ***	1.76 ^{ns}	1.84 ^{ns}	6.04 **
Number of leaves	4.66 *	28.21 ***	2.50 ^{ns}	0.52 ^{ns}	5.27 **
Number of branches	0.5 ^{ns}	3.68 ^{ns}	3.68 *	0.41 ^{ns}	0.41 ^{ns}
SPAD Value	42.74 ***	4.86 *	4.55 **	0.51 ^{ns}	0.65 ^{ns}
Fv/Fm Values	0.91 ^{ns}	20.26 ***	3.66 *	0.07 ^{ns}	2.86 *
Photosynthetic rate	0.92 ^{ns}	139.08 ***	26.27 ***	0.16 ^{ns}	16.41 ***
Stomatal conductance	0.59 ^{ns}	38.82 ***	9.24 ***	0.30 ^{ns}	5.212 ***
Transpiration rate	0.73 ^{ns}	83.89 ***	15.41 ***	0.01 ^{ns}	9.61 **
Intercellular CO_2 concentration	0.54 ^{ns}	2.86 ^{ns}	3.57 *	0.44 ^{ns}	6.42 **
PWUE	2.56 ^{ns}	1.60 ^{ns}	5.33 **	1.64 ^{ns}	7.22 ***
Leaf temperature	0.02 ^{ns}	30.98 ***	66.13 ***	0.94 ^{ns}	4.12 **
ΔT	0.10 ^{ns}	29.09 ***	43.34 ***	0.68 ^{ns}	3.97 **

Note * $p < 0.05$, ** $p < 0.01$, *** $p < 0.001$, ^{ns}, not significant and DF = degree of freedom.

3.3. Effect of Waterlogging on Photosynthetic Traits (Pn, E, gs, and PWUE)

The immediate effects of waterlogging on the photosynthetic traits, *Pn*, *E*, *gs*, *Ci*, and PWUE, of the two cassava cultivars at the early growth stage are presented in Figure 3A–E. The *Pn* values of the two waterlogged cultivars displayed significant differences between treatment durations ($p < 0.001$), treatment ($p < 0.001$), and the interaction between treatment duration (TrtD), and waterlogging ($p < 0.001$) (Table 1). However, the two cultivars did not differ between treatments (Table 1, Figure 3A). The *Pn* values of

https://doi.org/10.3390/su18010405

waterlogged Tokunoshima-white and Tokunoshima-yellow cultivars decreased by 75.1% and 60%, respectively, compared to WW cultivars at 4 DAT. The Tokunoshima-yellow cultivar exhibited the largest reduction in *Pn* values by 99.9% at 12 DAT compared with the Tokunoshima-white cultivar (99.2%) (Figure 3A).

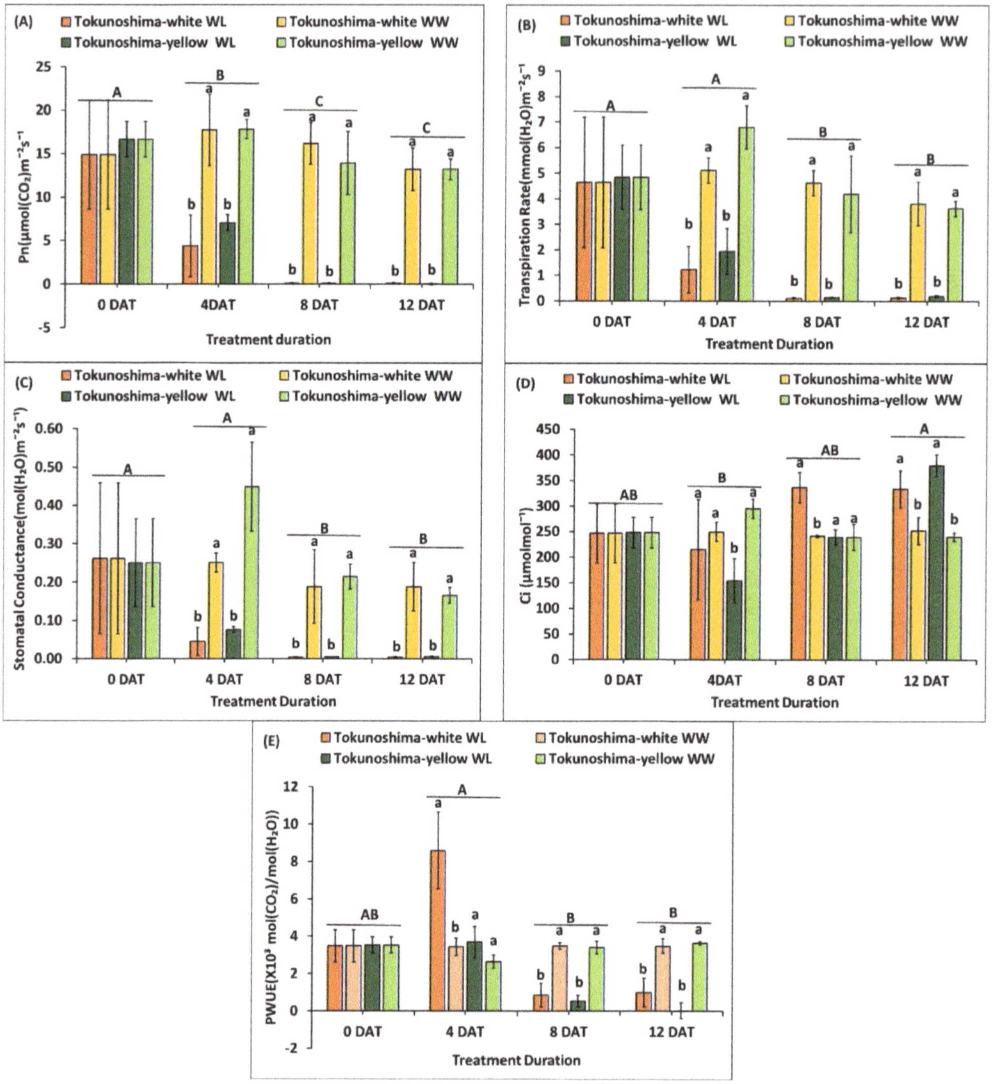

Figure 3. Effects of various waterlogging durations on photosynthetic rate (**A**), transpiration rate (**B**), stomatal conductance (**C**), intercellular CO_2 concentration (Ci) (**D**) and PWUE (**E**) at early growth stage of cassava cultivars. Uppercase letters above bars indicate significant differences between waterlogging durations, while those in lower case indicate differences among the waterlogging treatments. Bars are mean ± SD.

Similarly, the *E* values of the waterlogged cultivars displayed significant differences between TrtD ($p < 0.001$), waterlogging treatment ($p < 0.001$), and the interaction between duration and waterlogging treatment ($p < 0.001$) (Table 1). However, the cultivars

https://doi.org/10.3390/su18010405

did not differ among treatments (Table 1, Figure 3B). The *E* values of the waterlogged Tokunoshima-white and Tokunoshima-yellow cultivars decreased by 76% and 71%, respectively, compared to WW cultivars at 4 DAT. The Tokunoshima-white cultivar exhibited the largest reduction in *E* values by 96.8% at 12 DAT compared with the Tokunoshima-yellow cultivar (94.7%) (Figure 3B).

Additionally, the *gs* values of the waterlogged cultivars exhibited significant differences between TrtD ($p < 0.001$), waterlogging treatment ($p < 0.001$), and the interaction between duration and waterlogging treatment ($p < 0.001$) (Table 1). However, the cultivars did not differ among the treatments (Table 1, Figure 3C). The *gs* values of the waterlogged Tokunoshima-white and Tokunoshima-yellow cultivars decreased by 82% and 82.8%, respectively, compared to WW cultivars at 4 DAT. The Tokunoshima-white cultivar exhibited the largest reduction in *gs* values by 97.8% at 12 DAT compared with the Tokunoshima-yellow cultivar (96.2%) (Figure 3C).

The *Ci* values of the waterlogged cultivars displayed a significant difference between TrtD ($p < 0.05$) and the interaction between TrtD and waterlogging treatment ($p < 0.01$) (Table 1). However, the waterlogging treatment did not have a significant impact on *Ci*, and the cultivars did not display differences among treatments (Table 1, Figure 3D). The *Ci* values of the waterlogged Tokunoshima-white and Tokunoshima-yellow cultivars decreased by 13.8% and 47.9%, respectively, compared to WW cultivars at 4 DAT, which corresponded to their largest reduction in *Ci* values (Figure 3D).

The PWUE values of the waterlogged cultivars displayed a significant difference between TrtD ($p < 0.01$) and the interaction between TrtD and waterlogging treatment ($p < 0.001$) (Table 1). However, the waterlogging treatment did not display a significant difference, and the cultivars did not differ among treatments (Table 1, Figure 3E). The *PWUE* values of waterlogged Tokunoshima-white and Tokunoshima-yellow cultivars increased by 60.1% and 28.1%, respectively, compared to WW cultivars at 4 DAT. The Tokunoshima-yellow cultivar exhibited the largest reduction in PWUE values by 99.1% at 12 DAT compared with the Tokunoshima-white cultivar (71.8%) (Figure 3E).

These results revealed that the effects of WL treatment on the *Pn*, *E*, *gs*, *Ci*, and *PWUE* values of two cassava cultivars at the early growth stage increased with the waterlogging duration (Figure 3A–E).

3.4. Effect of WL Treatment on Photosynthetic Traits (SPAD and Fv/Fm)

The immediate effects of WL treatment on *Fv*/*Fm* and *SPAD* values of cassava cultivars at the early growth stage are presented in Figure 4A,B. The Fv/Fm values of the two cultivars were significantly affected by WL treatment ($p < 0.001$) and TrtD ($p < 0.05$) (Table 1). Additionally, the interaction between TrtD and waterlogging treatment significantly affected *Fv*/*Fm* ($p < 0.05$) (Table 1). The cultivars did not differ between treatments (Table 1, Figure 4A). The Fv/Fm values of the WL Tokunoshima-white and Tokunoshima-yellow cultivars decreased by 0% and 4.6%, respectively, compared to those of WW cultivars at 4 DAT. The Tokunoshima-white cultivar exhibited the largest reduction in *Fv*/*Fm* values by 7.6% at 12 DAT compared with the Tokunoshima-yellow cultivar (3.4%) (Figure 4A).

The *SPAD* values of the two WL cultivars displayed significant differences between TrtD ($p < 0.01$), waterlogging treatment ($p < 0.05$), and cultivars ($p < 0.001$) (Table 1, Figure 4B). However, the interaction between TrtD and WL treatment did not statistically differ from the WW treatment (Table 1, Figure 4B). The *SPAD* values of WL Tokunoshima-white and Tokunoshima-yellow cultivars decreased by 2.8% and 10%, respectively, compared to WW cultivars at 4 DAT. The Tokunoshima-yellow cultivar exhibited the largest reduction in *SPAD* values by 14.4% at 8 DAT compared with the Tokunoshima-

https://doi.org/10.3390/su18010405

white cultivar (10.3%). The effects of WL treatment on *Fv/Fm* and *SPAD* values of the two cassava cultivars at the early growth stage increases with the TrtD (Figure 4A,B).

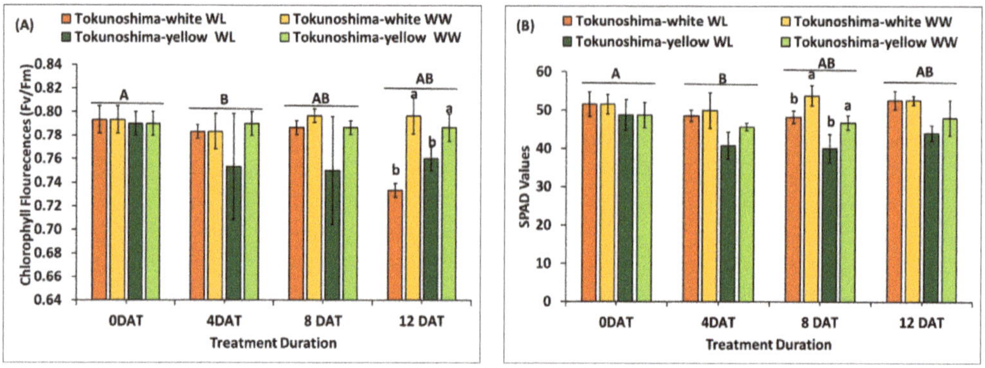

Figure 4. Effects of various waterlogging durations on Fv/Fm (**A**), SPAD values (**B**), at early growth stage of cassava cultivars. Uppercase letters above bars indicate significant differences between waterlogging durations, while those in lower case indicate differences among the waterlogging treatments. Bars are mean ± SD.

3.5. Effect of WL Treatment on Morphological Traits (Plant Height, Number of Branches, and Number of Leaves)

The immediate effects of various WL treatments on plant height, number of branches, and number of leaves of cassava cultivars at the early growth stage are presented in Figure 5A–C. The plant height of the two cultivars was significantly affected by WL treatment at 4 DAT ($p < 0.001$), 8 DAT ($p < 0.03$), and 12 DAT ($p < 0.01$), respectively (Table 1). However, the TrtD did not significantly affect plant height (Table 1). In contrast, the interaction of TrtD and waterlogging treatment had a significant effect ($p < 0.01$) (Table 1). The plant heights of WL Tokunoshima-white and Tokunoshima-yellow cultivars decreased by 40.5% and 25.5% at 4 DAT, respectively, compared to WW cultivars. The Tokunoshima-white cultivar exhibited the largest reduction in plant height of 51.1% compared to 40.5% for the Tokunoshima-yellow cultivar at 12 DAT compared to WW conditions (Figure 5A).

The number of branches per plant was not significantly affected by WL treatment at all treatment durations or by the interaction between treatment duration and waterlogging treatment (Table 1, Figure 5B). The cultivars did not display any difference between treatments (Table 1, Figure 5B). However, the treatment duration did not have a significant effect ($p < 0.05$) compared with WW cultivars (Table 1, Figure 5B). The number of branches per plant of WL Tokunoshima-white and Tokunoshima-yellow cultivars decreased by 14.3% and 25% at 4 DAT, respectively, compared to WW cultivars. Tokunoshima-yellow cultivar exhibited the largest reduction in the number of branches by 25% at 4 DAT compared with Tokunoshima-white cultivar with 14.3% compared with WW-treated cultivars (Figure 5B).

Finally, the number of leaves per plant of both cultivars was significantly ($p < 0.001$) affected by WL treatment at 8 and 12 DAT but was not significant at 4 DAT. However, the effect of treatment duration on the number of leaves per plant was not statistically significant (Table 1). The effect of the interaction of treatment duration and condition had a significant effect ($p < 0.01$), and cultivars were significantly different ($p < 0.05$) (Table 1). The number of leaves per plant of WL Tokunoshima-white and Tokunoshima-yellow cultivars decreased by 26.4% and 6%, respectively, compared to those of WW cultivars. The Tokunoshima-white cultivar exhibited the largest reduction in the number of leaves

https://doi.org/10.3390/su18010405

per plant with 56.3% at 12 DAT compared with 50% for the Tokunoshima-yellow cultivar (Figure 5C).

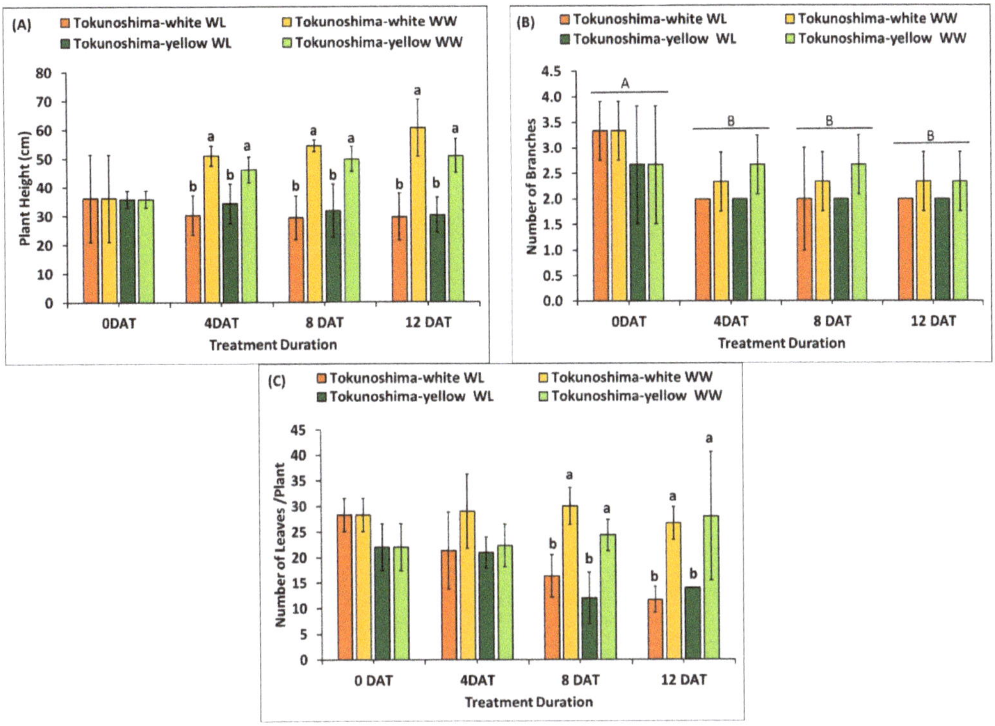

Figure 5. Effects of various waterlogging durations on plant height (**A**), number of branches (**B**), and number of leaves (**C**) at early growth stage of cassava cultivars. Uppercase letters above bars indicate significant differences between waterlogging durations, while lower case indicate differences among the waterlogging treatments. Bars are mean ± SD.

The WL treatment had a nonsignificant effect on the number of branches and leaves at 4 DAT for the two cassava cultivars at the early growth stage; however, it had a significant effect on plant height at 4, 8, and 12 DAT. The damage of WL treatment and treatment duration on plant height and number of leaves increased with the treatment duration (Figure 5A–C).

3.6. Relationships Between Morphological Parameters, Leaf Gas Exchange, and Photosynthetic Parameters

Figure 6 presents the results of Pearson's correlation matrix analysis of morphological parameters, leaf gas exchange, and photosynthetic parameters of the two cassava cultivars under WW (Figure 6b) and WL (Figure 6a) conditions. The results of the WL treatment revealed highly significant positive correlations between *Pn* and *gs* ($p < 0.001$, r = 0.94), *E* ($p < 0.001$, r = 0.98), *PWUE* ($p < 0.001$, r = 0.72). However, *Pn* was highly significantly correlated with leaf temperature ($p < 0.001$, r = −0.93), ΔT ($p < 0.001$, r = −0.91) but only significantly correlated with *Ci* ($p < 0.05$, r = −0.46). Net photosynthesis also showed non significance correlation with *SPAD* (r = 0.05) and *Fv/Fm* (r = 0.38) (Figure 6a).

https://doi.org/10.3390/su18010405

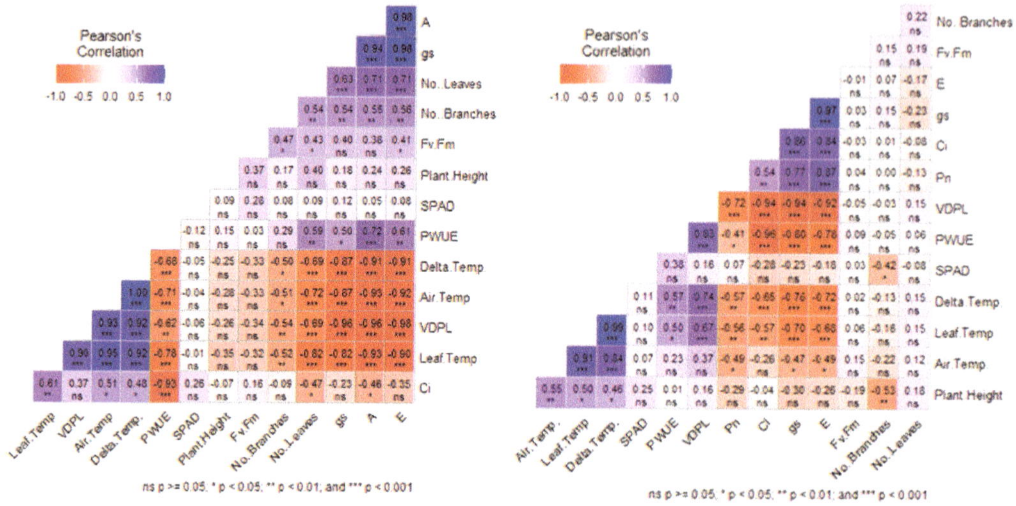

(**a**) WL (**b**) WW

Figure 6. Pearson's correlation matrix of the changes in morphological parameters, leaf gas exchange, and photosynthetic-related parameters of the two cassava cultivars under well-watered (**b**) and waterlogging (**a**) treatments. The number of biological replicates used was 3 plants/treatment.

In contrast, under WW treatment, the results revealed significant moderate positive correlation between *Pn* and *gs* ($p < 0.001$, r = 0.77), E ($p < 0.001$, r = 0.87), but significant weak negative correlated with *PWUE* ($p < 0.05$, r = −0.41). In addition, *Pn* showed significant weak negative correlation with leaf temperature ($p < 0.01$, r = −0.56) and ΔT ($p < 0.01$, r = −0.57), but higher significant weak positive correlation with *Ci* ($p < 0.01$, r = 0.54). However, *Pn* was not significantly correlated with *SPAD* (r = 0.07) and *Fv*/*Fm* (r = 0.04) (Figure 6b).

In addition, stomatal conductance was highly significantly negatively correlated with leaf temperature ($p < 0.001$, r = −0.82) and ΔT ($p < 0.001$, r = −0.87) as well as transpiration rate with leaf temperature ($p < 0.001$, r = −0.90 and ΔT ($p < 0.001$, r = −0.91) under waterlogging treatment (Figure 6a). Contrary to well-watered treatment, stomatal conductance was significantly moderately negatively correlated with leaf temperature ($p < 0.001$, r = −0.70) and ΔT ($p < 0.001$, r = −0.76) as well as transpiration rate with leaf temperature ($p < 0.001$,r = −0.68), and ΔT ($p < 0.001$, r = −0.72) (Figure 6b).

4. Discussion

When root growth and function are impaired by waterlogging, shoot growth and function are also restricted. In this study, waterlogging effects on morphological parameters, including plant height, number of branches, and number of leaves, were evaluated. At 4 DAT, reductions in the number of branches and leaves were not statistically significant, whereas plant height was significantly affected. With longer waterlogging durations (8 and 12 DAT), significant reductions were observed in plant height and number of leaves (Figure 5A–C). The severity of damage to cassava cultivars during early growth increased with waterlogging duration, which is consistent with findings by [32], who reported that prolonged waterlogging intensified stress effects in maize. These results indicate that 12 days of waterlogging significantly reduced cassava plant height but had no effect on the number of branches. The number of leaves began to decline from 8 DAT onward. The reduction in plant height and leaf number after 12 days of waterlogging can be attributed to decreased cell proliferation and expansion, as well as delayed leaf unfolding, which were shown to be inhibited by water stress [33]. "Similarly, Ledent J. (2002) reported that prolonged drought reduced leaf number and size while enhancing the retention of already expanded leaves [34]", and Suárez L. and V. Mederos (2011) noted that leaf production and longevity are varietal traits strongly influenced by environmental conditions [35]. In cassava, leaf loss is considered a strategy to conserve water [36,37]. Furthermore, Aina O.O. et al. (2007) demonstrated that plant height is sensitive to water deficit, with reductions of up to 47% under drought in greenhouse and field conditions [38]. These findings highlight the heightened sensitivity of cassava to excess water, reflected primarily in reductions in plant height and leaf number. Therefore, plant height and leaf number, rather than branch number, may serve as reliable parameters for evaluating waterlogging tolerance in cassava during the vegetative growth stage.

To evaluate the photosynthetic ability of cassava plants under waterlogging, chlorophyll content and Fv/Fm were measured. *SPAD* values, which indicate relative chlorophyll content, were used to assess chlorophyll degradation. Results revealed significant differences in *SPAD* values among treatment durations ($p < 0.01$), waterlogging treatment ($p < 0.05$), and cultivars ($p < 0.001$) (Table 1, Figure 4B). Changes in soil chemical properties under waterlogging may have influenced *SPAD* values, potentially explaining the observed responses. Notably, cassava can maintain higher PNUE under nitrogen-deficient conditions induced by waterlogging. This finding is consistent with [39], who reported that cassava can acclimate to nitrogen starvation, which is an adaptation linked to its evolution in environments with inherently low nitrogen availability.

Fv/Fm was used as a reliable indicator of photo-inhibitory impairment in plants under environmental stresses, including waterlogging. "According to Long S. et al. (1994), the Fv/Fm ratio is a key parameter for detecting damage to photosystem II (PSII) and identifying the occurrence of photoinhibition [40]". In this study, Fv/Fm values exhibited no significant differences between treatments, except at 12 DAT (Figure 4A). The decline in Fv/Fm value at 12 DAT may reflect the sensitivity of the photosynthetic apparatus to waterlogging. Low oxygen concentrations are known to trigger hypoxia-related metabolism, including the induction involved in anaerobic fermentation, sugar utilization, and antioxidant defense [41,42]. The stability of PSII at 0, 4, and 8 DAT suggests that oxidative stress was limited during early waterlogging, which may be attributed to effective ROS scavenging and the absence of chlorophyll degradation, as supported by *SPAD* data (Figure 4B). Nishizawa A.Y. et al. (2008) reported that ROS scavengers reduce oxidative damage under abiotic stress [43] "while Cao et al. (2022) found that 6 days of waterlogging significantly increased the activities of superoxide dismutase, catalase, and peroxidase in cassava leaves and roots [44]". Collectively, these findings suggest that cassava may possess mechanisms that allow metabolism adjustment and PSII protection under short-term anaerobic conditions. However, further studies are needed to elucidate the specific mechanisms underlying this tolerance.

Under waterlogging, root hypoxia reduces water uptake, and stomatal closure limits transpiration, which often results in increased leaf temperature due to reduced evaporative cooling [13]. In this study, leaf temperature and ΔT in two cassava cultivars increased under waterlogging at 4, 8, and 12 DAT compared to well-watered plants (Figure 2A,B). In addition, stomatal conductance was significantly negatively correlated with leaf temperature ($p < 0.001$, r = -0.82) and ΔT ($p < 0.001$, r = -0.87) as well as transpiration rate with leaf temperature ($p < 0.001$, r = -0.90 and ΔT ($p < 0.001$, r = -0.91) under waterlogging treatment (Figure 6a). Increased leaf temperature and ΔT under waterlogging caused physiological stress due to impaired transpiration and stomatal regulation (Figure 3B,C). This result was consistent with Pan, Jiang et al. (2019) who reported that, "leaf temperature

https://doi.org/10.3390/su18010405

was higher in the sensitive genotypes than that in the tolerant genotypes after waterlogging [45]". The increased leaf temperature (Figure 2A) and decreased stomatal conductance and transpiration rate (Figure 3B,C) under waterlogging suggest that waterlogged plants regulate stomatal aperture to conserve relative water content under stress. A similar finding was reported by Utsumi Y. et al. (2019), "who suggested that the stomatal conductance and transpiration rate in leaves of acetic acid-treated plants decreased due to higher leaf temperatures compared to control plants to maintain relative water content and avoid drought [46] " Surendar, K.K et al. (2013) "found increases of 2–3 °C in leaf temperature under water deficit relative to control conditions [47]". Our results are further supported by Siddique, M. et al. (2000), "who demonstrated that water stress substantially decreased leaf water potential, relative water content, and transpiration rate, with a concomitant increase in leaf temperature [48]". According to Nelson and Bugbee, (2002), "ΔT can reduce the influence of air temperature changes on leaf temperatures based on the monitoring of leaf water status [49]."

We also observed that the leaf gas exchange parameters of cassava were affected by waterlogging. A significant decrease in *Pn*, *E*, and *gs* was observed at 4 DAT (Figure 3A–C), indicating that the channel for water and CO_2 exchange was restricted. According to Ahmed, S. et al. (2002), "the earliest response of plant species to waterlogging is stomatal closure, which limits gas exchange and decreases *Pn* and *E* in many plant species [50]", including tomato (*Solanum lycopersicum* L.) [51], cassava [17], sorghum [52], and wheat (*Triticum aestivum* L.) [53]. This data suggests that the reduced *gs* is responsible for the decreased *Pn* and *E* in the evaluated cassava cultivars during early growth under waterlogging. Bansal and Srivastava (2015), reported that 6 days of waterlogging reduced the chlorophyll content of leaves in pigeon pea [54]. A similar observation was reported in sorghum after 12 days of waterlogging [52]. During photosynthesis, chlorophyll serves as a crucial pigment to absorb light energy [55], demonstrating that *SPAD* values are highly correlated with the leaf chlorophyll content. At 4 DAT, a significant decrease in SPAD values was observed in cassava cultivars under waterlogging (Figure 4B). However, no significant correlation was found between *Pn* and *SPAD* values (Figure 6a,b). Thus, reduced photosynthesis during the early growth of waterlogged cassava cultivars could be attributed to increased leaf temperatures and decreased *gs* and *E*. In this study, the effect of waterlogging on *gs* and *E* increased with increased treatment duration, resulting in the same trend as *Pn* (Figure 3A–C). The leaf temperature of WL cassava cultivars was significantly higher than that of WW cultivars at 4, 8, and 12 DAT (Figure 2A). These findings aligned with similar leaf temperature-based study on differential response of growth and photosynthesis in diverse cotton genotypes under hypoxia stress by [45], "who reported that, leaf temperature was higher in the sensitive genotypes than that in the tolerant genotypes after waterlogging", which means that the tolerant plants could maintain a stable metabolic environment to avoid anoxia stress damage in the roots.

Furthermore, Pearson's correlation matrix analysis revealed that leaf temperatures and ΔT were negatively correlated with *Pn* (r = −0.93, $p < 0.001$; r = −0.91, $p < 0.001$), *gs* (r = −0.82, $p < 0.001$; r = −0.87, $p < 0.001$), and *E* (r = −0.90, $p < 0.001$; r = −0.91, $p < 0.001$, respectively, under waterlogging treatment compared to *Pn* (r = −0.56, $p < 0.01$; r = −0.57, $p < 0.01$), *gs* (r = −0.70, $p < 0.001$; r = −0.76, $p < 0.001$), and *E* (r = −0.68, $p < 0.001$; r = −0.72, $p < 0.001$, respectively, under well-watered treatment (Figure 6a,b). This result indicates that increased leaf temperatures and ΔT caused by hypoxia conditions in the plant rhizosphere might have created unfavorable conditions for plants under waterlogged treatment compared to well-watered treatment, which resulted in strong a correlation with the WL treatment, because the abnormal leaf temperature and ΔT will disturb the physiological metabolism and subsequently lead to decreased leaf gas

https://doi.org/10.3390/su18010405

exchange parameters (*Pn*, *gs*, and *E*). These results align with Wang, W. et al. (2019), "who reported significant relationships among ΔT, RWC, and gas exchange parameters (*E* and *Pn*) [13]". Additionally, the results aligned with Djanaguiraman, M. et al. (2011) and Luan et al. (2018), "who reported that increased leaf temperature caused by water deficit creates unfavorable conditions for plants; the abnormal leaf temperature will disturb the physiological metabolism, destroy the chloroplast structure, and decrease *Pn* [56,57]". Surendar, K.K. et al. (2013) also "found a significant negative correlation between leaf temperature and photosynthetic rate [47]". Based on our findings, we deduce that ΔT is closely related to leaf water content and gas exchange parameters, representing a real-time response supported by highly significant correlations. Therefore, ΔT was observed as a reliable index for assessing the leaf gas exchange and performance of cassava leaves under waterlogging. However, more studies with diverse cassava genotypes and environment are needed to confirm these findings.

Results on PWUE at 4 DAT under waterlogging (Figure 3E) indicate a strategy to maintain carbon gain while minimizing water loss. PWUE results at 4 DAT suggest that waterlogged cassava plants retained relative water content by regulating stomatal aperture to reduce water loss while sustaining some photosynthetic activity. This finding aligns with [58], who reported that under drought conditions, stomatal closure reduces transpiration (*E*) and increases *PWUE*, especially when photosynthesis remains more active than transpiration. This boost in *PWUE* (Figure 3E) often reflects a strategy to maintain carbon gain while minimizing water loss. Therefore, the statistically significant decrease in *Pn*, *E*, and *gs* of waterlogged cassava cultivars at 4 DAT and the negative correlation between *Pn*, *gs*, *E*, and PWUE and leaf temperature and ΔT (Figure 6a,b) indicate that as leaves become hotter than the air, these physiological functions are still actively responding, although in a stressed or compensatory manner. The strong negative correlation between *Pn*, *E*, *gs*, and *PWUE* under waterlogging suggests a survival or compensatory mechanism, maintaining carbon gain despite harsh hypoxia conditions. Increased ΔT often correlates with reduced leaf RWC, photosynthetic performance, and transpiration, especially in waterlogging-sensitive cultivars [13]. Thus, ΔT could serve as a useful real-time, nondestructive indicator of plant water and photosynthetic status.

5. Conclusions

Plant responses to waterlogging must be monitored in an accurate and timely manner to mitigate the adverse effects of waterlogging, especially on cassava production. The main objective of this study was to identify the relationship between cassava leaf temperatures and ΔT and leaf gas exchange parameters, as well as to understand the effects of waterlogging on leaf temperature. Key results indicated that waterlogging stress markedly reduced photosynthetic rate, stomatal conductance, and transpiration rate in cassava, while increasing leaf temperature and ΔT led to impaired stomatal regulation and reduced evaporative cooling. Strong negative correlations between ΔT and photosynthetic parameters (*Pn*, *E*, *gs*, and *PWUE*) further suggested that ΔT can serve as a reliable, nondestructive indicator of cassava's physiological responses under hypoxic conditions. Maintaining cooler canopies may therefore contribute to improved tolerance and survival under waterlogged conditions. ΔT shows promise as an indicator for rapid screening and selection of waterlogging-tolerant cassava genotypes. However, validation across a wider range of cassava cultivars and environments, and integration with additional parameters, such as leaf RWC, free water content, and anatomical traits, is recommended to strengthen its application in breeding programs. In conclusion, ΔT offers a simple and effective tool to guide agronomic practices, breeding, and crop improvement strategies for enhancing cassava resilience under waterlogging stress.

https://doi.org/10.3390/su18010405

Author Contributions: L.A.: Conceptualization, methodology, data collection, data curation, formal analysis, writing, editing, proofreading, and experiment management. T.N.: Helped with data collection. A.T. and I.S.: Helped with data collection and proofreading. P.S. and R.C.: Proofreading and editing. J.-I.S.: Conceptualization, methodology, experiment administration, supervision, reviewing, editing, and proofreading. All authors have read and agreed to the published version of the manuscript.

Funding: This study was funded by the United Graduate School of Agricultural Sciences, Kagoshima University, Kagoshima City, Agricultural Studies Networks for Food Security (Agri-Net) scholarship program of the Japan International Cooperation Agency (JICA).

Institutional Review Board Statement: Not applicable.

Informed Consent Statement: Not applicable.

Data Availability Statement: The data that supports the findings of this study are available in Figshare at https://doi.org/10.6084/m9.figshare.30362716 (accessed on 15 October 2025).

Acknowledgments: We would like to thank the United Graduate School of Agricultural Sciences, Kagoshima University, Kagoshima, Japan, for providing funding for this study and the Agricultural Studies Networks for Food Security (Agri-Net) scholarship program of the Japan International Cooperation Agency (JICA) for providing a scholarship to the first author that allowed him to undertake this study, as well as Kizuna farm in Tokunoshima, Kagoshima prefecture, Japan, for provision of the experimental materials (Tokunoshima-white and Tokunoshima-yellow cassava cultivars).

Conflicts of Interest: The authors declare no conflicts of interest.

References

1. Stromquist, N.P. *World Development Report 2019: The Changing Nature of Work: By the World Bank*; World Bank: Washington, DC, USA, 2019; 151p, ISBN 978-1-4648-1342-9. [CrossRef]
2. Nguyen, T.L.T.; Gheewala, S.H.; Garivait, S. Full Chain Energy Analysis of Fuel Ethanol from Cassava in Thailand. *Environ. Sci. Technol.* **2007**, *41*, 4135–4142. [CrossRef] [PubMed]
3. Carsky, R.; Walker, P.; Hauser, S.; Dashiell, K.; Dixon, A. Response of Selected Crop Associations to Groundwater Table Depth in an Inland Valley. *Field Crop. Res.* **1993**, *34*, 1–13. [CrossRef]
4. Andriesse, W.; Fresco, L.O.; Dujvenbooden, N.V.; Windeijer, A.P.N. Multiscale Characterization of Inland Valley Agro-Ecosystems in West Africa. *Neth. J. Agric. Sci.* **1994**, *42*, 159–179.
5. Ekanayake, I.J.; Dixon, A.G.O.; Asiedu, R.; Izac, A.-M.N. Improved-Cassava-for-Inland-Valley-Agro-Ecosystems. In Proceedings of the 5th Symposium of the International Society for Tropical Root Crops-African Branch (ISTR-AB), Kampala, Uganda, 22–28 November 1992.
6. More, S.J.; Ravi, V.; Raju, S. The Quest for high Yielding Drought Tolerant Cassava Variety. *J. Pharmacogn. Phytochem.* **2020**, *SP6*, 433–439.
7. Zhai, P.; Pörtner, H.; Roberts, D.; Skea, J.; Shukla, P.; Pirani, A. Global Warming of 1.5 C. An IPCC Special Report on the impacts of global warming of 1.5 C above pre-Industrial Levels and Related Global Greenhouse Gas Emission Pathways, in the Context of Strengthening the Global Response to the Threat of Climate Change, Sustainable Development, and Efforts to Eradicate Poverty. In *Sustainable Development, and Efforts to Eradicate Poverty*; Intergovernmental Panel on Climate Change: Geneva, Switzerland, 2018; p. 32.
8. Ponnamperuma, F.N. Effects of Flooding on Soils. In *Flooding and Plant Growth*; Kozlowski, T.T., Ed.; Academic Press: New York, NY, USA, 1984; pp. 9–45.
9. Koslowski, T.T.; Pallardy, S.G. Effects of Flooding on Water, Carbohydrate and Mineral Relations. In *Flooding and Plant Growth*; Koslowski, T.T., Ed.; Academic Press Inc.: Orlando, FL, USA, 1984.
10. Voesenek, L.A.C.J.; Blom, C.W.P.M. Blom, Growth Responses of Rumex Species in Relation to Submergence and Ethylene. *Plant Cell Environ.* **1989**, *12*, 433–439. [CrossRef]
11. Setter, T.L.; Waters, I.; Sharma, S.K.; Singh, K.N.; Kulshreshtha, N.; Yaduvanshi, N.P.S.; Ram, P.C.; Singh, B.N.; Rane, J.; McDonald, G.; et al. Review of Wheat Improvement for Waterlogging Tolerance in Australia and India: The Importance of Anaerobiosis and Element Toxicities Associated with Different Soils. *Ann. Bot.* **2009**, *103*, 221–235. [CrossRef]
12. Popescu, M.; Popescu, G.C. Diurnal Changes in Leaf Photosynthesis and Relative Water Content of Grapevine. *Curr. Trends Nat. Sci.* **2014**, *3*, 74–81.

13. Zhang, R.; Zhou, Y.; Yue, Z.; Chen, X.; Cao, X.; Ai, X.; Jiang, B.; Xing, Y. The Leaf-Air Temperature Difference Reflects the Variation in Water Status and Photosynthesis of Sorghum under Waterlogged Conditions. *PLoS ONE* **2019**, *14*, e0219209. [CrossRef]
14. Feng, X.; Zhou, G. Relationship of Leaf Water Content with Photosynthesis and Soil Water Content in Summer Maize. *Acta Ecol. Sin.* **2018**, *38*, 177–185. [CrossRef]
15. Giorio, P.; Sorrentino, G.; D'andria, R. Stomatal Behaviour, Leaf Water Status and Photosynthetic Response in Field-Grown Olive Trees Under Water Deficit. *Environ. Exp. Bot.* **1999**, *42*, 95–104. [CrossRef]
16. Egilla, J.N.; Davies, F.T., Jr.; Boutton, T.W. Drought Stress Influences Leaf Water Content, Photosynthesis, and Water-Use Efficiency of Hibiscus Rosa-Sinensis at Three Potassium Concentrations. *Photosynthetica* **2005**, *43*, 135–140. [CrossRef]
17. Aquilino, L.; Nayatami, K.L.; Tamu, A.; Soe, I.; Sakagami, J.I. Photosynthetic Function Analysis under Rhizosphere Anaerobic Conditions in Early-Stage Cassava. *Photosynth. Res.* **2025**, *163*, 42. [CrossRef]
18. Blum, A.; Mayer, J.; Gozlan, G. Infrared Thermal Sensing of Plant Canopies as a Screening Technique for Dehydration Avoidance in Wheat. *Field Crop. Res.* **1982**, *5*, 137–146. [CrossRef]
19. Brahmesh, R.B.; Kiran, B.O.; Somanagouda, B.P.; Ashwathama, V.H. Canopy Temperature in Sorghum under Drought Stress: Influence of Gas-Exchange Parameters. *J. Cereal Res.* **2022**, *14*, 81–85. [CrossRef]
20. Leopold, A.C.; Sun, W.Q.; Bernal-Lugo, I. The Glassy State in Seeds: Analysis and Function. *Seed Sci. Res.* **1994**, *4*, 267–274. [CrossRef]
21. Liu, Y.; Shen, S.; Pan, Y.; Zhang, L.; Lin, S.; Xue, Y. Diurnal Variation in Leaf-air Temperature Difference and Leaf Temperature Difference and the Hybrid Difference in Maize under Different Drought Stress. *J. China Agric. Univ.* **2014**, *19*, 13–21. [CrossRef]
22. Dhıllon, R.; Rojo, F.; Roach, J.; Upadhyaya, S.; Delwıche, M.A. A Continuous Leaf Monitoring System for Precision Irrigation Management in Orchard Crops. *Tarım Makinaları Bilim. Derg.* **2014**, *10*, 267–272.
23. Rajarajan, K.; Ganesamurthy, K.; Raveendran, M.; Jeyakumar, P.; Yuvaraja, A.; Sampath, P.; Prathima, P.T.; Senthilraja, C. Differential Responses of Sorghum Genotypes to Drought Stress Revealed by Physio-Chemical and Transcriptional Analysis. *Mol. Biol. Rep.* **2021**, *48*, 2453–2462. [CrossRef] [PubMed]
24. Pineda, M.; Barón, M.; Pérez-Bueno, M.-L. Thermal Imaging for Plant Stress Detection and Phenotyping. *Remote. Sens.* **2020**, *13*, 68. [CrossRef]
25. Jones, H.G. Use of Thermography for Quantitative Studies of Spatial and Temporal Variation of Stomatal Conductance over Leaf Surfaces. *Plant Cell Environ.* **1999**, *22*, 1043–1055. [CrossRef]
26. Anderegg, J.; Aasen, H.; Perich, G.; Roth, L.; Walter, A.; Hund, A. Temporal Trends in Canopy Temperature and Greenness are Potential Indicators of Late-Season Drought Avoidance and Functional Stay-Green in Wheat. *Field Crop. Res.* **2021**, *274*, 108311. [CrossRef]
27. Pradhan, A.; Aher, L.; Hegde, V.; Jangid, K.K.; Rane, J. Cooler Canopy Leverages Sorghum Adaptation to Drought and Heat Stress. *Sci. Rep.* **2022**, *12*, 4603. [CrossRef] [PubMed]
28. Balota, M.; Payne, W.A.; Evett, S.R.; Peters, T.R. Morphological and Physiological Traits Associated with Canopy Temperature Depression in Three Closely Related Wheat Lines. *Crop. Sci.* **2008**, *48*, 1897–1910. [CrossRef]
29. Ginkel, M.V.; Reynolds, M.P.; Trethowan, R.; Hernandez, E. (Eds.) Complementing the Breeder's Eye with Canopy Temperature Measurements. In *International Symposium on Wheat Yield Potential: Challenges to International Wheat Breeding*; International Maize and Wheat Improvement Center: El Batán, Mexico, 2006.
30. Yabuta, S.; Fukuta, T.; Tamaru, S.; Goto, K. The Productivity of Cassava (*Manihot esculenta* Crantz) in Kagoshima, Japan, Which Belongs to the Temperate Zone. *Agronomy* **2021**, *11*, 2021. [CrossRef]
31. Jones, H.G. Use of infrared thermometry for estimation of stomatal conductance as apossible aid to irrigation scheduling. *Agric. For. Meteorol.* **1999**, *95*, 139–149. [CrossRef]
32. Huang, C.; Gao, Y.; Qin, A.; Liu, Z.; Zhao, B.; Ning, D.; Ma, S.; Duan, A.; Liu, Z. Effects of waterlogging at different stages and durations on maize growth and grain yields. *Agric. Water Manag.* **2022**, *261*, 107334. [CrossRef]
33. Alves, A.A.C.; Setter, T.L. Response of cassava leaf area expansion to water deficit: Cell proliferation, cell expansion and delayed development. *Ann. Bot.* **2004**, *94*, 605–613. [CrossRef]
34. Ledent, J. *Deficit Hidrico y Crecimiento de las Plantas: Respuestas al Deficit Hidrico: Comportamiento Morfofisiologico/Modelado Del Crecimiento De Las Plantas*; Fundacion PROINPA: Cochabamba, Bolivia; Centro Internacional de la Papa (CIP): Lima, Peru; Proyecto Papa Andina: Lima, Peru, 2002; 79p, Available online: https://agris.fao.org/search/en/providers/122619/records/647248dd53aa8c896304ef1b (accessed on 15 November 2008).
35. Suárez, L.; Mederos, V. Apuntes sobre el cultivo de la yuca (*Manihot esculenta* Crantz). Tendencias actuales. *Cultiv. Trop.* **2011**, *32*, 27–35.
36. León, P.R.; Pérez Macias, M.; Gutiérrez Trocel, M.; Rodríguez Izquierdo, A.; Fuenmayor Campos, F.; Marín Rodriguez, C. Caracterización ecofisiológica de cuatro clones de yuca del banco de germoplasma del INIA-CENIAP. *Agron. Trop.* **2014**, *64*, 97–105.

37. Vandegeer, R.; Miller, R.E.; Bain, M.; Gleadow, R.M.; Cavagnaro, T.R. Drought Adversely Affects Tuber Development and Nutritional Quality of the Staple Crop Cassava (*Manihot Esculenta* Crantz). *Funct. Plant Biol.* **2012**, *40*, 195–200. [CrossRef]
38. Aina, O.O.; Dixon, A.G.O.; Akinrinde, E.A. Effect of Soil Moisture Stress on Growth and Yield of Cassava in Nigeria. *Pak. J. Biol. Sci.* **2007**, *10*, 3085–3090. [CrossRef]
39. Cruz, J.; Mosquim, P.; Pelacani, C.; Araújo, W.; DaMatta, F. Photosynthesis Impairment in Cassava Leaves in Response to Nitrogen Deficiency. *Plant Soil* **2003**, *257*, 417–423. [CrossRef]
40. Long, S.; Humphries, S.; Falkowski, P.G. Photoinhibition of photosynthesis innature. *Annu. Rev. Plant Physiol. Plant Mol. Biol.* **1994**, *45*, 633–662. [CrossRef]
41. Blokhina, O.; Virolainen, E.; Fagerstedt, K.V. Antioxidants, Oxidative Damage and Oxygen Deprivation Stress: A Review. *Ann. Bot.* **2003**, *91*, 179–194. [CrossRef]
42. Chang, W.W.; Huang, L.; Shen, M.; Webster, C.; Burlingame, A.L.; Roberts, J.K. Patterns of Protein Synthesis and Tolerance of Anoxia in Root Tips of Maize Seedlings Acclimated to a Low-Oxygen Environment, and Identification of Proteins by Mass Spectrometry. *Plant Physiol.* **2000**, *122*, 295–317. [CrossRef] [PubMed]
43. Nishizawa, A.; Yabuta, Y.; Shigeoka, S. Galactinol and raffinose constitute a novel function to protect plants from oxidative damage. *Plant Physiol.* **2008**, *147*, 1251–1263. [CrossRef] [PubMed]
44. Chen, Y.; Cao, M.; Zheng, L.; Li, J.; Mao, Y. Transcriptomic Profiling Suggests Candidate Molecular Responses to Waterlogging in Cassava. *PLoS ONE* **2022**, *17*, e0261086.
45. Pan, R.; Jiang, W.; Wang, Q.; Xu, L.; Shabala, S.; Zhang, W.Y. Differential Response of Growth and Photosynthesis in Diverse Cotton Genotypes under Hypoxia Stress. *Photosynthetica* **2019**, *57*, 772–779. [CrossRef]
46. Utsumi, Y.; Utsumi, C.; Tanaka, M.; Van Ha, C.; Takahashi, S.; Matsui, A.; Matsunaga, T.M.; Matsunaga, S.; Kanno, Y.; Seo, M.; et al. Acetic Acid Treatment Enhances Drought Avoidance in Cassava (*Manihot esculenta* Crantz). *Front. Plant Sci.* **2019**, *10*, 521. [CrossRef]
47. Surendar, K.K.; Devi, D.D.; Ravi, I.; Jeyakumar, P.; Velayudham, K. Effect of Water Stress on Leaf Temperature, Transpiration Rate, Stomatal Diffusive Resistance and Yield of Banana. *Plant Gene Trait.* **2013**, *4*, 43–47. [CrossRef]
48. Siddique, M.R.B.; Hamid, A.; Islam, M.S. Drought stress effects on water relations of wheat. *Bot. Bull. Acad. Sin.* **2000**, *41*, 35–39.
49. Nelson, J.; Bugbee, B. Analysis of Environmental Effects on Leaf Temperature under Sunlight, High Pressure Sodium and Light Emitting Diodes. *PLoS ONE* **2015**, *10*, e0138930. [CrossRef]
50. Ahmed, S.; Nawata, E.; Hosokawa, M.; Domae, Y.; Sakuratani, T. Alterations in Photosynthesis and Some Antioxidant Enzymatic Activities of Mungbean Subjected to Waterlogging. *Plant Sci.* **2002**, *163*, 117–123. [CrossRef]
51. Else, M.A.; Tiekstra, A.E.; Croker, S.J.; Davies, W.J.; Jackson, M.B. Stomatal Closure in Flooded Tomato Plants Involves Abscisic Acid and a Chemically Unidentified Anti-Transpirant in Xylem Sap. *Plant Physiol.* **1996**, *112*, 239–247. [CrossRef]
52. Zheng, C.; Jiang, D.; Liu, F.; Dai, T.; Jing, Q.; Cao, W. Changes in Photosynthesis, Chloroplast Ultrastructure, and Antioxidant Metabolism in Leaves of Sorghum under Waterlogging Stress. *Photosynthetica* **2019**, *57*, 1076–1083. [CrossRef]
53. Zheng, C.; Jiang, D.; Liu, F.; Dai, T.; Jing, Q.; Cao, W. Effects of Salt and Waterlogging Stresses and their Combination on Leaf Photosynthesis, Chloroplast ATP Synthesis, and Antioxidant Capacity in Wheat. *Plant Sci.* **2009**, *176*, 575–582. [CrossRef] [PubMed]
54. Bansal, R.; Srivastava, J.P. Effect of Waterlogging on Photosynthetic and Biochemical Parameters in Pigeonpea. *Russ. J. Plant Physiol.* **2015**, *62*, 322–327. [CrossRef]
55. Kawamitsu, Y. Effects of Nitrogen Supply on Growth Characteristics and Leaf Photosynthesis in Sugarcane. *Sci Bull. Coll. Agr. Univ. Ryukyus* **1999**, *46*, 1–14.
56. Djanaguiraman, M.; Prasad, P.V.V.; Boyle, D.L.; Schapaugh, W.T. High-Temperature Stress and Soybean Leaves: Leaf Anatomy and Photosynthesis. *Crop. Sci.* **2011**, *51*, 2125–2131. [CrossRef]
57. Luan, H.; Guo, B.; Pan, Y.; Lv, C.; Shen, H.; Xu, R. Morpho-Anatomical and Physiological Responses to Waterlogging Stress in Different Barley (*Hordeum vulgare* L.) Genotypes. *Plant Growth Regul.* **2018**, *85*, 399–409. [CrossRef]
58. Santos, C.M.D.; Endres, L.; Ferreira, V.M.; Silva, J.V.; Rolim, E.V.; Filho, H.C.W. Photosynthetic Capacity and Water Use Efficiency in *Ricinus communis* (L.) under Drought Stress in Semi-Humid and Semi-Arid Areas. *Acad. Bras. Cienc.* **2017**, *89*, 3015–3029. [CrossRef] [PubMed]

https://doi.org/10.3390/su18010405

sustainability

MDPI

Article

Toward Sustainable Crop Monitoring: An RGB-Based Non-Destructive System for Predicting Chlorophyll Content in Peanut Leaves

Kui Ge *, Huan Li, Xinqi Fan, Yixuan Wang, Juan Zhao, Jiatong Huang and Changcheng Tian *

School of Food and Biological Engineering, Bengbu University, Bengbu 233030, China; huanli@bbc.edu.cn (H.L.);
* Correspondence: gekui@bbc.edu.cn (K.G.); tiancc2006@bbc.edu.cn (C.T.)

Abstract

Accurate assessment of plant photosynthetic responses under drought and high-temperature stress is critical for understanding crop resilience. Chlorophyll content is a key indicator of photosynthetic efficiency, but conventional methods are destructive and time-consuming. Here, we developed a non-destructive detection system that captures Red (R), Green (G), and Blue (B) values from peanut (*Arachis hypogaea* L.) leaves and predicts chlorophyll content using machine learning. We optimized sensor distance (3–6 mm) and found 3 mm provided the most reliable RGB readings. Among Bayesian ridge and linear regression models, linear regression performed best (coefficient of determination $R^2 = 0.93$), yielding a robust predictive formula: chlorophyll = $-0.0308 \times [2 \times G - R - B] + 4.386$. Integration of this formula into the detection system enabled real-time estimation of chlorophyll as a proxy for photosynthetic status and stress response. By enabling low-cost, non-destructive and rapid chlorophyll monitoring, this framework can help support resource-efficient crop monitoring and high-throughput screening for stress-resilient cultivars, with potential relevance to sustainable production in water-limited environments.

Keywords: peanut; plant stress; color sensing; resource-efficient monitoring; chlorophyll prediction

1. Introduction

Peanut (*Arachis hypogaea* L.), a key leguminous crop, is renowned for its high-quality edible oil and rich protein content. In regions such as Asia and Africa, where populations rely heavily on plant-based proteins, peanut serves as a vital dietary component [1]. However, approximately 70% of global peanut production takes place in arid and semi-arid environments, where drought and heat stress are widespread and significantly reduce crop yields [2,3]. These environmental constraints not only limit water availability but also intensify thermal stress, leading to marked declines in peanut growth, pod development, and seed quality. Among these abiotic stresses, drought is the most detrimental [4–6].

To survive and adapt under drought conditions, plants have developed a variety of morpho-physiological and molecular mechanisms [7,8]. One of the earliest and most visible responses to environmental stress, particularly water deficit, is the alteration of leaf color, closely associated with changes in chlorophyll content [9–11]. As the central pigment in photosynthesis, chlorophyll is directly linked to plant productivity, stress adaptation, and overall vitality [6]. Recent studies have shown that chlorophyll content often exhibits more pronounced decline under abiotic stresses (e.g., low temperature, high humidity, soil pollution) than carotenoids, making chlorophyll a useful proxy indicator of environmental stress [12–14].

Reliable estimation of leaf chlorophyll is essential for assessing plant responses under environmental stress, especially in large-scale genotypic screening. Conventional methods typically involve solvent-based pigment extraction followed by spectrophotometric or chromatographic analysis [15,16]. While accurate, these techniques are destructive, labor-intensive, and unsuitable for repeated or high-throughput measurements [9,17]. They also require specialized equipment and cannot track temporal changes in the same leaf, limiting their field applicability [18]. To address these limitations, non-destructive approaches such as optical sensing and chlorophyll meters have gained popularity. Devices like the SPAD-502 (The Soil–Plant Analysis Development (SPAD) unit, developed by Minolta (Osaka, Japan), is a hand-held, dual-wavelength chlorophyll meter (models 501 and 502) used in plant physiology for rapid, non-destructive field measurement of chlorophyll content) estimate chlorophyll indirectly by measuring light transmittance at specific wavelengths and have shown strong correlation with actual chlorophyll content [19–21]. However, the high cost of such instruments (ranging from 2000 to 3000 USD) limits their accessibility and widespread adoption, particularly in resource-constrained environments. In contrast, recent advancements in artificial intelligence (AI) and machine learning (ML) present promising alternatives. These approaches enable the integration of low-cost imaging devices with automated data analysis pipelines, facilitating rapid, scalable, and cost-effective solutions for addressing complex challenges in agricultural research across diverse production systems [22–25].

In recent years, proximal optical sensors, particularly low-cost RGB imaging devices have emerged as valuable tools for monitoring plant physiological status [26,27]. These sensors enable real-time, non-destructive measurements and are highly portable, making them well suited for high-throughput phenotyping across diverse environmental conditions. For instance, consumer-grade RGB cameras have been successfully employed to capture subtle stress-induced changes in leaf color and morphology, providing insights into physiological responses to water deficit or nutrient limitations [28,29]. Furthermore, integrated thermal-RGB imaging systems have demonstrated enhanced sensitivity for detecting water stress by combining canopy temperature data with visible color features, allowing for earlier and more reliable stress detection compared with RGB data alone [30,31]. Additionally, proximal spectral-thermal fusion platforms have been implemented in precision agriculture to monitor plant water status, underscoring the practical advantages of sensor configurations that optimize cost, responsiveness, and measurement sensitivity [32,33].

RGB-based chlorophyll estimation has progressed toward standardized acquisition and learning-based modeling for practical low-cost phenotyping [22,34] (Table S1). Leaf-scale smartphone/contact RGB imaging with compact machine learning/deep learning predictors has shown promising performance, but robustness depends strongly on illumination; recent work demonstrates the need for illumination-aware calibration or multi-light training [22,35]. Complementary efforts include handheld RGB-sensor devices with controlled light for rapid chlorophyll a/b and total chlorophyll quantification, as well as improved RGB-derived indices and recent reviews that summarize RGB pipelines and emphasize calibration/transferability as key limitations [34,36,37]. Crop-specific demonstrations (e.g., tomato) further support applicability under appropriate calibration [38].

In peanut, drought stress frequently suppresses photosynthesis and leads to a marked decline in leaf chlorophyll, making chlorophyll a sensitive indicator of stress severity [6]. Given that peanut is a major oilseed crop predominantly cultivated in arid and semi-arid regions, developing an RGB-based chlorophyll estimation system could provide a scalable, low-cost complement to expensive SPAD meters and destructive laboratory assays,

https://doi.org/10.3390/su18021001

enabling rapid monitoring and high-throughput screening under drought conditions. Despite recent progress in RGB-based chlorophyll/SPAD estimation, practical use for peanut drought monitoring still has challenges. Many models are trained on other crops and specific imaging setups, so their performance may not transfer well to peanut because leaf optical properties differ among crops and cultivars, and RGB signals can be sensitive to lighting and measurement geometry. In addition, methods based on uncontrolled field photography or complex models may show reduced repeatability and may be less suitable for low-cost, on-device use. Moreover, studies that directly validate RGB-based predictions against destructive biochemical assays under drought-stress conditions in peanut are still limited.

To help address these issues, we developed a low-cost (the parts cost of the system developed in this study is approximately 22 USD (Table S3)) detection system to estimate chlorophyll content in peanut leaves using RGB-derived color traits, together with a standardized measurement setup intended to reduce lighting-related variation. We built predictive models linking RGB features to chlorophyll content and evaluated them under simulated drought stress by comparing model outputs with conventional biochemical assays. Within the tested conditions, the proposed leaf-color-based system shows potential as a practical tool for rapid chlorophyll monitoring in peanut, and it may be adaptable to other crops after calibration, supporting high-throughput phenotyping and drought-stress assessment.

Therefore, this study aimed to:

(a) develop a low-cost leaf-color-based detection system for estimating chlorophyll content in peanut leaves;
(b) construct and validate predictive models linking RGB-derived color features with chlorophyll content under drought-stress conditions;
(c) evaluate the system's accuracy and applicability by comparing model predictions with conventional biochemical assays to assess its potential for high-throughput stress monitoring.

2. Materials and Methods

2.1. Hardware and Software Components of the Color Detection System

The chlorophyll content detection system includes two components: a hardware system and a software system. The hardware system consists of an Arduino UNO microprocessor (Core processor model, ATmega328P; SIMMTECH Co., Ltd., Suzhou, China) for the whole circuit and a color acquisition system (Color detection module, GY-33 TCS34725; SIMMTECH Co., Ltd.) consisting of a microcontroller, a "TCS34725" chip, and two white light LEDs, which are used to light the object being measured. When the light reflected or transmitted by the object enters the sensor, it passes through a filter, and the light intensities for the R, G, and B color channels are separately captured. The hardware also includes a wireless communication system (Wireless communication system modules, ESP8266-WIFI; SIMMTECH Co., Ltd.), a display system (OLED liquid crystal display module, SSD1306; SIMMTECH Co., Ltd.), and an alarm system (Active buzzer module, MH-FMD; SIMMTECH Co., Ltd.). The RGB value range for normal peanut leaves is predefined in the Arduino UNO microprocessor. When the RGB values detected by the color sensor and input into the processor exceed the predefined range, the alarm system will be triggered. The software system includes Arduino IDE (Arduino IDE version 1.8.19, the software programming environment of Arduino UNO) and the Bafa Cloud Platform (https://bemfa.com/). The specific structure is illustrated in Figure 1.

https://doi.org/10.3390/su18021001

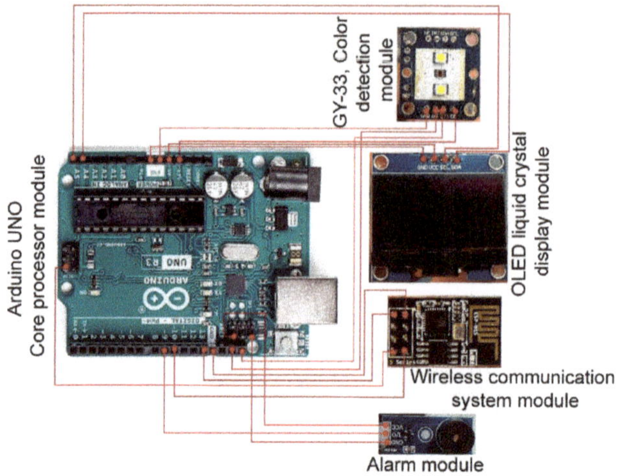

Figure 1. Overview of the RGB-based chlorophyll content detection system.

The chlorophyll content detection system uses Arduino UNO as a microprocessor. Arduino UNO controls the realization of each function through programming in Arduino IDE, including controlling the color acquisition system to obtain the plant leaf R, G, and B values. The chlorophyll content is calculated separately based on the R, G, and B values and the formulas. The resulting values for chlorophyll content are transmitted to the display system for real-time display using the Inter-Integrated Circuit (IIC) method; at the same time, the values for chlorophyll content in the leaf are uploaded to the Bafa Cloud Platform in real time through a wireless communication system to realize remote plant monitoring.

The color acquisition module was the GY-33, which integrates the TCS34725 color light-to-digital converter and on-board white LEDs (Table S2). The sensor provides four channels (red, green, blue, and clear; RGBC) with digital output and includes an IR-blocking filter to reduce infrared contamination in color sensing. The TCS34725 supports programmable analog gain (1×/4×/16×/60×) and programmable integration time in 2.4-ms steps (integration time = 2.4 ms × (256 − ATIME)), and communicates via a digital interface (I^2C; fast-mode compatible). In this study, the GY-33 module outputs 8-bit RGB values (0–255) computed by its on-board MCU after white balancing.

Before data collection, the GY-33 module was white-balanced using the vendor PC software (ATK-XCOM, version 2.3) by sending the command [0 × A5 + 0 × BB + 0 × 60] with a white reference under the built-in LEDs. This white-balance setting is stored by the module and retained after power cycling. During measurements, a light-shielding structure and a fixed sensor-to-leaf geometry were used to minimize ambient-light interference and reduce variability caused by viewing angle or distance. For each leaf sample, RGB readings were collected at multiple positions and the mean value was used for subsequent analysis. The RGB values used in model development were the module outputs after white balance (not raw sensor register counts), and the regression index [2G − R − B] was computed from these RGB values.

2.2. Light-Shielding Structure and Measurement Geometry for RGB Acquisition

To minimize the influence of variable ambient illumination on RGB measurements, a custom light-shielding structure was used during data acquisition (Supplementary Figure S2). The target leaf region was positioned inside a semi-enclosed measurement cavity that surrounds the sensor's field of view. During measurement, the cavity limits the entry of external light from lateral and overhead directions, so that the reflected signal recorded by the sensor is dominated by the module's built-in LED illumination rather than room or laboratory lighting. In addition, the shield provides a fixed opening and a constrained sensor–leaf geometry, which helps reduce variability caused by changes in angle, distance, or positioning between measurements. The key dimensions of the light shield are shown in Supplementary Figure S2; the gap parameter (*d*) can be adjusted according to leaf thickness to ensure consistent contact/spacing and repeatable positioning.

2.3. Plant Growth and Treatment Conditions

The peanut (*Arachis hypogaea* L.) cultivars 'Yue you 7', 'Yue you 13', 'Zhong hua 16', and 'Pu hua 36' were provided by the Crop Research Institute, Guangdong Academy of Agricultural Sciences, China. The seeds (*Medicago sativa* L., Zhong mu 1; *Arabidopsis thaliana* (L.) Heynh., Columbia-0; *Stevia rebaudiana* Hemsl.) were soaked in water for 12 h, placed on moist filter paper, and transferred to a growth chamber (Guang Qi, GHP-160, Shanghai, China) with a cycle of 16 h light from fluorescent and incandescent lamps (200 μmol $m^{-2}s^{-1}$) at 26 °C, followed by 8 h darkness at 22 °C. The plants were cultivated as previously described [39]. To simulate the effects of drought, the peanut taproot was treated with 20% (*w*/*v*) polyethylene glycol 6000 (PEG 6000) (Sinopharm Chemical Reagent Co., Ltd., Cat No. 30151728, Shanghai, China) for 24 h or 48 h. Whole peanut plants were subjected to a high-temperature treatment at 45 °C in a plant incubator.

2.4. Determination of Chlorophyll Content

Chlorophyll was extracted as follows. Approximately 0.2 g of fresh peanut leaf tissue was weighed and immersed in 6 mL of a 1:1 (*v*/*v*) acetone–ethanol mixture. The samples were kept in darkness for 24 h to allow thorough pigment diffusion into the solvent while minimizing light-induced chlorophyll degradation. After 24 h, the extract was filtered, and the residual tissue was re-extracted with an additional 2 mL of the same solvent in darkness for another 12 h to recover remaining pigments. The two extracts were combined, and absorbance was measured at 645 nm and 663 nm using a Multi-scan GO Microplate Photometer (Thermo Fisher Scientific Inc., Waltham, MA, USA), with the acetone–ethanol (1:1, *v*/*v*) mixture used as the blank. Chlorophyll content was calculated as follows: Chlorophyll content (mg/g FW) = $[[20.21 \times OD645 + 8.02 \times OD663] \times V]/[1000 \times W]$, where OD645 and OD663 are the absorbance values at 645 nm and 663 nm, respectively; V is the extract volume, and W is the fresh weight of leaf tissue (g).

2.5. Acquisition of Actual RGB Values for Peanut Leaves

To obtain the actual RGB values of peanut leaves, leaves were collected from plants at the four-leaf stage, and RGB values were measured using a Hunter Lab Agera (A) colorimeter (Shanion Creative Inc., Shanghai, China) at 460 nm, 140 Lux (parameters consistent with those used by the chlorophyll content detection system's color sensor).

2.6. Predictive Modeling of Physiological Parameters in Peanut

Regression modeling was performed in Python using scikit-learn. The input variables were RGB values (R, G, B) output by the GY-33 module after the white-balance procedure described in Section 2.1. For each sample, an RGB-derived index was computed as [2G − R − B]. The response variable was the biochemically measured chlorophyll content (mg/g FW). Candidate RGB combinations were screened by Pearson correlation analysis, and [2G − R − B] was used as the predictor for the regression models as described in Section 3.2.

Five regression algorithms were evaluated: Linear Regression, Bayesian Ridge, Elastic Net, SVR, and Gradient Boosting Regressor (scikit-learn). The dataset was split into training and test sets at a 70:30 ratio. Model fitting was performed on the training set. Five-fold

https://doi.org/10.3390/su18021001

cross-validation was conducted within the training set. The held-out test set was used for final evaluation.

Model performance was quantified using explained variance (EV), mean absolute error (MAE), mean squared error (MSE), root mean squared error (RMSE), and coefficient of determination (R^2). Linear regression has no tunable hyperparameters. For Bayesian ridge regression, Elastic Net, SVR, and Gradient Boosting Regressor, the default scikit-learn parameter settings were used, and the software version and all relevant parameters are reported in the revised manuscript for reproducibility. Where a random state argument was available, it was fixed.

2.7. Data Analysis

Pearson's correlation coefficient analysis, significance testing, and paired t-tests were conducted using IBM SPSS Statistics 25 software. The predictive model of the in vivo chlorophyll content from peanut leaf color eigenvalues was established using machine learning regression algorithms in Python (Python 3.11.0). Plots were generated in Origin Pro 2025 (64-bit) SR1 (Origin Pro Student Version). Quantitative data were expressed as mean $\pm$ SD. The statistical significance of experimental data was assessed by Student t-test using IBM SPSS Statistics 25 software.

3. Results

3.1. Evaluation and Optimization of Our Chlorophyll Content Detection System Based on the R, G, and B Values of Peanut Leaves

To evaluate the accuracy of the leaf color detection system developed in this study, we determined its optimal working distance by measuring the R, G, and B values of peanut leaves from seedlings at the four-leaf stage at distances of 3, 4, 5, or 6 mm from the color sensor (Figure 2a). At 3 mm, the R, G, and B values obtained by the color detection system were tightly grouped and closely aligned with the actual values estimated from the colorimeter (referred to as 'Actual Color, AC'). As the distance increased, however, the R, G, and B values measured by the system gradually dropped and deviated from the true values (Figures 2b and S1a). To validate this finding, we selected two visually distinct peanut leaves, labeled 'a' and 'b' (Figure 2c), and used the color detection system to measure their R, G, and B values. The values measured with our detection system (at 3 mm) were close to the values estimated using the colorimeter (Figure 2d). We also measured the chlorophyll content of leaves 'a' and 'b' via biochemical assays, revealing notable differences in their chlorophyll levels (Figure 2e). These findings indicate that the chlorophyll content detection system developed in this study can accurately capture the R, G, and B values of peanut leaves and that leaves with different RGB profiles have distinct chlorophyll content.

3.2. Prediction of Peanut Leaf Chlorophyll Content Using RGB-Derived Index [$2 \times G - R - B$]

To establish a mathematical relationship to predict the chlorophyll content of peanut leaves from their RGB profiles, we used our chlorophyll content detection system to measure the R, G, and B values from 147 leaf samples collected from peanut plants grown under normal or drought conditions. We then detected their chlorophyll content. Following the approach described by Xing Liu et al. [6], we calculated the Pearson's correlation coefficient between measured chlorophyll content and combinations of R, G, and B values (Figure 3a). We obtained the highest absolute correlation coefficients between chlorophyll content and [2 × G − R − B] values, with coefficient of −0.957. A scatterplot of chlorophyll content as a function of [2 × G − R − B] value confirmed the linear relationship among these parameters (Figure 3b). These results demonstrate that the [2 × G − R − B] index is a reliable predictor of chlorophyll content in peanut leaves.

https://doi.org/10.3390/su18021001

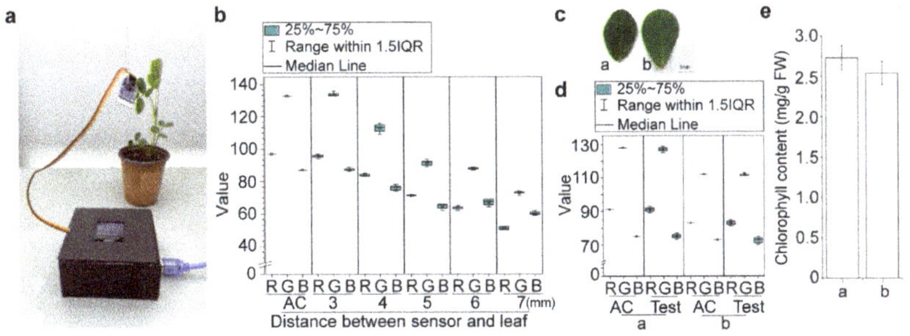

Figure 2. Measurements of RGB values and chlorophyll content in the leaves of peanut seedlings at the four-leaf stage using RGB sensors and biochemical methods. (**a**) Photograph of the chlorophyll content detection system used to acquire the red (R), green (G), and blue (B) values from four different sections of peanut leaves. (**b**) Box plots illustrating the RGB values measured from the leaves of peanut seedlings at the four-leaf stage, obtained at distances of 3, 4, 5, 6, or 7 mm from the leaf using RGB sensors. As reference, the RGB values of the same leaf measured with a colorimeter (AC, Actual Color) are shown (n = 10 the number of measurements taken for each leaf at each distance.). IQR, interquartile range. (**c**) Photograph of the two peanut leaves from seedlings at the four-leaf stage whose RGB values were measured with our detection system. a, left leaf; b, right leaf. (**d**) Box plots showing the R, G, and B values of the two leaves a and b in (**c**), measured using a colorimeter (AC, Actual Color) (colorimeter-estimated values) and RGB sensors at a distance of 3 mm from the leaves (n = 6 technical replicate measurements). (**e**) Chlorophyll content of the two leaves labeled a and b in (**c**), measured using biochemical assays (n = 6 technical replicate measurements).

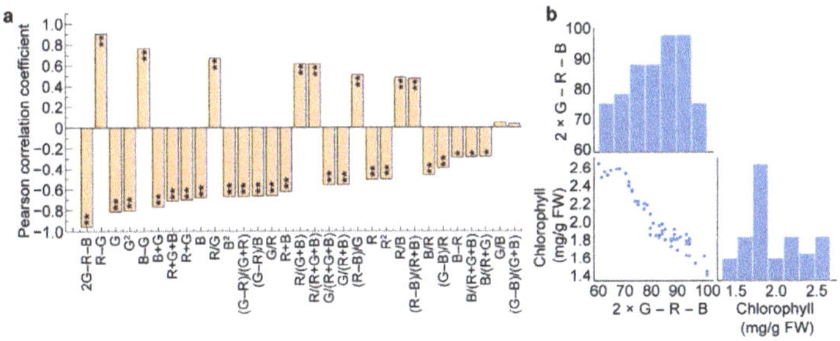

Figure 3. Analysis of chlorophyll contents in leaves from peanut seedlings at the four-leaf stage using RGB values. (**a**) Pairwise Pearson's correlation coefficients between combinations of leaf RGB values, obtained with our detection system, and the leaf chlorophyll content measured by biochemical assays at 0 h (control), 24 h, and 48 h into drought treatment from seedlings at the four-leaf stage. The coefficients were calculated using SPSS software. Asterisks indicate significant differences as determined by a Student's t-test: * $p < 0.05$; ** $p < 0.001$. (RGB values and chlorophyll content were measured in 147 leaf samples); (**b**) Scatterplots showing the pairwise correlation between the indicated RGB values [$2 \times G - R - B$] and chlorophyll content in the same leaves. The histograms along the diagonal show the data distribution for each set of values.

3.3. Establishment of a Mathematical Model for Predicting Chlorophyll Content in Peanut Leaves Using RGB Values and Machine Learning Regression

We employed five machine learning regression algorithms for model development, dividing the datasets for chlorophyll content, [$2 \times G - R - B$] value into a training set and a testing set at a 70:30 ratio (102:45 leaves). The Bayesian ridge and linear regression models resulted in the highest explained variance (EV) and coefficient of determination (R^2) score, both reaching 0.94 and 0.93, respectively. Additionally, the Bayesian ridge and linear regression models exhibited lower mean absolute error (MAE), and root mean squared error (RMSE) values than the other three regression models, suggesting their superior predictive performance (Figure 4a). Notably, the chlorophyll content predicted from the Bayesian ridge and linear regression models were close to the true measured values, indicating better fitting performance (Figure 4b). In internal five-fold cross validation, the linear regression model emerged as the most effective, producing the smallest relative variation coefficient of determination (R^2) score (Figure 4c). The linear regression model performed better in the analyses and can generate prediction formulas for real-time display. By contrast, the other models are more complex and do not readily produce formulas for real-time, uniform calculations. To visualize overall prediction performance on the held-out test set, we plotted predicted versus biochemically measured chlorophyll values, showing high agreement (Figure 4d). A Bland–Altman style difference-versus-mean plot further indicated a small bias (0.017 mg g^{-1} FW) and no strong concentration-dependent error (Figure 4e). We therefore selected the linear regression model for predicting chlorophyll content, with the formula: Chlorophyll = [−0.0308 × [2 × G − R − B] + 4.386], using the [2 × G − R − B] value as predictors of chlorophyll content.

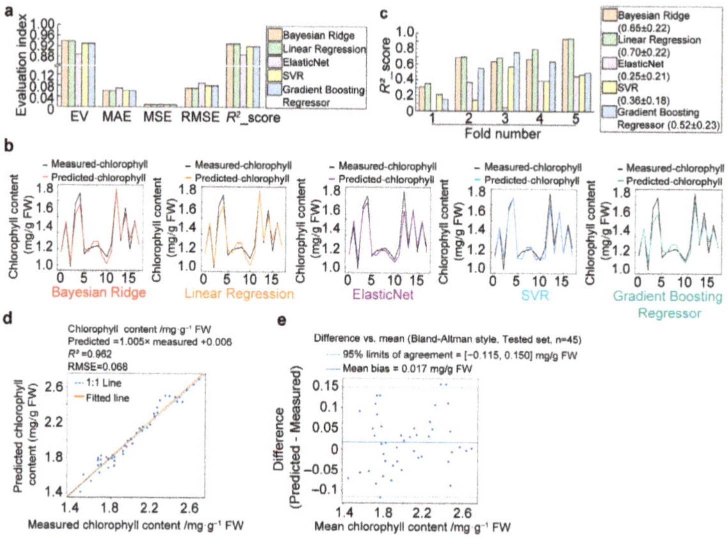

Figure 4. Analysis and prediction of chlorophyll content in leaves from peanut seedlings at the four-leaf stage using RGB values and machine learning models. (**a**) Evaluation indices from the use of each of five machine learning regression algorithms (Linear regression, Bayesian ridge, Elastic Net, SVR, and Gradient Boosting Regressor) using 70% of the data from the chlorophyll content [2 × G – R – B] value as the training set. The explained variance (EV), mean absolute error (MAE), mean squared error (MSE), root mean squared error (RMSE), and coefficient of determination (R^2) values for each model were calculated. (**b**) Predicted and ground truth values for chlorophyll content by the five machine learning regression algorithms (Linear regression, Bayesian ridge, Elastic Net, SVR, and Gradient Boosting Regressor) models, using the [2 × G − R − B] value from 17 individual leaf samples, representing 30% of all leaf samples in the test dataset. Black lines, chlorophyll content obtained from biochemical experiments. (**c**) Five-fold cross-validation of the coefficient of determination (R^2) scores obtained from the outputs of each of the five training models. The data are the average of R^2 scores ± SE. (**d**) Predicted vs. measured chlorophyll content on the held-out test set. (**e**) Agreement and error analysis using a Bland–Altman style difference-versus-mean plot (test set).

3.4. Assessment of the Rapid Detection System for Measuring Chlorophyll Content in the Leaves of Peanut Plants Exposed to Stress or Different Peanut Cultivars

To further evaluate the applicability of the chlorophyll content detection system across different scenarios, we predicted the chlorophyll content of leaves from peanut plants

https://doi.org/10.3390/su18021001

(Predicted-chlorophyll) subjected to drought or high-temperature stress. In parallel, we measured the chlorophyll content via biochemical assays (Measured-chlorophyll). The chlorophyll content of leaves from peanut plants experiencing drought or high-temperature conditions was lower than that of the control (Figure 5a); these results are consistent with previous studies showing that drought and high-temperature stress inhibit chlorophyll biosynthesis and accelerate chlorophyll degradation, resulting in decreased leaf chlorophyll content [40,41]. The rapid detection system yielded similar results to biochemical assays for chlorophyll content across all conditions. Similarly, we assessed the chlorophyll content of leaves from three cultivated peanut varieties. We predicted the highest chlorophyll levels in Yue you 13, followed by Pu hua 36 and Zhong hua 16 (Figure 5b). These chlorophyll content aligned with prior findings on the photosynthetic capacity of these varieties [42]. Again, the rapid detection system and biochemical assays produced comparable results.

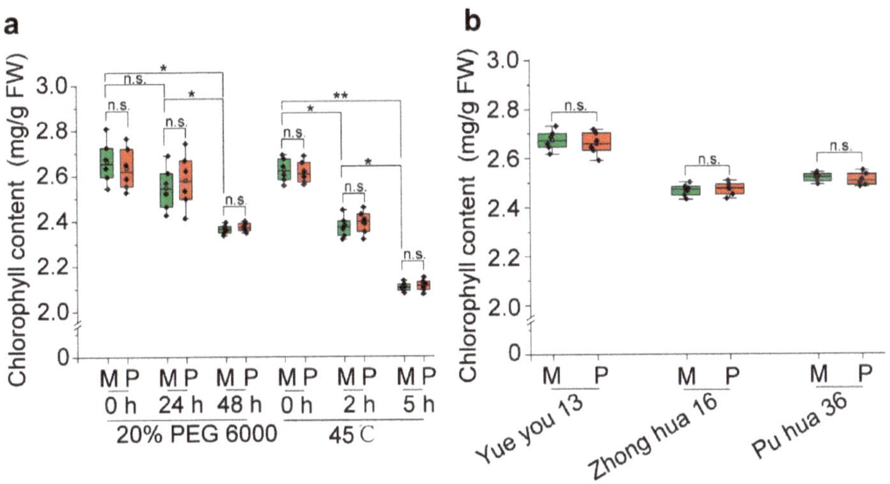

Figure 5. Analysis of chlorophyll content in leaves from peanut seedlings at the four-leaf stage under different conditions and in different cultivars. (**a**) Chlorophyll content in the leaves of peanut seedlings at the four-leaf stage, estimated by the chlorophyll content detection system (Predicted-chlorophyll, P) or biochemical assays (Measured-chlorophyll, M) under the following conditions: 20% (w/v) PEG6000 treatment for 0 h, 24 h, or 48 h, and high-temperature stress treatment at 45 °C for 0 h, 2 h, or 5 h. (**b**) Chlorophyll content in the leaves of peanut seedlings at the four-leaf stage from cultivars Yue you 13, Zhong hua 16, and Pu hua 36 grown under normal conditions, estimated by the chlorophyll content detection system or biochemical assays. Asterisks indicate significant differences, as determined by a Student t-test; * $p < 0.05$; ** $p < 0.001$; n.s.: not significant.

4. Discussion

In this study, an Arduino-based color detection system was developed to acquire peanut leaf RGB values and to estimate chlorophyll content using lightweight regression models. The measurement setup was optimized experimentally, and a sensor-leaf distance of 3 mm produced the most stable RGB readings and the closest agreement with reference color measurements (Figures 2 and S1). Using biochemically quantified chlorophyll as reference measurements, correlation analysis identified [2G − R − B] as the most informative predictors for chlorophyll within the collected datasets (Figure 3). Model comparison across five regression algorithms supported selecting linear regression as a deployable option that balances predictive performance with an explicit analytic equation suitable for on-device implementation (Figure 4). Within the tested controlled treatments and genotypes, predicted values followed trends consistent with biochemical measurements

https://doi.org/10.3390/su18021001

(Figures 5 and S3). Thus, the RGB sensor in the chlorophyll content detection system designed in this study provides a low-cost and repeatable approach for acquiring peanut leaf RGB profiles under controlled illumination and fixed geometry.

As key participants in photosynthesis, leaves significantly contribute to plant growth and development. Changes in leaf coloration reflect changes in pigment contents, which are closely associated with plant nutrient and water status [6,41,43–45]. Abiotic stress induced symptoms often manifest in distinct patterns in plants, which can be effectively predicted using color sensing [46,47]. Although multispectral and hyperspectral sensors, including those in the NIR range, offer deeper insights into vegetation, such as vegetation health, chlorophyll content, and nutrient deficiencies, RGB-based sensors provide a more cost-effective solution for detecting stress indicators [48–50]. RGB sensing provides a low-cost option for capturing visible-color traits with minimal computational requirements, which is advantageous for portable and embedded devices [51,52]. In this study, RGB acquisition depended on controlled illumination and fixed geometry: the optimized 3 mm distance reduced ambient-light effects and improved repeatability of the measured RGB values (Figures 2b and S1). These results support using simple RGB features as inputs for rapid estimation under controlled measurement conditions.

Additionally, a device was designed to shield against ambient light interference, addressing the sensitivity of RGB sensors to external illumination (Figure S2). In practice, transferability and long-term stability may be improved by controlling illumination geometry, calibrating different units with a standard reference target, and periodically re-checking the reference to monitor drift and re-calibrate as needed. In field applications, measurement data can be transferred to a computer via USB or wireless networks or stored locally for later analysis [50]. In this study, a wireless communication system (ESP8266-WIFI), a display system (OLED) were integrated into the chlorophyll content detection system, enabling on-site data recording or remote transmission (Figure 1). The selection of an RGB sensor as the core component of the chlorophyll content detection system provides a practical and repeatable way to capture leaf color characteristics, enabling quantitative estimation of chlorophyll content under controlled conditions.

To keep the predictor simple and interpretable, we used a single RGB-derived index, [2G − R − B], as the input feature (Figure 3). This index can be viewed as a visible-band descriptor of leaf greenness. It is closely related to the core idea used in the Excess Green family and other RGB greenness indices that assign more weight to the green channel while reducing the contribution of red and blue, and such indices have been used in RGB-based plant phenotyping and vegetation color analyses as practical descriptors of leaf greenness and color variation [34,50]. Using one index also allows the final model to be written as an explicit linear equation, which is convenient for integration into the Arduino-based device for real-time calculation and display (Figure 4). In the current study, 147 sets of leaf RGB profiles were used to develop predictive models for estimating chlorophyll content in peanut leaves. Despite the limited number of data points, the predicted chlorophyll content was comparable across all plant and conditions, validating the accuracy and reliability of the rapid detection system (Figures 5 and S3). However, the models were developed and evaluated mainly under controlled experimental settings; further validation using larger and more diverse datasets, including field-grown plants and broader environmental conditions, will be necessary to fully assess generalization. In addition, PEG-induced osmotic stress and constant 45 °C chamber heat are controlled proxies that do not fully reproduce field drought and heat, which involve progressive soil drying, heterogeneous root–soil interactions, fluctuating radiation and vapor pressure deficit, and diurnal temperature cycles [53]. Therefore, applying the current models beyond

https://doi.org/10.3390/su18021001

the tested controlled settings will require re-calibration and validation using field-grown plants and broader environmental variability.

While our system was developed and validated using peanut leaves, the underlying principle of predicting chlorophyll content from RGB color metrics has broad potential across other crops [26]. Non-destructive RGB-based approaches have been successfully applied in Arabidopsis, rice, and Hami melon, showing strong concordance with biochemical assays and SPAD measurements [21,54]. These studies indicate that, although the optimal color indices and regression formulas may differ among species, the general strategy can be transferable after appropriate calibration. However, the present study was evaluated under controlled laboratory conditions with a fixed measurement configuration (built-in LEDs, light shielding, and fixed sensor–leaf geometry) (Figures 5 and S2). A key limitation is that RGB-based greenness indices can be sensitive to illumination; changes in ambient light intensity and spectrum may shift measured RGB values and affect prediction performance, as noted in prior image-based/SPAD prediction and RGB phenotyping workflows [34,35]. Another limitation is the lack of benchmarking against a conventional SPAD meter, which we plan to include in subsequent validation studies to assess comparability across instruments and operating conditions. In addition, leaf surface conditions (e.g., gloss, dust, or water droplets) and variations in measurement angle or contact pressure can introduce variability that may reduce transferability across environments. Therefore, although the current device uses on-board LEDs and a light shield to reduce ambient-light interference (Figure S2), the model has not yet been validated for open-field use. For outdoor deployment, additional calibration and validation across variable sunlight and field microclimates will be required, and a simple reference-based procedure (e.g., periodic checking with a white/gray target) could be introduced to support calibration under irregular lighting conditions. Future work will expand the dataset using field-grown plants and test robustness across environments, following common practice in sensor-based plant phenotyping and proximal/remote sensing applications.

Beyond methodological performance, the proposed system has practical sustainability implications. Its low component cost and non-destructive workflow can reduce dependence on consumables and solvent-based extraction when frequent measurements are needed, and it enables higher-frequency monitoring to support timely, resource-efficient management under drought and heat. In addition, the same framework can facilitate high-throughput phenotyping for selecting stress-resilient genotypes, which is relevant to sustaining productivity in water-limited production regions. Overall, this study shows that combining RGB color sensing with regression modeling can support a practical, non-destructive estimation of leaf chlorophyll for peanut under the measurement configuration tested here. The proposed workflow—standardized RGB acquisition, reference biochemical measurement, and simple model deployment—may provide a useful basis for developing similar rapid estimation tools in other plant species, provided that appropriate calibration and validation are conducted for the new species and environments.

5. Conclusions

In summary, this study developed a low-cost, non-destructive RGB-based sensing system to estimate chlorophyll content in peanut leaves under a standardized measurement configuration. By pairing leaf RGB readings acquired with an integrated LED illumination and light-shielding structure with biochemically quantified chlorophyll as the reference, we identified an interpretable RGB-derived index [2G − R − B] that strongly correlates with chlorophyll content and established a simple linear regression model (R^2 = 0.93). The resulting explicit equation, chlorophyll = [−0.0308 × [2G − R − B] + 4.386], can be implemented on-device to support real-time chlorophyll estimation. Within the tested scope, the

https://doi.org/10.3390/su18021001

predicted chlorophyll values showed trends consistent with destructive biochemical assays. These results indicate that the proposed system can serve as a practical complementary tool for rapid screening and high-frequency monitoring of chlorophyll dynamics when destructive assays are impractical and conventional meters are not readily accessible.

Several limitations should be noted. The models were developed and evaluated under controlled laboratory conditions with a fixed sensor-leaf geometry, and RGB-based indices can be sensitive to illumination, leaf surface properties, and measurement handling; therefore, extension to other growth stages, field environments, or additional species will require recalibration and independent validation. In addition, we did not benchmark the device against a conventional SPAD meter in this study, which limits direct cross-instrument comparability for routine field users. Future work will focus on expanding the dataset across developmental stages and environments, conducting outdoor/field validation under variable sunlight, developing simple reference-based calibration procedures, and performing cross-instrument benchmarking (including SPAD) to improve robustness and transferability. In practice, integrating the sensor into a clip-like holder could further improve crop-specific adaptability by standardizing contact/spacing across leaves with different thickness, lamina size, or dense trichomes, thereby supporting more scalable measurements across crops. Overall, by directly linking low-cost RGB traits to chlorophyll content with a deployable, interpretable on-device equation, this study provides a practical framework that complements destructive biochemical assays for rapid photosynthetic status monitoring in peanut under standardized measurement conditions.

Supplementary Materials: The following supporting information can be downloaded at: https://www.mdpi.com/article/10.3390/su18021001/s1, Figure S1: RGB values measured by the chlorophyll content detection system using an RGB sensor at distances of 3, 4, 5, 6, or 7 mm, and RGB values measured using a colorimeter (AC, Actual Color) (n = 10 measurements taken for each leaf at each distance.). IQR, interquartile range. Figure S2: Light-shielding chamber for RGB acquisition in the chlorophyll detection system. Views from multiple angles with key dimensions (mm). Panel 'a' corresponds to the setup shown in Figure 2a. The units are in mm. Figure S3: Cross-species evaluation of the RGB-based chlorophyll detection system. (a) Representative plants of *Medicago sativa*, *Arabidopsis thaliana*, and *Stevia rebaudiana* (light-green) and *Stevia rebaudiana* (normal) used for leaf chlorophyll measurements; (b) Leaf chlorophyll content of the above species under normal conditions, estimated by the RGB-based chlorophyll content detection system (Predicted, P) and measured by biochemical assays (Measured, M). Asterisks indicate significant differences, as determined by a Student *t*-test; n.s., not significant. Table S1: Quantitative RGB chlorophyll comparison. Table S2: GY33-TCS34725 specifications. Table S3: Hardware costs of RGB-based chlorophyll predicted system.

Author Contributions: Conceptualization, K.G. and C.T.; methodology, H.L., X.F.; software, Y.W.; validation, J.Z., J.H.; formal analysis, K.G., H.L.; investigation, K.G., H.L.; resources, K.G.; data curation, K.G.; writing—original draft preparation, K.G.; writing—review and editing, K.G., C.T.; visualization, Y.W.; supervision, K.G. and C.T.; project administration, K.G.; funding acquisition, Y.W. All authors have read and agreed to the published version of the manuscript.

Funding: This research was funded by the Talent Introduction Project of Bengbu University, grant number 2025GQD014, and the Higher Education Research Project (Youth Program) of the Education Department of Anhui Province, grant 2025AHGXZK40614, to Kui Ge; and the Innovation and Entrepreneurship Training Project of College Student of Bengbu University (S202511305084 to Xinqi Fan and Yixuan Wang; S202511305085 to Juan Zhao and Jiatong Huang).

Institutional Review Board Statement: Not applicable.

Informed Consent Statement: Not applicable.

Data Availability Statement: Data will be made available on request.

https://doi.org/10.3390/su18021001

Conflicts of Interest: The authors declare no conflicts of interest.

Abbreviations

The following abbreviations are used in this manuscript:

R	Red
G	Green
B	Blue
R^2	Coefficient of determination
SPAD	Soil–Plant Analysis Development
USD	United States Dollar
AI	Artificial intelligence
ML	Machine learning
SVR	Support vector machine regression
EV	Explained variance
MAE	Mean absolute error
MSE	Mean squared error
RMSE	Root mean squared error
AC	Actual Color
M	Measured
P	Predicted

References

1. Li, W.; Yoo, E.; Lee, S.; Sung, J.; Noh, H.J.; Hwang, S.J.; Desta, K.T.; Lee, G.-A. Seed Weight and Genotype Influence the Total Oil Content and Fatty Acid Composition of Peanut Seeds. *Foods* **2022**, *11*, 3463. [CrossRef]
2. Pokhrel, S.; Kharel, P.; Pandey, S.; Botton, S.; Nugraha, G.T.; Holbrook, C.; Ozias-Akins, P. Understanding the impacts of drought on peanuts (*Arachis hypogaea* L.): Exploring physio-genetic mechanisms to develop drought-resilient peanut cultivars. *Front. Genet.* **2025**, *15*, 1492434. [CrossRef]
3. Reddy, T.; Reddy, V.; Anbumozhi, V. Physiological responses of groundnut (*Arachis hypogea* L.) to drought stress and its amelioration: A critical review. *Plant Growth Regul.* **2003**, *41*, 75–88. [CrossRef]
4. Hamidou, F.; Ratnakumar, P.; Halilou, O.; Mponda, O.; Kapewa, T.; Monyo, E.; Faye, I.; Ntare, B.; Nigam, S.; Upadhyaya, H.; et al. Selection of intermittent drought tolerant lines across years and locations in the reference collection of ground-nut (*Arachis hypogaea* L.). *Field Crops Res.* **2012**, *126*, 189–199. [CrossRef]
5. Attia, Z.; Pogoda, C.S.; Reinert, S.; Kane, N.C.; Hulke, B.S. Breeding for sustainable oilseed crop yield and quality in a changing climate. *Theor. Appl. Genet.* **2021**, *134*, 1817–1827. [CrossRef] [PubMed]
6. Liu, X.; Li, L.; Li, M.; Su, L.; Lian, S.; Zhang, B.; Li, X.; Ge, K.; Li, L. AhGLK1 affects chlorophyll biosynthesis and photosynthesis in peanut leaves during recovery from drought. *Sci. Rep.* **2018**, *8*, 2250. [CrossRef]
7. Wei, Y.S.; Javed, T.; Liu, T.T.; Ali, A.; Gao, S.J. Mechanisms of abscisic acid (ABA)-mediated plant defense responses: An updated review. *Plant Stress* **2025**, *15*, 100724. [CrossRef]
8. Liu, S.; Li, M.; Su, L.; Ge, K.; Li, L.; Li, X.; Liu, X.; Li, L. Negative feedback regulation of ABA biosynthesis in peanut (Arachis hypogaea): A transcription factor complex inhibits AhNCED1 expression during water stress. *Sci. Rep.* **2016**, *6*, 37943. [CrossRef]
9. Ren, J.; Guo, P.; Zhao, X.; Ma, X.; Ai, X.; Wang, J.; Zou, H.; Yu, H. Differential photosynthetic responses to drought stress in peanut varieties: Insights from transcriptome profiling and JIP-Test analysis. *BMC Plant Biol.* **2025**, *25*, 957. [CrossRef] [PubMed]
10. Shafi, S.; Zaffar, A.; Riyaz, I.; Zargar, S.M.; Najeeb, S.; Sofi, P.A. A non-destructive, low cost and high throughput colorimetric method for chlorophyll estimation in rice (*Oryza sativa* L.). *Discov. Plants* **2024**, *1*, 2. [CrossRef]
11. Yan, W.; Lu, Y.; Guo, L.; Liu, Y.; Li, M.; Zhang, B.; Zhang, B.; Zhang, L.; Qin, D.; Huo, J. Effects of Drought Stress on Photosynthesis and Chlorophyll Fluorescence in Blue Honeysuckle. *Plants* **2024**, *13*, 2115. [CrossRef] [PubMed]
12. Amin, B.; Atif, M.J.; Kandegama, W.; Nasar, J.; Alam, P.; Fang, Z.; Cheng, Z. Low temperature and high humidity affect dynamics of chlorophyll biosynthesis and secondary metabolites in Cucumber. *BMC Plant Biol.* **2024**, *24*, 903. [CrossRef]
13. Xu, P.; Yu, J.; Ma, R.; Ji, Y.; Hu, Q.; Mao, Y.; Ding, C.; Li, Z.; Ge, S.; Deng, W.-W.; et al. Chlorophyll and Carotenoid Metabolism Varies with Growth Temperatures among Tea Genotypes with Different Leaf Colors in *Camellia sinensis*. *Int. J. Mol. Sci.* **2024**, *25*, 10772. [CrossRef]
14. Tomaškinová, J.; Tomaškin, J.; Drimal, M.; Bellido, J. The Impact of Abiotic Environmental Stressors on Fluorescence and Chlorophyll Content in *Glycine max* (L.) Merrill. *Agronomy* **2025**, *15*, 263. [CrossRef]

https://doi.org/10.3390/su18021001

15. Porra, R.J.; Thompson, W.A.; Kriedemann, P.E. Determination of accurate extinction coefficients and simultaneous equations for assaying chlorophylls a and b extracted with four different solvents: Verification of the concentration of chlorophyll standards by atomic absorption spectroscopy. *Biochim. Biophys. Acta BBA Bioenerg.* **1989**, *975*, 384–394. [CrossRef]
16. Couture, J.J.; Singh, A.; Rubert-Nason, K.F.; Serbin, S.P.; Lindroth, R.L.; Townsend, P.A. Spectroscopic determination of ecologically relevant plant secondary metabolites. *Methods Ecol. Evol.* **2016**, *7*, 1402–1412. [CrossRef]
17. Wang, Z.J.; Wang, J.H.; Liu, L.Y.; Huang, W.J.; Zhao, C.J.; Wang, C.Z. Prediction of grain protein content in winter wheat (*Triticum aestivum* L.) using plant pigment ratio (PPR). *Field Crops Res.* **2004**, *90*, 311–321. [CrossRef]
18. Li, Y.; Sun, Y.; Jiang, J.; Liu, J. Spectroscopic determination of leaf chlorophyll content and color for genetic selection on *Sassafras tzumu*. *Plant Methods* **2019**, *15*, 73. [CrossRef]
19. Gholizadeh, A.; Amin, M.S.M.; Anuar, A.R.; Aimrun, W. Evaluation of SPAD chlorophyll meter in two different rice growth stages and its temporal variability. *Eur. J. Sci. Res.* **2009**, *37*, 591–598.
20. Shah, S.H.; Houborg, R.; McCabe, M.F. Response of chlorophyll, carotenoid and SPAD-502 measurement to salinity and nutrient stress in wheat (*Triticum aestivum* L.). *Agronomy* **2017**, *7*, 61. [CrossRef]
21. Liang, Y.; Urano, D.; Liao, K.-L.; Hedrick, T.L.; Gao, Y.; Jones, A.M. A nondestructive method to estimate the chlorophyll content of *Arabidopsis* seedlings. *Plant Methods* **2017**, *13*, 26. [CrossRef]
22. Barman, U.; Saikia, M.J. Smartphone contact imaging and 1-D CNN for leaf chlorophyll estimation in agriculture. *Agriculture* **2024**, *14*, 1262. [CrossRef]
23. Lopin, P.; Nawsang, P.; Laywisadkul, S.; Lopin, K.V. Evaluation of Low-Cost Multi-Spectral Sensors for Measuring Chlorophyll Levels Across Diverse Leaf Types. *Sensors* **2025**, *25*, 2198. [CrossRef]
24. Wacker, K.; Kim, C.; van Iersel, M.W.; Sidore, B.; Pham, T.; Haidekker, M.; Seymour, L.; Ferrarezi, R.S. Development of an Automated Low-Cost Multispectral Imaging System to Quantify Canopy Size and Pigmentation. *Sensors* **2024**, *24*, 5515. [CrossRef] [PubMed]
25. Singhal, G.; Choudhury, B.U.; Singh, N.; Goswami, J. An enhanced chlorophyll estimation model with a canopy structural trait in maize crops: Use of multi-spectral UAV images and machine learning algorithm. *Ecol. Inform.* **2024**, *83*, 102811. [CrossRef]
26. Wang, R.-F.; Qu, H.-R.; Su, W.-H. From sensors to insights: Technological trends in image-based high-throughput plant phenotyping. *Smart Agric. Technol.* **2025**, *12*, 101257. [CrossRef]
27. Anshori, M.F.; Dirpan, A.; Sitaresmi, T.; Rossi, R.; Farid, M.; Hairmansis, A.; Purwoko, B.; Suwarno, W.B.; Nugraha, Y. An overview of image-based phenotyping as an adaptive 4.0 technology for studying plant abiotic stress: A bibliometric and literature review. *Heliyon* **2023**, *9*, e21650. [CrossRef] [PubMed]
28. Niu, Y.; Xu, X.; Huang, W.; Li, J.; Li, S.; Zhao, N.; Li, B.; Xu, C.; Lu, S. RGB Imaging and Irrigation Management Reveal Water Stress Thresholds in Three Urban Shrubs in Northern China. *Plants* **2025**, *14*, 2253. [CrossRef]
29. Samsuzzaman; Islam, S.; Ali, M.R.; Dey, P.K.; Bicamumakuba, E.; Reza, M.N.; Chung, S.O. Quantification of Light, Photoperiod, Temperature, and Water Stress Symptoms Using Image Features for Smart Vegetable Seedling Production. *Horticulturae* **2025**, *11*, 1340. [CrossRef]
30. Chandel, N.S.; Rajwade, Y.A.; Dubey, K.; Chandel, A.K.; Subeesh, A.; Tiwari, M.K. Water Stress Identification of Winter Wheat Crop with State-of-the-Art AI Techniques and High-Resolution Thermal-RGB Imagery. *Plants* **2022**, *11*, 3344. [CrossRef]
31. Zhou, Z.; Diverres, G.; Kang, C.; Thapa, S.; Karkee, M.; Zhang, Q.; Keller, M. Ground-based thermal imaging for assessing crop water status in grapevines over a growing season. *Agronomy* **2022**, *12*, 322. [CrossRef]
32. Sánchez-Virosta, A.; Gómez-Candón, D.; Montoya, F.; Pérez-García, Y.; Jiménez, V.; Martínez-López, J.A.; González-Piqueras, J.; López-Urrea, R.; Sánchez, J.M. Remote and proximal sensing assessment of water status and its correlation with yield on almond orchards in Southeast Spain. *Precis. Agric.* **2025**, *26*, 92. [CrossRef]
33. Mertens, S.; Verbraeken, L.; Sprenger, H.; De Meyer, S.; Demuynck, K.; Cannoot, B.; Merchie, J.; De Block, J.; Vogel, J.T.; Bruce, W.; et al. Monitoring of drought stress and transpiration rate using proximal thermal and hyperspectral imaging in an indoor automated plant phenotyping platform. *Plant Methods* **2023**, *19*, 132. [CrossRef]
34. Kior, A.; Yudina, L.; Zolin, Y.; Sukhov, V.; Sukhova, E. RGB Imaging as a Tool for Remote Sensing of Characteristics of Terrestrial Plants: A Review. *Plants* **2024**, *13*, 1262. [CrossRef] [PubMed]
35. Kamiwaki, Y.; Fukuda, S. Effect of the Light Environment on Image-Based SPAD Value Prediction of Radish Leaves. *Algorithms* **2023**, *17*, 16. [CrossRef]
36. Agarwal, A.; Colwell, F.D.J.; Galvis, V.A.C.; Hill, T.R.; Boonham, N.; Prashar, A. Two-fold red excess (TREx): A simple and novel digital color index that enables non-invasive real-time monitoring of green-leaved as well as anthocyanin-rich crops. *Plant Methods* **2025**, *21*, 24. [CrossRef]
37. Pan, W.; Ma, H.; Wang, R.; Wang, H.; Wang, D.; Guo, W.; Guo, X. Development of a handheld chlorophyll content detector on wheat and maize leaves based on RGB sensor. *Front. Plant Sci.* **2025**, *16*, 1606413. [CrossRef]
38. Zhang, X.; Yu, H.; Yan, J.; Meng, X. Study on the Detection of Chlorophyll Content in Tomato Leaves Based on RGB Images. *Horticulturae* **2025**, *11*, 593. [CrossRef]

https://doi.org/10.3390/su18021001

39. Ge, K.; Liu, X.; Li, X.; Hu, B.; Li, L. Isolation of an *ABA Transporter-Like 1* Gene from *Arachis hypogaea* That Affects ABA Import and Reduces ABA Sensitivity in *Arabidopsis*. *Front. Plant Sci.* **2017**, *8*, 1150. [CrossRef]
40. Ma, Y.; Dias, M.C.; Freitas, H. Drought and Salinity Stress Responses and Microbe-Induced Tolerance in Plants. *Front. Plant Sci.* **2020**, *11*, 591911. [CrossRef]
41. He, K.; Xu, Y.; Ding, H.; Guo, Q.; Ci, D.; Zhang, J.; Qin, F.; Xu, M.; Zhang, G. The Impact of Short-Term Drought on the Photosynthetic Characteristics and Yield of Peanuts Grown in Saline Alkali Soil. *Plants* **2024**, *13*, 2920. [CrossRef]
42. Hao, X.; Zhang, J.; Liu, J.; Zang, X.; Zhu, S.; Tang, F. Evaluation of drought resistance in different peanut varieties. *J. Peanut Sci.* **2020**, *49*, 47–51. (In Chinese)
43. Ge, K.; Lv, Q.; Chen, S.; Guo, Z.; Peng, Y.; Chen, Y.; Sun, S.; Wang, X. The nodule-specific transcriptional repressor Top Hub 4 regulates nodule structure and nitrogen fixation capacity in soybean. *J. Genet. Genom.* **2024**, *51*, 96–99. [CrossRef]
44. Wang, M.; Lee, J.; Choi, B.; Park, Y.; Sim, H.J.; Kim, H.; Hwang, I. Physiological and Molecular Processes Associated with Long Duration of ABA Treatment. *Front. Plant Sci.* **2018**, *9*, 176. [CrossRef]
45. Menzies, I.J.; Youard, L.W.; Lord, J.M.; Carpenter, K.L.; Van Klink, J.W.; Perry, N.B.; Schaefer, H.M.; Gould, K.S. Leaf colour polymorphisms: A balance between plant defence and photosynthesis. *J. Ecol.* **2016**, *104*, 104–113. [CrossRef]
46. Zandi, A.; Hosseinirad, S.; Zadeh, H.K.; Tavakolian, K.; Cho, B.-K.; Vasefi, F.; Kim, M.S.; Tavakolian, P. A systematic review of multi-mode analytics for enhanced plant stress evaluation. *Front. Plant Sci.* **2025**, *16*, 1545025. [CrossRef] [PubMed]
47. Pérez-Patricio, M.; Osuna-Coutiño, J.A.d.J.; Ríos-Toledo, G.; Aguilar-González, A.; Camas-Anzueto, J.L.; Morales-Navarro, N.A.; Velázquez-González, J.R.; Cundapí-López, L.Á. Plant Stress Detection Using a Three-Dimensional Analysis from a Single RGB Image. *Sensors* **2024**, *24*, 7860. [CrossRef] [PubMed]
48. Tao, M.; Ma, X.; Huang, X.; Liu, C.; Deng, R.; Liang, K.; Qi, L. Smartphone-based detection of leaf color levels in rice plants. *Comput. Electron. Agric.* **2020**, *173*, 105431. [CrossRef]
49. Ponti, M.P. Segmentation of low-cost remote sensing images combining vegetation indices and mean shift. *IEEE Geosci. Remote Sens. Lett.* **2012**, *10*, 67–70. [CrossRef]
50. Bodor-Pesti, P.; Nguyen, L.L.P.; Nguyen, T.B.; Dam, M.S.; Taranyi, D.; Baranyai, L. LeafLaminaMap: Exploring Leaf Color Patterns Using RGB Color Indices. *AgriEngineering* **2025**, *7*, 39. [CrossRef]
51. Coelho, F.G.T.; Maciel, G.M.; Siquieroli, A.C.S.; Gallis, R.B.d.A.; de Oliveira, C.S.; Ribeiro, A.L.A.; Pereira, L.M. High-throughput phenotyping for the evaluation of agronomic potential and root quality in tropical carrot using RGB sensors. *Agriculture* **2024**, *14*, 710. [CrossRef]
52. Furbank, R.T.; Jimenez-Berni, J.A.; George-Jaeggli, B.; Potgieter, A.B.; Deery, D.M. Field crop phenomics: Enabling breeding for radiation use efficiency and biomass in cereal crops. *New Phytol.* **2019**, *223*, 1714–1727. [CrossRef] [PubMed]
53. Zhang, M.; Wang, L.-F.; Zhang, K.; Liu, F.-Z.; Wan, Y.-S. Drought-induced responses of organic osmolytes and proline metabolism during pre-flowering stage in leaves of peanut (*Arachis hypogaea* L.). *J. Integr. Agric.* **2017**, *16*, 2197–2205. [CrossRef]
54. Baek, S.-H.; Park, K.-H.; Jeon, J.-S.; Kwak, T.-Y. A Novel Method for Calibration of Digital Soil Images Captured under Irregular Lighting Conditions. *Sensors* **2022**, *23*, 296. [CrossRef]

https://doi.org/10.3390/su18021001

sustainability

MDPI

Article

How Will Smart Technology Support SDG 12? An Empirical Study on Sustainability in Indian Agricultural Operations †

Usha Ramanathan [1,2] and Ramakrishnan Ramanathan [1,3,*]

1 College of Business Administration, University of Sharjah, Sharjah 27272, United Arab Emirates; uramanathan@sharjah.ac.ae
2 Nottingham Business School, Nottingham Trent University, 50 Shakespeare Street, Nottingham NG1 4FQ, UK
3 Business and Management Research Institute, University of Bedfordshire Business School, Luton LU1 3JU, UK
* Correspondence: rramanathan@sharjah.ac.ae; Tel.: +971-6-5050579

† This article is a revised and expanded version of a paper entitled "Diffusion of smart technology in agricultural sector of rural India", which was presented at the European Operations Management (EurOMA) Conference, Leuwen, Belgium, 4–7 July 2023.

Abstract

India is one of the fastest growing economies with significant potential for the use of smart farming operations. Although agriculture is a major sector, implementation of smart technologies in the agriculture sector has not progressed in India. We use a mixed-methods approach to develop knowledge on the factors determining this slow adoption of smart technology and develop strategies for large-scale adoption in the Indian agriculture sector. First, qualitative interviews are used to understand the factors behind the slow diffusion of smart technology in the agriculture sector. Based on the responses, we link the results of the qualitative study from the agri-sector to the well-known Diffusion of Innovations (DoI) theory. We then develop a framework for applying Fuzzy-set Qualitative Comparative Analysis (fsQCA) to analyze the impact of multiple causal factors. We apply our research findings to help achieve SDG 12 in the agriculture sector. Our findings indicate individual factors on their own may influence adoption, but some reasonable combinations of factors (e.g., a combination of technology, knowhow, experience, benefits-operation, and finance and reliability) could also result in the large-scale adoption of smart technologies in improving Indian agricultural operations. By doing so, we provide a contextual empirical configurational test of the DoI theory in the Indian smart agricultural context.

Keywords: agri-technology; technology diffusion; smart technology; agriculture sector; qualitative comparative analysis; SDG 12

1. Introduction

In the second decade of the 21st century, several nations around the globe are excelling in adapting smart technology in agriculture [1–4]. Especially in developing nations, such as India, smart IoT technology is being used widely in agricultural supply chains [5]. The share of agriculture and allied sectors in India's national GDP was 20.19% in 2021 [6]; hence, any effort to improve the efficiency of agricultural operations using smart technology will have many benefits for the Indian economy. In this research, we consider India to explore smart technology's diffusion into the agricultural sector.

India is one of the top producers of agricultural goods in the world. There are many small family holders involved in farming, and it is their main source of income. The prevalence of small and marginal farmers contributes to growing a wide range of fruits and vegetables, depending on various climatic conditions throughout the year. Despite its enormous potential, Indian agriculture is hampered by several difficulties [7]. Farming crises such as water shortage, lower productivity, and natural calamities are often reported in the news headlines. Several studies have reported that the adoption of modern technologies could help overcome these challenges [7].

The sustainable development goals (SDG) proposed by the United Nations aim to achieve zero hunger (SDG 2) and responsible consumption and production (SDG 12) around the globe. According to the United Nations description of SDG 12, high-income countries contribute a larger environmental footprint compared to low-income countries. In this era of technology, every nation aims to adopt smart technology in every field of production and the service sector [1]. This diffusion of technology will bring progress in production and services [8]. The resulting higher productivities will support the socio-economic position of the local population, which in turn can support the SDGs, including SDG 12.

In spite of several success stories on the contribution of smart technology, such as the Internet of Things sensors, Big Data, cloud computing, and artificial intelligence, farmers in India have differing perspectives on the use of smart technology in their everyday activities [9]. Here, we use the term smart technology as a comprehensive term for several modern technologies, including Internet of Things (IoT), sensors, blockchain, temperature-controlled logistics with monitoring options, connected devices, smart farming, and precision farming. Although we did not follow a systematic literature review, our search terms included the smart technology terms included in this paragraph and two more terms namely "agriculture" and "India".

Mohan et al. [7] discussed the role of smart technology in food storage and distribution to avoid post-harvest waste in the Indian context. They highlighted the role of temperature-controlled logistics with simple smart technology, to reduce the wastage of fresh food. Danese at al. [10] used blockchain technology in wine supply chains to prevent counterfeiting. In this article, the authors used multiple case studies to identify a set of key variables, but no interaction effect was discussed.

A recent article by Wang et al. [11] developed a technology-based model, namely the YOLO-BLBE model using multi-scale Retinex with a color restoration method, to identify the maturity of blueberry fruits. On the other hand, Mohan et al. [7] suggested a postharvest loss management using a simple temperature control option. Several business studies have reported the potential of smart agriculture in the Indian scenario, but to the best of our knowledge, there is no comprehensive academic study that discusses the factors behind smart technology adoption, as well as which combinations of these factors would support the large-scale application of smart technology to boost productivity in Indian agriculture.

This study fills these two research gaps (i) via a qualitative study to understand the factors behind smart technology adoption and (ii) by exploring whether these factors acting alone or in combination will help large-scale adoption. Accordingly, the main aim of this research is to understand how smart technologies affect agricultural operations in India. The main research question is as follows: How does smart technology diffusion make a positive change in the agricultural sector of India? The following are the associated research questions.

RQ 1: What is the status of adoption of smart technology in the Indian agriculture sector?

RQ 2: What are the primary factors contributing to the adoption or non-adoption of smart technology?

RQ 3: To what extent are the previous theories on technology adoption frameworks meaningful in explaining the behavior of Indian farmers in adopting smart technologies in this era?

https://doi.org/10.3390/su18031344

RQ 4: Which factor(s) will expedite the adoption of smart technology in the agriculture sector?

RQ 5: How can the adoption of smart technology in the agriculture sector help to achieve SDG 12?

We employed a mixed methods approach to answer these research questions. First, in Phase 1, we conducted a set of interviews in a rural area of Southern India, specifically in the state of Karnataka. We designed the interviews without linking to any previous theoretical frameworks to ensure that we receive unbiased responses from the farmers. This approach helped us to understand the current scenario and factors conducive to smart technology adoption in the agriculture sector. This, in turn, helped to answer the first two RQs. Interestingly, a thorough analysis of the qualitative responses helped us to link the responses to the tenets of the well-known Diffusion of Innovations (DoI) theory [12]. Further, in this process we identified DoI as the theoretical background, which addresses RQ3. The DoI theory helps to approach smart technology adoption from the perspectives of success factors and barriers such as the relative advantage and complexity of the new technology. The interviews shed light on the factors responsible for successful adoption or otherwise. We then build on the knowledge of these factors to understand whether these factors on their own or in combination could help in successful adoption. For this purpose, the well-known Qualitative Comparative Analysis (QCA) is employed. We use fuzzy ideas and, hence, we use the fuzzy-set version (called fsQCA). This fsQCA analysis introduces possible solutions to the research questions—RQ4 and RQ5.

Thus, this study makes multiple contributions to the literature. First, this is one of the first studies to collect empirical data on the adoption of smart technology in Indian agriculture from the theoretical lens of the DoI theory. Second, it not only synthesizes the factors based on the results of the interviews but extends the results further using fsQCA. By doing so, it provides a contextual empirical configurational test of the DoI theory in the Indian smart farming context. While traditional innovation studies focus only on the specific characteristics stipulated in the DoI theory towards successful adoption, we try to determine the combinations of some factors that could support successful adoption. To our knowledge, this is the first study that provides an empirical test for the applicability of the DoI theory in Indian smart farming context.

This paper is structured as follows. Section 2 provides a background study including a literature review of the Indian agriculture scenario and Fuzzy-Set Qualitative Comparative Analysis (FsQCA). The qualitative research methodology is elaborated in Section 3. This also explains the research design and two phases of data analysis. Section 4 discusses the results of the analysis of the interview data from Phase 1 and the application of fuzzy-set QCA (fsQCA) in Phase 2. This section highlights the theories identified from the qualitative study. Our findings of the analyses, presented in Section 4, are discussed further in Section 5. This section not only contextualizes our findings for the Indian agriculture context but also provides an empirical configurational test of the DoI theory. Section 6 provides a summary including contributions to practice, especially to SDGs, and also highlights the future research opportunities.

2. Background Study and Methodology

India's expertise in smart technology has created several new avenues for farmers to make the agriculture production process simple and reliable. Despite having the resources in place, the adoption of smart technology for the purpose of agriculture sector (hereafter, this will be mentioned as the agri-sector) is following a slow phase of diffusion in India and around the globe. This research focuses on the critical factors that limit farmers' adoption of smart technologies using empirical research. We also explore the ways to more quickly diffuse this smart technology within the community to bring young farmers into the agribusiness.

In view of answering the proposed research questions, we have conducted a background study in the southern part of India, and we have devised a research method accordingly.

We have used a qualitative case study approach in Phase 1, as this is the best way to explore the less studied area of research in the context of Indian agriculture, in comparison to developed nations [13]. The case study data are used to validate the smart-farming acceptance model and also to address the research questions considered in this study. We have performed qualitative data analysis from the interview transcripts to arrive at initial observations on technology use in the Indian agri-sector.

We conducted the case study in three stages as shown in Figure 1. In the first stage, we performed initial observation field visits to understand the current practices of agri-food production. In Stage two, we interviewed four groups of farmers, totaling 20 individual farm-holders. A previous version of this part has been presented at the EurOMA 2023 conference [14]. In Stage three, we collated the interview data and analyzed the data. The results of the data analysis were discussed with the focus groups to raw thematic conclusions. Following these thematic conclusions, we further analyzed the data using fsQCA to derive variables and their interactions in smart technology adoption decisions.

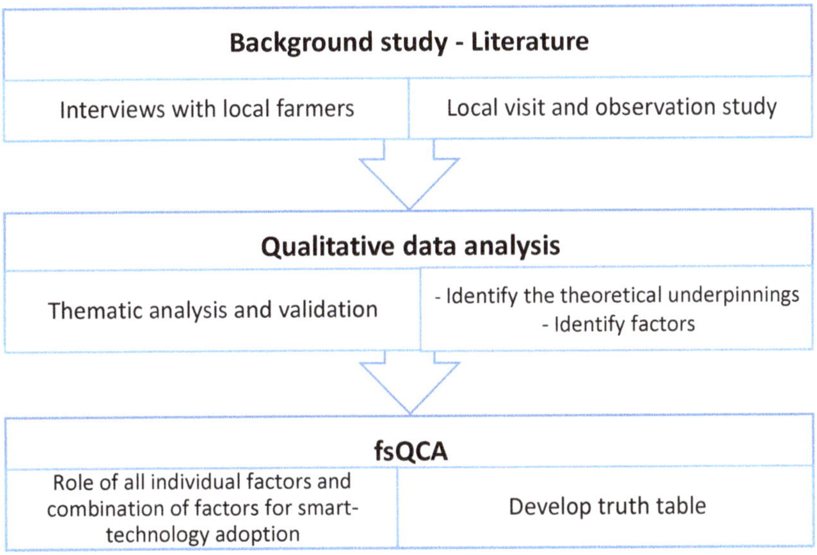

Figure 1. Research design.

In this section, we present the literature on the Indian agriculture sector and introduce the Fuzzy-set Qualitative Comparative Analysis (fsQCA) approach.

2.1. Studies on the Indian Agriculture Scenario

The growing global population demands a high volume of food production to feed human beings and household animals. Small and marginal farmers, with around an 85% share, still dominate the number of holdings at the national level [15]. Since the invention of smartphones with abundant Wi-Fi-internet connectivity, the use of IoT in Indian agri-sector has gained a rapid advantage [16]. The level of readiness for technology differs for each sector in various geographical locations [17]. The Indian agri-sector uses

https://doi.org/10.3390/su18031344

IoT in various forms, namely, precision farming, smart irrigation, greenhouse automation, and predictive analytics by well-established technology companies such as mKRISHI from Tata Consultancy Services, Fasal, WayCool, Farm again, and many more [16].

Smart Farming is one of the recent innovations of IoT technology that emphasizes the use of information and communication technology (ICT) in agricultural farm management with the support of cloud computing and data analysis [1–6]. The technologies include Internet of Things sensors, drones, blockchain, computer vision, machine learning, deep learning, image recognition, artificial intelligence, and more [17]. As Artificial Intelligence (AI) and Big Data are becoming an integral part of our life and businesses around the globe [18,19], adopting these technologies in the food sector and agriculture will enrich the productivity of food products [17].

There are some interesting studies on the adoption of new smart technology in Indian agriculture. General reviews of technology in Indian agriculture are available [20]. Jarial [18] has provided a review of such studies focusing on a description of various technologies, their potential, and their issues and challenges. The study has provided a list of leading Indian companies engaged in various smart technology development. While this paper provides a good overview of smart technologies in Indian agriculture, it does not look at specific factors affecting individual farmers. Pal et al. [21] describe an empirical study on the adoption of a specific smart technology, called Laser Land Levelling, in some parts of Indian agriculture. Thus, there are some studies that have focused on the application of smart agriculture in Indian context. However, a detailed review shows that there is no holistic attempt to look at the factors that promote or inhibit the application of smart agriculture in the Indian context. We fill this gap by drawing on the tenets of the Diffusion of Innovations theory. This theory is discussed in the next section.

Some small/medium farm holders and several large-scale farmers are interested in adapting Internet of Things (IoT) in smart farming to take their yield to the next level. Although agricultural operations look simple from the outset, the agri-sector has inherited complexity, which needs an innovative approach to achieve dynamic solutions [22]. However, some initial investment is required to have smart IoT in place, and the small amount of maintenance/service cost for using the smart technology is not perceived as a positive investment but as a recurring expenditure. Local government and councils are supporting farmers to adopt smart technology by showcasing technology demonstration projects and through some financial support such as grants and small incentives for using the technology. Now, several young entrepreneurs and farmers realize the potential of using the smart technology in agriculture, replacing human labor [7]. We explore the diffusion of innovation in the agri-sector of India to be able to make suggestions for stakeholders and policy makers in this prominent field of agriculture, in the era of technology. Then, we use Qualitative Comparative Analysis (QCA) to analyze the results of the qualitative study; we briefly describe this methodology in the next sub-section.

2.2. *Fuzzy-Set Qualitative Comparative Analysis (fsQCA)*

The idea of QCA was introduced by Ragin [23] as a tool for detecting causal complexity in an outcome of interest across different scientific fields. QCA is based on set theoretic ideas and the basic operations of Boolean algebra (i.e., union, intersection, and negation of sets). Based on empirical observations of the occurrence of the outcome of interest and the presence or absence of causal factors linked to the outcome, QCA is used in an outcome's membership score in complex situations (indicated by the presence or absence of one or more causal factors). Saridakis et al. [24] highlight that QCA uses Boolean methods to assess whether a complex causal condition is necessary and/or sufficient for the outcome of interest to occur.

https://doi.org/10.3390/su18031344

If empirical observations indicate that a particular factor is present in all observations of the presence of the outcome, then the factor could be considered a necessary factor. Thus, a necessary factor (but not a sufficient factor) is indicated if the factor is always present for the outcome to occur, but this factor alone is not enough. In set theory language, the outcome set will be the subset of the set of the necessary factor. That is, 100% of the outcome set will be the subset of the necessary factor.

In contrast, a sufficient factor (but not a necessary factor) is indicated if the factor itself can produce the outcome, even though there are other factors that may also produce the outcome even if this factor is not present. In set theory language, the set of a sufficient factor will be a subset of outcome set. That is, 100% of the sufficient factor will be the subset of outcome set.

However, based on limited empirical observations, a factor may not be a fully necessary or a fully sufficient factor, as there might be a small number of empirical observations that can deviate from expected patterns. Hence, we need to establish some thresholds (less than 100%) to claim that a factor is a necessary factor or sufficient factor. In QCA, consistency and coverage are used to assess whether a factor can be called necessary or sufficient using some threshold values. These two measures range from 0 to 1. Consistency measures the degree to which a causal factor (or a combination of factors) leads to the outcome. Coverage measures how many cases in the observations with the presence of positive outcome that also have the presence of the particular factor (or a combination). The recommended threshold for necessary consistency is 0.9 [24] when analyzing individual factors. In general, the higher the consistency score is, the lower the solution coverage will be [23], and hence, some compromise is necessary in choosing appropriate threshold values, especially when considering a combination of factors. In this case, an acceptable threshold for consistency (above 0.70) and coverage between 0.25 and 0.65 have been suggested.

The truth table is considered the main part of a QCA. If there are k factors in consideration in the analysis, then there are 2^k possible combinations of the factors. Based on empirical observations, QCA identifies all possible 2^k combinations of the factors and the corresponding presence or absence of the outcome and estimates the corresponding consistency score. If the consistency score is 1, then the combination of factors represented by the corresponding row is always present when the outcome is present. If the consistency score is less than but close to 1, then there might be some observations when the outcome is absent, even though the outcome is present for most of the corresponding combinations of factors. In QCA, a measure called "raw consistency" is usually used to interpret the sufficiency condition. Thus, if a combination of factors has a raw consistency of 0.9 or more in indicating the presence of the outcome, then the combination of factors is considered sufficient for predicting the outcome. Combinations with high consistency scores signify that these combinations almost always lead to the given outcome condition [25,26].

Please note that when the membership values of factors and coverage are crisp (either 0 or 1), then the analysis will be QCA. However, it is possible to provide fuzzy membership functionality. If some or all the factors and outcome are specified using fuzzy membership, then the analysis will be called fsQCA.

3. Research Design and Data Collection

3.1. Data Collection

The present study was carried out in Karnataka state in India during 2021. The state of Karnataka was selected purposefully because it is one among the top states in India promoting the adoption of smart farming technologies in rural India through various initiatives. Karnataka is one of the established agri-tech start-up ecosystems in the country, and it has the highest number of agri-tech start-ups. Hence it was assumed that there

https://doi.org/10.3390/su18031344

would be a high rate of diffusion of smart farming technologies in the state, which would support and facilitate conducting this study.

3.2. Interview Question Design

Our qualitative study was designed as an exploratory study, without any preconception to link to any previous theoretical frameworks. This is because there are relatively few studies in the Indian context that link existing technology adoption theories. This approach has helped us ensure that we receive unbiased responses from the farmers. Our interview instrument consisted of 10 open-ended questions to elicit rich responses from farmers, and the questions are listed in Table 1. These interview questions are related to RQs 1–5.

Table 1. Interview instrument and their links to research questions and theory.

Interview Questions	Research Questions Linked to the Interview Questions	Theory Linked to the Interview Questions
1. Tell us about crop, technology in your farm and region. 2. Tell us about the agriculture technologies deployed in your farm, its application and usage.	RQ 1: What is the status of adoption of smart technology in the Indian agriculture sector?	
3. How did you come across this technology and what made you adopt the technology. 4. From how long have you been using this technology? What are the limitations or hurdles you have faced related to technology? 5. What are the advantages of this method over the conventional method?	RQ 2: What are the primary factors contributing to the adoption or non-adoption of smart technology?	
6. To what extent has it affected the labor/manpower/maintenance cost? 7. What are the financial benefits that you have gained through this method? 8. What are the factors that may be inhibiting technology adoption by other farmers? 9. Do you share the knowledge and influence other farmers to adopt smart farming technologies? If yes, share your experience with us?	RQ 4: Which factor(s) will expedite the adoption of smart technology in the agriculture sector?	RQ 3: To what extent are the previous theories on technology adoption frameworks meaningful in explaining the behavior of Indian farmers in adopting smart technologies of technology era?
10. What is your overall opinion on Smart Farming Technology? Closing: Please tell us more- about anything: your experience of technology in agriculture, agriculture productivity, performance, and job satisfaction.	RQ 5: How can adoption of smart technology in the agriculture sector help achieve SDG 12?	

3.3. Sample and Data Collection

Respondents who were willing to provide the required information voluntarily across the rural side of the state of Karnataka, India, were selected for the research study. Following initial field visits and observation study, the respondents were categorized into four segments, namely, full technology adoption farmers, partial technology adoption farmers, willing to adopt but not yet adopted, and farmers least interested in technology adoption.

Full technology adoption: This means they understand fully and accept, adopt the technology, use it productively and intuitively, and look out for further enhancements. This segment will basically help us understand what influenced them to fully adopt the technology and how can this help other segments of farmers to adopt the technology.

https://doi.org/10.3390/su18031344

Partial technology adoption: This means they do not fully rely on the technology or have some hesitation to invest fully into the technology, but they have some form of technological adoption. The partially adopted segment is the extension of segment 1, which will give a clear understanding of factors limiting full adoption. We collected the farmers' opinion on ground level problems while adopting the technology.

Willing to adopt but not yet adopted: This means they are interested to learn, invest and use the technology, but they have not adopted technology yet due to various factors. This is mainly related to farmers with a dilemma regarding technology adoption. 'Willing to adopt' will be the focus group, to understand the factors that are stopping them from adopting the technology and how to help this segment of farmers to start adopting technology.

Not willing to adopt: This group includes those least interested in technological adoption or people having resistance to change. This segment basically helps us understand the gap and difference between segment 1 and 4 (i.e., fully technology adoption farmers and those not willing to adopt). It also helps us in understanding the ground level problems for an unwillingness to adopt technology.

Five respondents from each category were selected using a stratified random sampling technique, making the total sample size 20 (Table 2). With due consideration to the scope of the study, a systematically structured interview schedule was developed after a pilot study. For data collection, the personal interview technique was used. All the interviews were recorded in raw audio format with consent from the responding farmer. The technology use of the farmers varied widely from local providers for simple sensors to established service providers for fully automated Smart Irrigation Systems. The farmers from two segments, namely 'willing to adopt' and 'not willing to adopt', did not provide much information on technology, and this is reflected in the empty cells in Table 2.

Table 2. Respondents' profile and technology use.

Segment	Sl	Farmer Code	Activity	Technology
Fully Adopted	1	S1F1	Floriculture (Sindhu Flora)	Smart Irrigation System
	2	S1F2	Aquaponics	Smart sensors for maintaining water quality, greenhouse parameters, and fish management
	3	S1F3	Poultry	Environmental control poultry farm
	4	S1F4	Pomegranate, Dragon Fruit, Guava	Fassal Smart Sensors and Bayer
	5	S1F5	Pomegranate	Fassal Smart Sensors and Apps
Partially Adopted	6	S2F1	Nursery (Ekalavya Hightech Nursery)	Sensors
	7	S2F2	Pomegranate	Fassal Smart Sensors
	8	S2F3	Pomegranate	Fassal Smart Sensors
	9	S2F4	Aquaponics	Beegle IoT Sensor
	10	S2F5	Capsicum, Tomato, and Cucumber	Smartflow Irrigation
Willing to adopt	11	S3F1	Biofloc Fish Farming	-
	12	S3F2	Mulberry and Cattle Farming	-
	13	S3F3	Millets and Mulberry	Government funded basic sensors such as temperature sensors
	14	S3F4	Sericulture	-
	15	S3F5	Pomegranate and Other Fruit Crops	Government funded basic sensors such as temperature sensors
Not Willing to Adopt	16	S4F1	Agroforestry	-
	17	S4F2	Mulberry	-
	18	S4F3	Ragi	-
	19	S4F4	Rose, Papaya, Guava	-
	20	S4F5	Maize, Beans, Eggplant, Kohl	-

https://doi.org/10.3390/su18031344

4. Data Analysis and Findings

4.1. Case Study: Interview Data

As shown in Table 2, the 'full technology adoption farmers' from Segment 1 use an array of technologies namely Fassal Smart sensors, environmentally controlled poultry farms and fish farms, and high-level sensors to monitor the water quality and greenhouse parameters. These are investment-intensive farms having medium- and large-sized agriculture lands. In Segment 2, 'partial technology adoption farmers' provide high-value agriculture production such as aquaponics, poultry, and smart irrigation systems, including smart flow systems. These farmers are small and medium size, mainly family growers having a sizable portion of their agriculture fields fitted with sensors and other smart technology (low- to medium-range equipment) for seamless operations in their fields. These farmers are benefited by availing themselves of the local agri-tech companies sale offers and incentives provided by the local government. Segment 3 are family growers with a medium- or small-sized limited area of land. These farmers are influenced by success stories of the technology use in neighboring fields. Eventually in the next few years these farmers will embrace smart agriculture technologies. Category 4 farmers are above 50 years of age, are not familiar with smart technology, and are reluctant to use technology. This is due to 'resistance to change behavior' and fear of embracing technology without having knowledge of the ease of use.

A transcript of all the interviews was prepared to support the analysis. The interview protocol covered the following areas: product variety (crop), use of technology, knowledge of smart technology for agri use, limitations, advantages of using technology in agriculture over conventional methods, impact of technology on employability and manual labor, factors that drive use/non-use of technology, overall performance, and knowledge dissemination. Some of the respondents from Segment 4 expressed distrust in the technology, and some were not willing to talk about the technology use. We analyzed the interview data using the content analysis for thematic consideration (see Table 3). We used repeated words such as technology, interest, benefit, trust, future, agriculture yield, smart IoT, and sensors. We also used several other key words in this process, in line with our research questions.

Table 3. Thematic analysis of interview data.

Fully Adopted Technology	Partially Adopted Technology	Willing to Adopt Technology	No Adoption
Business: High value crop, poultry, and aquaponics.	Nursery farming for vegetable crops, pomegranate, onion, vegetables, and fruits.	Fishing, animal husbandry, millets, long-life crops, and pomegranate.	Perennial crops, trees such as guava.
Location: 60% urban and 40% rural farmers. Medium to large farm holders.	Mostly rural area but 60% closer to technology hub of Bangalore. Medium size land holders.	Outskirts of Bangalore (semi urban).	Mostly rural.
Use of technology and period: Smart irrigation, weather app, humidity sensors, pH measurement, environmentally controlled sheds for chickens. Handheld device for luminous intensity. 1–3 years in use.	Auto fertilization, rainwater harvesting, temperature, humidity and moisture sensors. 1–8 years in use.	Need technology. To avoid spreading of disease. To spread chemicals. To avoid labor shortage. To reduce water use.	Not using technology for routine life (no smart phones).
Knowledge on smart agri-tech: Through education, awareness programs, social media news and training.	Through fellow farmers and social media.	Limited knowledge but willing to consider if the investment requirement is affordable.	Not much knowledge on technology.

https://doi.org/10.3390/su18031344

Table 3. *Cont.*

Fully Adopted Technology	Partially Adopted Technology	Willing to Adopt Technology	No Adoption
Benefits and advantages: 24/7 remote monitoring, less manpower, comfort of farm operations through mobile app, optimal use of resources such as waste, fertilizer, and feed. High yield and high financial benefit.	To monitor humidity and temperature for sapling germination. Irrigation and feeding are based on the sensor monitor and timely (only when needed). Yield is increased by at least 20%.	Looking for technology which updates data immediately without delay. Willing to adopt the existing technologies, provided subsidies are provided.	Not ready to consider potential benefits of adoption.
Barriers: Investment need, lack of high level of support from local government.	Initial investment and lack of support. Lack of awareness about technologies.	Not believing technology fully, as there are several loopholes in the existing technologies. Need accurate data on water parameters, prediction of diseases, and mechanized harvesters.	Fear of adoption in terms of investment requirement and actual ease of use.
Willingness to share knowledge and influence other farmers: Enthusiastic to showcase the success of technology adoption. Willing to set up similar technology in other farms.	Highly willing to expand, provided financial support is offered.	Yes ready to share the information after installing IoT sensors and equipment.	Not yet to the point of opening discussion.

We have made a detailed analysis of the interview data to identify the underlying factors for smart technology adoption. In this effort, we also tried to identify the link with the existing theories. Excerpts of the interview transcripts are given below:

Segment 1 farmers expressed their full support for technology adoption in the agriculture field, F2 mentioned—"...So smart farming is the need of the day because if we do not adopt technologies, the next generation is never going to enter farming. For them to come into farming we need to make it more certain."

F4 mentioned—"...Especially in Pomegranate, where the plant's 50% growth depends on irrigation. So, I would like to suggest farmers to install this technology, keep track on updates from their app and plan for future of their farm".

Statements from farmers F2 and F4 clearly expressed interests of 21st century farmers in considering smart farming as the first option in contrast to a conventional farming approach. Young technology entrepreneurs are willing to consider smart farming as the same as any technology-related job, as it does not demand as much physical effort for farmers as in previous generations.

Segment 2 farmers who use technology partially for their routine agriculture field operations have limited investment and use automation for essential operations. One of the farmers F1 in this group strongly expressed his view on technology use—"After installing these sensors we are able take right decision on when and how much to water".

Farmer F5 said—"I first saw this technology when I visited my friend's plot. He had adopted this automated drip system for 1 acre land. Seeing this, I decided to take up the fertigation system in my 0.5-acre polyhouse."

As the size of the farms are not huge, and the investment is not heavily required for simple smart technologies such as sensors for temperature and humidity monitoring, these smart agri-technologies have been adopted by a large group of farmers in this category. Real opinions of F1 and F5 evidenced the role of technology for small farms with a minimum investment. It also expressed the possibility of having a partial technology system instead

https://doi.org/10.3390/su18031344

of having a full adoption with high investment. This flexibility in adoption may help young farmers to choose agriculture as their choice of job or business in the future.

Three young farmers from Segment 3 expressed their willingness to adopt technology and also showed their concerns. The following quotes will be linked to the theory in the next section.

> "It would be great if such support is provided. Now they are already giving weather-based insurance and few assistances. We are making use of them. It would be great if they update about these to farmers." "It will be helpful and will yield better results. It is necessary. But to cost factor people are hesitating. If costs are managed people would adopt it.". "...Yes, there is a good future. Many farmers having larger farms could be benefitted by such technologies".

Segment 3 farmers have two different viewpoints on adoption. The first point is their lack of investment capability, and the other important point is questioning the use of technology and its need in small farms.

Based on the interview transcripts and data analysis, our preliminary results are given below: While educational background and initial success were the main drivers of the full adoption of technology in Segment 1, partial success (not realizing the full potential of technology adoption) was the reason for the partial adoption in Segment 2. The difference in production performance of Segments 1 and 2 are mainly due to the variety of crop and soil conditions. This difference in adoption by Segment 3 could be easily altered if the mindset of the farmers is tuned through appropriate training programs and demonstration of technologies locally within the community.

Finally, a document comprising the comparative study of each segment for each of the questions was prepared using the transcript. All qualitative responses were closely compared and analyzed to obtain the key findings. Before reporting the key findings, we discussed the preliminary results with the respondents in a focus group dissemination activity. This is further listed in Table 4. The farmers and their fully IoT-fitted agricultural fields from Segment 1 can be a catalyst for diffusion of technology in the local farming community, and they can co-create value for all potential new generation farmers. This could also create a distributed demand for crops and produce in relation to seasonal cultivations. Segment 2 farmers can integrate themselves with Segment 1 farmers by actively participating in the dissemination programs. While Segment 3 intends to adopt technology, the group of farmers from Segments 1 and 2 can encourage and support these farmers through knowledge dissemination. Social media and smart phones can play a role in taking technology to the next level in the Indian agri-sector. Segment 4 farmers, who are exhausted due to strenuous agricultural activities in the past, do not have any further interest in spending money or acquiring new knowledge of adopting technology. Family members of this group can be trained and supported. Training programs for young family members can attract the next generation of farmers in the existence of technology and ease of use. In addition, fertile agriculture land (not currently in use) can be made available for young farmers within a legal framework. Local government organizations, councils, and technology companies can work together to achieve this objective (refer to Table 4).

Table 4. Preliminary results on the adoption of IoT technology in the Indian agri-sector.

Technology Adoption	Age	Farm Size (L: Large; M: Medium; S: Small)	Source of Investment	Technology Awareness/Knowledge/Ease of Use	Potential Strategy to Wider Adoption of Technology
Technology adoption: full	20–40	M and L	Personal funds	High knowledge and can spread awareness	Co-create: volunteer and support other farmers through dissemination programs and showcase activities.

https://doi.org/10.3390/su18031344

Table 4. *Cont.*

Technology Adoption	Age	Farm Size (L: Large; M: Medium; S: Small)	Source of Investment	Technology Awareness/Knowledge/Ease of Use	Potential Strategy to Wider Adoption of Technology
Technology adoption: partial	20–50	S, M, and L	Personal funds and government support	High knowledge but yet to experience the benefits fully	Be part of the social change in terms of technology adoption.
Willing to adopt technology	20–50	S and M	Seeking investment	Awareness and knowledge are reasonably good but yet to experience the use and benefits	Support and be supported by the fellow members of the society.
No adoption	Above 50	S and M	Government can support and educate	No interest in technology use in agriculture	Need to train young family members. Available agriculture land can be made accessible for young farmers within a legal framework.

4.2. Analysis of the Results of the Qualitative Study and Linking to the DoI Theory

The results presented in Tables 3 and 4 yield interesting insights on the factors that influence the adoption of smart technology in the Indian agriculture sector.

First, there is evidence that the sample of farmers that we approached have differing levels of adoption—full adoption, partial adoption, willingness to adopt, and no adoption. This is consistent with technology adoption theories such as the Diffusion of Innovations theory by Rogers [12].

The second insight is that those farmers who have fully or partially adopted the technology have shown some positive results such the ability to measure important variables such as humidity, pH, luminous intensity, etc. This has helped these farmers to simplify some traditional operations. For example, farmers traditionally relied on their own perceptions of humidity before the use of smart technology, but with sensors they are more confident on the level of humidity. This has helped to confirm that the farmers who have adopted smart technologies are able to improve the observability of complex agriculture data, to register a relative advantage compared to other farmers. Similarly, the data from sensors have helped them to reduce the level of complexity in agricultural operations. These three tenets, namely observability, relative advantage, and complexity, are primarily employed in the Diffusion of Innovations theory. Thus, the responses of farmers seem to concur with similar theoretical properties of the DoI theory. Several farmers have found the output from sensors to be reliable, confirming the reliability of smart technologies.

By trialing smart technology, those who have either fully adopted or partially adopted it are able to see the benefits of 24×7 monitoring, a reduction in manpower, and the reduced use of resources. Thus, there is evidence that the technology can be trialed by farmers before a larger level of adoption. This resonates well with the tenet of trialability in the DoI theory. Given that farmers who partially adopted the technology first are able increase the level of adoption gradually signifies high levels of compatibility of the technology with existing farming operations.

A mapping of the observed characteristics from the qualitative study with the tenets of the DoI theory is shown in Figure 2. The interview data analysis highlighted the points that coincide with the attributes of DoI theory namely relative advantage, complexity, compatibility, trialability, and observability (please refer to the quotes from the interview data in Section 4.1).

https://doi.org/10.3390/su18031344

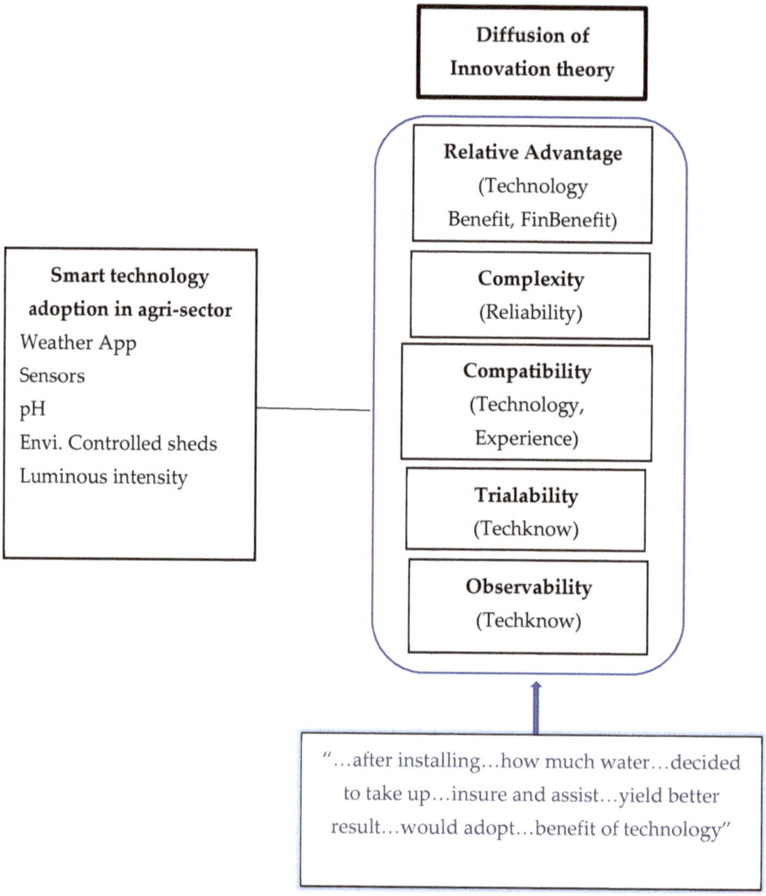

Figure 2. Smart farming adoption model: mapping of the themes identified from the qualitative study with the DoI theory. Source: created by the authors.

4.3. Theoretical Observations from the Case Study

The Diffusion of Innovations (DoI) theory [12] is one of the most widely applied theories in the prediction of the level technology adoption in private organizations [27]. The basic idea of this theory is that certain characteristics of the innovation can affect the rate of adoption [28]. The DoI theory has been used to establish the interface of e-commerce with the marketing and operational functions of SMEs [5]. A recent article by Wang et al. [29] considered DoI for a construction project by linking it with organizational innovation. DoI is able to be adapted to different fields, as long as the innovation is the one of the key elements. Specifically, the following characteristics have been suggested in this theory.

- Relative Advantage: the extent to which users or organizations believe that the new innovation is better than the previous practices.
- Complexity: the extent to which the new innovation is perceived by users as relatively difficult to understand and use.
- Compatibility: the degree to which the new innovation is perceived as consistent with the existing values, past values, and needs of potential adopters.

https://doi.org/10.3390/su18031344

- Trialability: the extent to which users or organizations believe that there are chances for the new innovation to be experienced before deciding whether to adopt it or not.
- Observability: the degree to which the benefits of the new innovations are observable to the users and others.

In this research, the primary research instrument, the qualitative interviews, are based on these innovation characteristics. Thus, the DoI theory has helped to understand what made the adopters of smart agriculture to implement the innovation fully and also to understand why others are reluctant to adopt or slow to adopt. Additional information (refer to Tables 1–3) has helped us to understand the reasons for adoption or non-adoption.

During the course of our research, we observed that, although the DoI theory, introduced by Rogers in 1962, has been widely applied in various innovation contexts, there has been little effort to extend this theory in different disciplines. Following Roger's Diffusion of Innovations theory, a few researchers improved the DoI concept in multi-dimensions, fitting it to the context of study [30]. Efforts have been made to determine whether these factors are all necessary or a few combinations are sufficient (e.g., Moore and Benbasat [30]. In the current paper, we focus more on whether the identified factors alone are sufficient to encourage adoption or whether we should consider a combination of factors. These combinations of factors, if identified, could help in extending DoI theory. Hence, we have attempted a fuzzy set QCA analysis in Phase 2 after identifying the factors of technology diffusion in Phase 1.

4.4. Qualitative Comparative Analysis

The first step in this Phase 2 is to identify important success factors from Phase 1. Based on the interview analysis presented in Section 3, we have identified six factors for the adoption of smart technology in Indian agriculture. These factors are based on the characteristics of DoI theory but further built on additional insights generated from the interview analysis.

Technology: This factor captures the totality of several DoI characteristics, such as the relative advantage and compatibility, of the new smart agriculture technology. Our interview analysis shows that these characteristics play a critical role in encouraging adoption. For example, farmers who have already fully or partly adopted the technology and those who are planning to introduce these technologies are convinced of these characteristics, namely the advantage of use and compatibility. Those who have not adopted so far with no plans to adopt in the near future seem unconvinced by the new technologies.

We consider the adoption of current technology in routine operations of farm activities. Here, the smart technology for agriculture includes both simple technology such as sensors and an automated watering system and advanced technology such as data-driven decisions-making using apps and the use of multi-sensors in the agri-fields, logistics, and distribution centers.

Technical knowledge (Techknow): This refers to the knowledge of the user in handling the technology used within their agri-businesses. This factor captures the observability and trialability characteristics specified in the DoI theory. If farmers are able to observe the results, have the technical knowledge to apply (trial) the technology, and are convinced of the new technology, then they have a higher intention to adopt the new technology. As Table 3 shows, the reason for farmers' partial adoption of technology is the lack of accuracy observed in the results.

Reliability: This is related to the users' experience, based on consistent results and the benefits achieved through the IoT technology in their agri-business. The level of reliability required seems to influence the decision on technology adoption. This factor is related the complexity characteristic specified in the DoI theory. For situations where the application

https://doi.org/10.3390/su18031344

of the new smart agriculture is perceived as complex, its reliability is affected. This factor is critical in that those who have partly or fully adopted the technology perceive investment as an important need. Investment requirements seem to be an important barrier for the other two categories (wiling to adopt and no adoption).

Experience: In the agri-sector, family experience plays a vital role for day-to-day activities. However, in this research we consider that the experience of the user in handling the technology for their routine operations is vital for decision-making on the adoption of technology. These data are captured through our interviews with the participants.

Benefit: The interview data brought several benefits for our consideration, such as the low labor cost, easy solution, and quick update. The actual benefit of using the technology could be realized in various stages of the supply chain such as less manpower, production, logistics, distribution, and other operations. Here, we consider benefit as the benefits to the farmers of using smart technology.

Financial benefit (Finbenefit): The financial outcome from agriculture operations is crucial to continue farming activities. For several small landholders, farming is the only lifeline for the entire family. We consider Finbenefit as the financial benefit realized by the user when using the smart technology in their agri-businesses.

We have used the above six factors as our scale items to measure the technology adoption level in the agri-sector.

Saridakis [24] has introduced a clear guide for qualitative comparative analysis (QCA); the authors have highlighted the importance of relating the theory with actual practices in QCA (see Table 5).

Table 5. An estimate of the success factor and adoption of smart technologies in Indian agriculture.

Farmer Number	Adoption	Technology	Techknow	Experience	Benefit	Reliability	FinBenefit
1	1	0.7	0.8	1	0.8	1	1
2	1	0.7	0.9	0.8	0.8	0.8	0.5
3	1	0.8	0.8	0.7	0.6	0.8	0.8
4	1	0.85	0.9	0.65	0.6	0.8	1
5	1	0.85	0.8	0.6	0.8	0.5	0.8
6	0.75	0.6	1	1	0.8	0.6	1
7	0.75	0.7	0.7	0.6	0.8	0.6	x
8	0.75	0.7	0.6	0.6	0.8	x	1
9	0.75	0.85	0.6	0.6	0.8	0.8	0.8
10	0.75	0.85	0.6	0.65	0.7	0.6	0.8
11	0.5	0.4	0.9	0.6	0.8	0.2	0.8
12	0.5	0.4	0.9	0.5	0.8	0.4	0.8
13	0.5	0.4	0.9	0.5	0.8	0.2	0.4
14	0.5	0.4	0.7	0.6	0.1	0.2	0.8
15	0.5	0.4	0.8	0.5	0.8	0.4	0.8
16	0.05	0.1	0.3	0	0	0	0
17	0.05	0	0	0	0	0	0
18	0.05	0.05	0	0	0	0	0
19	0.05	0	0	0	0	0	0
20	0.05	0	0	0	0	0	0

Application of FSQCA to Predict the Successful Adoption of Smart Farming in Indian Agriculture

In this section, we apply fsQCA to the data derived from the interviews. We have first used the results of the interviews to (i) identify various success factors for adoption and then (ii) to develop membership functions. The resulting table is shown below (Table 5). The responses of the interviewees highlighted their level of importance for specific variables: adoption, technology, knowledge, experience, benefit, reliability, and financial benefits. These are expressed in their usage of words. To accomplish this exercise, we used insights

https://doi.org/10.3390/su18031344

from trained local representatives who supported the interview process. For example, if a farmer mentions highly appreciating the smart technology, a membership score of 0.8 was assigned, while the membership of 1 was used if a farmer is confident of fully adopting the smart technology. If no response was provided, we represented this through 'x'. This case is treated equivalent to non-adoption with a membership value of zero.

The methodology of fsQCA has been applied to the data in Table 5. The details are described below.

As indicated in Section 2.2, the first step in fsQCA is to verify whether the factors on their own are necessary. We used the fsQCA software version 4.1 developed by the original QCA team for our analysis. The results are shown in Table 6.

Table 6. Necessary condition analysis of the presence of individual success factors to ensure adoption.

Factor	Necessity Consistency Score	Coverage
Technology	0.974359	0.822511
Techknow	0.819672	0.865801
Experience	0.954545	0.818182
Benefit	0.870370	0.813853
Reliability	0.993671	0.679654
FinBenefit	0.840708	0.822511

Table 6 indicates that some factors have high consistency above 0.9. However, the corresponding coverages are not high. Setting a combination of high consistency (above 0.9) and high coverage (above 0.9), we conclude that there is no individual factor that meets the necessary condition for successful adoption.

The next step is to develop a truth table for the $2^6 = 64$ combinations of factors. This analysis was conducted using the software. After reducing the results, Table 7 shows the combinations of factors that lead to high consistency scores.

Table 7. Final reduced truth table (Note: 1 represents the presence of the factor, and 0 represents its absence).

Pathway (Combination) Number	Technology	Techknow	Experience	Benefit	Reliability	FinBenefit	Adoption	Consistency	Coverage
1	1	1	1	1	1	0	1	1	0.588744
2	1	1	1	1	0	1	1	1	0.614719
3	0	1	1	1	0	1	1	0.973684	0.662338
4	0	1	1	0	0	1	1	0.964286	0.709957
5	1	1	1	1	1	1	1	1	0.519480

Thus, all the above combinations yield a consistency above 0.7 and coverage above the bottom threshold of 0.25. Thus, these are the combinations that could predict the successful adoption of smart technologies in Indian agriculture.

Please note that technical knowledge and experience occur in all these pathways (combinations). Thus, our analysis places the highest importance on these two factors. Pathway # 1 indicates that the presence of all but the financial benefit could lead to the successful adoption of smart technologies. Pathway # 2 shows that all but the presence of reliability may also help in successful adoption. Pathways 3 and 4 show that if technology is not present, then its reliability is not important in fostering adoption. Finally, the last

https://doi.org/10.3390/su18031344

pathway # 5 shows that the presence of all the six factors will always result in the successful adoption of smart technologies in Indian agriculture.

We believe that the results of the fsQCA have the potential to extend the DoI theory. First, we observed that individual factors are not sufficient to predict the successful adoption of smart agriculture in Indian scenario. This is somewhat counterintuitive to the tenets of DoI theory, which claims that some of the characteristics are important for successful adoption. However, we did find evidence of the successful application of the theory when we consider certain combinations of these factors rather than individual factors. Thus, we propose that the DoI theory can be extended to consider combinations of characteristics in explaining adoption in addition to the stress on individual characteristics.

5. Further Discussion

In this section, we revisit our research questions and summarize the answers on the basis of the analysis presented earlier.

RQ1: What is the status of adoption of smart technology in the Indian agriculture sector?

RQ 2: What are the primary factors contributing to the adoption or non-adoption of smart technology?

These two research questions (RQ1 and RQ2) have been answered through the observation field visits and interview data analysis. RQ3 related to the theory was established from the qualitative interview data analysis. In this paper, the Diffusion of Innovations theory is identified as a suitable theoretical framework.

Our next research question on factors that will expedite the adoption of smart technology in the agriculture sector, used the fsQCA approach to identify six factors from the qualitative data: technology, technology knowledge, experience, benefit, reliability, and financial benefit. Five pathways have been discussed with combinations of factors (Table 7). In Pathway 1, the combination of all factors except FinBenefit has been considered as technology adoption with a coverage of 58%; in Pathway 2, the combination of factors except reliability provides a positive adoption with a coverage of 61%. In the third pathway, technology and reliability do not play a role, but the adoption will take place with a coverage of 66%. In the fourth pathway, the combination of factors that lead successful adoption (coverage 70%) of technology includes technology knowledge, experience, and financial benefit. Interestingly, the final pathway includes all six factors for adoption with a coverage of 51%. These results show the highest consistency for pathways 1, 2, and 5. This implies that all factors are important for adoption. However, some combinations will work better together: although the technology is not in place currently, the farmers will eventually adopt the smart technology if they have realized the benefits with good technology knowledge. Financial benefit will also be positive in this case.

Further, if we look at Table 6, two factors (Techknow and Experience) occur constantly in all the pathways. Thus, we find that these two are critical factors for successful adoption. Two more factors, FinBenefit and Benefit, occur in four of the five pathways and thus can be considered the next critical factors.

Thus, the fsQCA shows not only the combinations of factors that would lead to successful adoption, but it also provides a way of understanding the criticality of the factors in encouraging adoption. We have found TechKnow and Experience are extremely critical for adoption. In other words, for Indian farmers to consider adopting smart agriculture technology, they should have adequate experience and a level of technical knowledge. The expectation of benefits, financial or otherwise, is the next set of critical factors.

https://doi.org/10.3390/su18031344

6. Conclusions, Contribution, Limitations, and Future Research

This research focused on understanding how smart technologies support Indian agriculture and join world leading efforts in this direction (e.g., Nikolić et al. [1]). It included a detailed interview with practitioners who use smart farming technology at various stages of operations. A total of five factors identified from the analysis of interview data have been used to conduct fsQCA. While all five factors, on their own, can help influence the successful adoption of smart technologies, the fsQCA further identified that five different combinations of factors could also influence adoption. Further analysis of the combinations of factors revealed that technical knowledge and experience are the most critical factors for facilitating adoption, followed by benefits (financial and other).

In this research article, we reflected on our empirical study considering a practice-based research question: How does technology diffusion make a positive change in the agricultural sector of India? In this attempt, we identified some individual factors and specific primary factors contributing to full and partial adoption of the technology for routine operations in the agri-sector: financial investment and technological knowledge. While technology diffusion is imperative among young professionals, farmers using traditional approaches are reluctant to change due to age and ignorance of technical aspects. We identified from two phases of data analysis that the Indian agriculture sector will benefit from a combination of factors. Technology diffusion is possible in every agri-business, irrespective of the size of the operation, by involving technology professionals. Family landholders can make use of the younger generations with a technology background to achieve responsible high production. Diffusion of technology in the field of agriculture is one of the best possible approaches to achieve success.

SDG 12, "Responsible Consumption and Production", can be made possible in a highly populated country like India by involving technology professionals in agri-sector operations. Here, digitalization supports the dynamic changes in production and agri-business sector [31]. Creating synergy between smart technology and traditional agri-sector can make a major change in food production and distribution. This will also enhance the economic and social value of the local society.

About 86.2% of the country's farmers are small farm holders who cultivate 47.3% of the country's arable land [12]. Fragmented landholdings result in poor bargaining power and reduced-price realization of agri-produce. Achieving economies of scale through technology adoption will attract the future generation of farmers. We suggest that a feasible approach to current farming in India is to adopt simple, cost-effective, and smart agri-technology. This can be achieved through systematic training and involvement. DoI in the agri-sector [12,30] is possible only if the full potential of the chosen technology reaches a wider population with reliable consistent benefits.

6.1. Theoretical Contribution

The literature on IoT technology for agriculture is growing steadily. Though DoI has been applied to several fields, not many academic articles highlight technology diffusion in the agriculture sector, specific to the Indian context [4]. After the qualitative interview data analysis, we identified a strong link with the DoI. This was further tested with fsQCA by analyzing the combination of factors and their influence on smart technology adoption. Two prominent factors, namely technical knowledge and experience, turned out to be the critical factors for all pathways.

6.2. Practical Contribution

This research contributes directly to practitioners intending to adopt technology (see Table 3). Several barriers that inhibit adoption have been identified. Age and technology

https://doi.org/10.3390/su18031344

use are highly related and limit the exploration of technology in a traditional agriculture set-up. In particular, farmers in rural India with less exposure to technology have resistance to change unless and until there is a push from young members of the family or friends. A higher level of the demonstration of technology adoption in the rural agriculture sector can influence the previous generation of farmers. At the same time, it is wise to think about the financial aspects of the agricultural production and its growth after the installation of smart IoT in small or medium farms. Testing farms and demonstration fields with live agri-technology facilities can help increase the adoption of technology.

The results of this research will contribute directly to farmers who are intending to consider smart farming. The dissemination of the results will help many young farmers to embrace smart farming and IoT technologies for routine processes in farming operations. Farmers who adopted and benefited through complete adoption can co-create and support societal welfare.

6.3. Research Propositions, Limitations, and Future Research

On the basis of the findings of this research, we highlight multiple propositions in the context of the use of modern technologies for smart agriculture:

RP1: an awareness program on smart farming will help remove the phobia of technology in agriculture.

RP2: a training program on smart farming will help provide hands-on experience to young agricultural professionals.

RP3: involving young professionals in smart farming may encourage several others to focus on the agri-sector.

RP4: the IT sector can include more and more career opportunities for smart-farming related technologies.

Future research can consider a longitudinal experimental study with specific cases, having observations in real-time on the adoption of technology and the behavior change along with the business performance. Our research included data from 20 interviews of individual farm holders, and hence, the results of this study are difficult to generalize. We consider this as the main limitation of our study. A larger sample size covering multiple parts of the country could be attempted. A large dataset can also help in refining the fuzzy membership estimates for the fsQCA analysis. More discussions with interviewees linking more directly to SDG12 could be attempted. A large-scale quantitative study could be attempted to confirm the above propositions. In addition, our data were from 2021 to 2023; more recent data will support new research on technology adoption in the agriculture sector. A detailed recording of the benefits and complexities of using smart technology will help farmers to choose their path of smart farming. Finally, the results will inspire more millennials to consider agriculture as their profession, which is currently not favorable in rural India.

Author Contributions: Conceptualization, U.R.; Methodology, U.R. and R.R.; Software, R.R.; Validation, U.R. and R.R.; Formal Analysis, U.R.; Investigation, U.R.; Resources, U.R.; Data Curation, U.R.; Writing—Original Draft Preparation, U.R.; Writing—Review and Editing, U.R. and R.R.; Visualization, U.R.; Supervision U.R.; Project Administration, U.R.; Funding Acquisition U.R. All authors have read and agreed to the published version of the manuscript.

Funding: This research was funded by the Nottingham Trent University Global Challenge Research Fund, 2021–2023, Funding number RD046.

Institutional Review Board Statement: The study was conducted after gaining ethical approval (ref RAMANATHAN 2019/283) from the Ethics Committee of Nottingham Trent University, UK, on 11 December 2019.

https://doi.org/10.3390/su18031344

Informed Consent Statement: Informed consent was obtained from all subjects involved in the study.

Data Availability Statement: The data presented in this study are available on request from the corresponding author.

Acknowledgments: An earlier version was presented at the EurOMA 2023 conference in Belgium. We would like to express our sincere thanks to the teams from Beegle Agritech and Agriproducts Pvt. Ltd., India, for supporting data collection.

Conflicts of Interest: The authors declare no conflicts of interest.

References

1. Nikolić, D.; Jovanović, S.; Jurišević, N.; Nikolić, N.; Radulović, J.; Radović, M.V.; Grujić, I. Sustainable Design in Agriculture—Energy Optimization of Solar Greenhouses with Renewable Energy Technologies. *Energies* **2025**, *18*, 416. [CrossRef]
2. Choudhary, V.; Guha, P.; Pau, G.; Mishra, S. An overview of smart agriculture using internet of things (IoT) and web services. *Environ. Sustain. Indic.* **2025**, *26*, 100607. [CrossRef]
3. Thilakarathne, N.N.; Bakar, M.S.A.; Abas, P.E.; Yassin, H. Internet of things enabled smart agriculture: Current status, latest advancements, challenges and countermeasures. *Heliyon* **2025**, *11*, e42136. [CrossRef]
4. Ibrahim, A.S.; Mohsen, S.; Selim, I.M.; Alroobaea, R.; Alsafyani, M.; Baqasah, A.M.; Eassa, M. AI-IoT based smart agriculture pivot for plant diseases detection and treatment. *Sci. Rep.* **2025**, *15*, 16576. [CrossRef]
5. Gupta, S.; Rikhtehgar Berenji, H.; Shukla, M.; Murthy, N.N. Opportunities in farming research from an operations management perspective. *Prod. Oper. Manag.* **2023**, *32*, 1577–1596. [CrossRef]
6. FICCI. Ushering in a New Growth Wave: From Artificial Intelligence to Agricultural Intelligence. In Proceedings of the International Virtual Conference on AI & Digital Applications in Agriculture, New Delhi, India, 8–9 October 2020.
7. Mohan, A.; Krishnan, R.; Arshinder, K.; Vandore, J.; Ramanathan, U. Management of postharvest losses and wastages in the Indian tomato supply chain—A temperature-controlled storage perspective. *Sustainability* **2023**, *15*, 1331. [CrossRef]
8. Ramanathan, R.; Ko, L.W.K.; Chen, H.; Ramanathan, U. Green characteristics of RFID technologies: An exploration in the UK logistics sector from innovation diffusion perspective. In *RFID Technology Integration for Business Performance Improvement*; Lee, I., Ed.; IGI Global: Hershey, PA, USA, 2014; pp. 158–180.
9. Ram, V.S.; Singh, K.; Sreshta, B. Chapter-15: Smart Farming—A case study from India. In *Digital Agricultural Ecosystem: Revolutionary Advancements in Agriculture*; Singh, K., Kolar, P., Eds.; John Wiley & Sons: Hoboken, NJ, USA, 2024; pp. 269–290.
10. Danese, P.; Mocellin, R.; Romano, P. Designing blockchain systems to prevent counterfeiting in wine supply chains: A multiple—Case study. *Int. J. Oper. Prod. Manag.* **2021**, *41*, 1–33. [CrossRef]
11. Wang, C.; Han, Q.; Li, J.; Li, C.; Zou, X. YOLO-BLBE: A Novel Model for Identifying Blueberry Fruits with Different Maturities Using the I-MSRCR Method. *Agronomy* **2024**, *14*, 658. [CrossRef]
12. Rogers, E.M. *Diffusion of Innovations*; Free Press: New York, NY, USA, 1983.
13. Ramanathan, U.; Ramanathan, R.; Adefisan, A.; Da Costa, T.; Cama-Moncunill, X.; Samriya, G. Adapting digital technologies to reduce food waste and improve operational efficiency of a frozen food company—The case of Yumchop Foods in the UK. *Sustainability* **2022**, *14*, 16614. [CrossRef]
14. Ramanathan, U.; Havyas, K.S. Diffusion of smart technology in agricultural sector of rural India. In Proceedings of the EurOMA Conference, Leuwen, Belgium, 4–7 July 2023.
15. Little, D. *Small Farmers in Indian Agriculture*; Foundation for Agrarian Studies: Bengaluru, India, 2017. Available online: https://fas.org.in/small-farmers-in-indian-agriculture/#:~:text=Almost%2050%20per%20cent%20of,162 (accessed on 22 November 2025).
16. Balakrishna, G.; Moparthi, N.R. Study report on Indian agriculture with IoT. *Int. J. Electr. Comput. Eng.* **2020**, *10*, 2322–2328. [CrossRef]
17. Ancarani, A.; Di Mauro, C.; Legenvre, H.; Cardella, M.S. Internet of things adoption: A typology of projects. *Int. J. Oper. Prod. Manag.* **2020**, *40*, 849–872. [CrossRef]
18. Jarial, S. Internet of Things application in Indian agriculture, challenges and effect on the extension advisory services—A review. *J. Agribus. Dev. Emerg. Econ.* **2022**, *13*, 505–519. [CrossRef]
19. Duan, Y.; Edwards, J.S.; Dwivedi, Y.K. Artificial intelligence for decision making in the era of Big Data: Evolution, challenges and research agenda. *Int. J. Inf. Manag.* **2019**, *48*, 63–71. [CrossRef]
20. Upendra, R.S.; Umesh, I.M.; Varma, R.R.; Basavaprasad, B. Technology in Indian agriculture—A review. *Indones. J. Electr. Eng. Comput. Sci.* **2020**, *20*, 1070–1077. [CrossRef]
21. Pal, B.D.; Kapoor, S.; Saroj, S.; Jat, M.L.; Kumar, Y.; Anantha, K.H. Adoption of climate-smart agriculture technology in drought-prone area of India–implications on farmers' livelihoods. *J. Agribus. Dev. Emerg. Econ.* **2022**, *12*, 824–848. [CrossRef]

https://doi.org/10.3390/su18031344

22. Malacina, I.; Lintukangas, K. On the edge: A multilevel perspective on innovation complexities and dynamic attractors in the supply network. *Int. J. Oper. Prod. Manag.* **2025**, *45*, 925–950. [CrossRef]
23. Ragin, C.C. *Fuzzy-Set Social Science*; University of Chicago Press: Chicago, IL, USA, 2020.
24. Saridakis, C.; Zaefarian, G.; Ganotakis, P.; Angelidou, S. A step-by-step guide of (fuzzy set) qualitative comparative analysis: From theory to practice via an implementation in a B2B context. *Ind. Mark. Manag.* **2022**, *107*, 92–107. [CrossRef]
25. Elliott, T. *Fuzzy Set Qualitative Comparative Analysis*; Research Notes; UCI Statistics Group: Irvine, CA, USA, 2013; pp. 1–6.
26. Woodside, A.G. Moving beyond multiple regression analysis to algorithms: Calling for adoption of a paradigm shift from symmetric to asymmetric thinking in data analysis and crafting theory. *J. Bus. Res.* **2013**, *66*, 463–472. [CrossRef]
27. Wang, Y.M.; Wang, Y.S.; Yang, Y.F. Understanding the determinants of RFID adoption in the manufacturing industry. *Technol. Forecast. Soc. Change* **2010**, *77*, 803–815. [CrossRef]
28. Zhang, L.; Wen, H.; Li, D.; Fu, Z.; Cui, S. E-learning adoption intention and its key influence factors based on innovation adoption theory. *Math. Comput. Model.* **2010**, *51*, 1428–1432. [CrossRef]
29. Wang, R.; Li, Q.; Chong, H.-Y. Untangling innovation diffusion in construction projects: A configural analysis of project attribute and corporate strategy. *Eng. Constr. Archit. Manag.* 2025, *ahead-of-print*. [CrossRef]
30. Moore, G.C.; Benbasat, I. Development of an Instrument to Measure the Perceptions of Adopting an Information Technology Innovation. *Inf. Syst. Res.* **1991**, *2*, 192–222. [CrossRef]
31. Yang, Z.; Hu, W.; Shao, J.; Shou, Y.; He, Q. How does digitalization alter the paradox of supply base concentration? The effects of digitalization intensity and breadth. *Int. J. Oper. Prod. Manag.* **2023**, *43*, 1690–1722. [CrossRef]

https://doi.org/10.3390/su18031344

MDPI AG
Grosspeteranlage 5
4052 Basel
Switzerland
Tel.: +41 61 683 77 34

Sustainability Editorial Office
E-mail: sustainability@mdpi.com
www.mdpi.com/journal/sustainability

www.ingramcontent.com/pod-product-compliance
Lightning Source LLC
Chambersburg PA
CBHW061813210926
51850CB00006B/691
* 9 7 8 3 7 2 5 8 7 1 4 2 1 *